Texts and
Monographs
in Physics

W. Beiglböck
M. Goldhaber
E. H. Lieb
W. Thirring

Series Editors

Robert D. Richtmyer

Principles of Advanced Mathematical Physics

Volume II

With 60 Figures

Springer-Verlag
Berlin Heidelberg

Robert D. Richtmyer

Department of Mathematics
University of Colorado
Boulder, Colorado 80309
USA

Editors:

Wolf Beiglböck

Institut für Angewandte Mathematik
Universität Heidelberg
Im Neuenheimer Feld 5
D-6900 Heidelberg 1
Federal Republic of Germany

Maurice Goldhaber

Department of Physics
Brookhaven National Laboratory
Associated Universities, Inc.
Upton, NY 11973
USA

Elliott H. Lieb

Department of Physics
Joseph Henry Laboratories
Princeton University
P.O. Box 708
Princeton, NJ 08540
USA

Walter Thirring

Institut für Theoretische Physik
der Universität Wien
Boltzmanngasse 5
A-1090 Wien
Austria

Library of Congress Cataloging in Publication Data

Richtmyer, Robert D
  Principles of advanced mathematical physics.

  (Texts and monographs in physics)
  Bibliography: v. 1, p.
  Includes indexes.
  1. Mathematical physics. I. Title.
QC20.R56   530.1'5   78-16494
                   (v. 2) AACR1

9 8 7 6 5 4 3 2 1

ISBN 978-3-642-51078-6   ISBN 978-3-642-51076-2 (eBook)
DOI 10.1007/ 978-3-642-51076-2

Contents

20   Group Representations I: Rotations and Spherical Harmonics   40

21   Group Representations II: General; Rigid Motions; Bessel Functions   62

22   Group Representations and Quantum Mechanics   80

30   Invariant Manifolds in the Taylor Problem   263

31   The Early Onset of Turbulence   276

References   313

Index   317

Preface to Volume II

The first eleven chapters in this volume, 18 through 28, contain material that was developed in the third year of the three-year mathematical physics sequence at the University of Colorado. The central concepts are groups, manifolds, and differential geometry. I wish to thank Professors Wesley Brittin and Russel Dubisch for extensive discussions of this material, and I wish to thank Professor Wolf Beiglböck for advice and suggestions on the overall plan and on the material on group representations.

The material in the last three chapters, related broadly to recent work in differentiable dynamical systems, has been discussed in special courses on hydrodynamic stability and seminars on mathematical physics. That material is somewhat less well organized than the older subjects, but has been included because it contains various concepts of great potential value in physical science.

Boulder, August 1981   Robert D. Richtmyer

CHAPTER 18

Elementary Group Theory

The group axioms; Abelian group; cyclic group; subgroup; order;
isomorphism; homomorphism; automorphism; permutation; symmetric
group; cycle; transposition; parity; alternating group; kernel of a
homomorphism; normal subgroup; simple group; conjugate elements; cosets;
Lagrange's theorem; factor group; law of homomorphism; translations; inner
automorphisms; Cayley's theorem; conjugate subgroups; simplicity of
$\mathscr{A}_5$; composition series; Jordan–Hölder theorem; generators and relations;
free group; free abelian group; the word problem; space and point groups;
direct and semidirect product; symmorphic space groups.

Prerequisite: Elementary algebra.

This chapter contains a survey of elementary group theory. For applications
in later chapters, the high point is the law of homomorphism and the concepts
associated with it.

18.1 The Group Axioms; Examples

A *group G* is any set or collection of elements $\{a, b, c, \ldots, x, y, z, \ldots\}$, finite
or infinite, together with a law of composition, denoted by $\circ$, such that:

   i. If a and b are any two elements of G, then $a \circ b$ is an element of G.
   ii. If a, b, and c are any three elements of G, then $(a \circ b) \circ c = a \circ (b \circ c)$
       (associative law).
   iii. If a and b are any two elements of G, then there exist unique elements
       x and y in G such that $a \circ x = b$ and $y \circ a = b$.

If the elements are numbers, matrices, quaternions, etc., the composition
$a \circ b$ may be either the sum or the product of a and b; in the examples below,
the word "under" is used to identify the law of composition. In the case of
mappings, transformations, rotations, permutations, etc., the law is understood as the usual law of composition; if a and b are transformations, then
$a \circ b$ is the transformation that results from performing b first, then a.

Note. In some books, the axiom iii above is replaced by the fully equivalent axiom that G contains a unique identity element e and that each element a of G has a unique inverse a^{-1}—see next section.

As a first example, let G be the set of all rotations in the plane: let R_φ denote the transformation in which a point x, y is moved to, or mapped onto, the point x', y', where

$$x' = x \cos \varphi - y \sin \varphi,$$
$$y' = x \sin \varphi + y \cos \varphi.$$

(18.1-1)

If the transformations R_{φ_1} and R_{φ_2} are performed in succession, the result is a rotation through the angle $\varphi_1 + \varphi_2$, i.e., it is the transformation $R_{\varphi_1 + \varphi_2}$. It is easily verified that the set $\{R_\varphi : 0 \leq \varphi < 2\pi\}$ of all such rotations satisfies the group axioms.

A rotation in 3 dimensions may be described by first choosing a direction through the origin and then performing a rotation through some angle about that direction as a fixed axis. It follows from Euler's theorem, proved in Section 19.2, below, that the resultant of two such transformations, performed in succession, is another such, i.e., is a rotation through *some* angle about *some* axis. [This seems evident (because everyone knows that it is true) until one tries to prove it.] In consequence, the set of all rotations in 3 dimensions is a group. The group of all rotations in n dimensions is denoted by SO(n), for reasons that will appear.

As a third example, consider the set of all rotations in 3 dimensions under which a cube, centered at the origin, is invariant (i.e., is mapped into a cube that coincides with the original cube). One can rotate the cube through 90°, 180°, or 270° about an axis through the midpoints of opposite faces, through 180° about an axis through the midpoints of opposite edges, or through 120° or 240° about an axis through opposite vertices. It is easily verified that these transformations (including the identity transformation) form a group of 24 elements. More generally, the set of all transformations of a specified kind (e.g., rotations, general linear transformations, rigid motions, conformal mappings) under which a given figure is invariant is a group, because the figure is clearly invariant under composition and inverses of such mappings. The rigid motions under which a crystal lattice is invariant constitute the *space group* of the crystal—see Section 18.13.

The set of all permutations of n objects is a group; such groups are discussed in Section 18.4.

Certain sets of real or complex numbers or quaternions are groups with respect to addition or multiplication, e.g., the set of all integers (positive, negative, and zero) under addition, the set of all positive real numbers under multiplication, the integers $0, 1, \ldots, m - 1$ under addition modulo m, or the set of all nonzero (real) quaternions under multiplication.

When addition is the rule of composition, $a \circ b$ is denoted by $a + b$, the inverse of a by $-a$, and the identity by 0. Often the little circle is omitted and the composition of two elements a and b is written simply as a product ab.

A finite group can be fully described by its *multiplication table*. For example, *Klein*'s 4-*group* V_4 is defined by

$$
\begin{array}{c|cccc}
 & e & a & b & c \\
\hline
e & e & a & b & c \\
a & a & e & c & b \\
b & b & c & e & a \\
c & c & b & a & e \\
\end{array}
$$

which means that $a \circ b = c$, etc. Each group element appears just once in each row and once in each column; furthermore, all rows are different and all columns are different. Any square arrangement of letters having this property is called a *latin square* (Euler). Any latin square defines an abstract group, provided that the multiplicative structure thus determined has an identity and satisfies the associative law.

Abstract group theory deals with the relations indicated in the multiplication table and completely ignores the inherent nature of the elements, a, b, etc. In contrast with calculus, real and complex analysis, differential equations, and other subjects in analysis (group theory belongs to *algebra*), numerical quantities hardly ever appear, except integers for the purposes of enumeration and counting.

The theory of groups plays a role in quantum mechanics, in the theory of spectra, in the analysis of classical dynamical systems, in the theory of automorphic functions, in the theory of algebraic equations, and so on.

18.2 Elementary Consequences of the Axioms; Further Definitions

The following laws are consequences of axioms i, ii, and iii of the preceding section:

Law of cancellation: If a, b, c are any elements of a group G, then

$$a \circ b = a \circ c \quad \text{implies} \quad b = c$$

and

$$b \circ a = c \circ a \quad \text{implies} \quad b = c.$$

Identity: In G there is a unique element e such that $a \circ e = e \circ a = a$ for all a in G.

Inverses: If a is any element of G, there exists in G a unique element a^{-1} such that $a \circ a^{-1} = a^{-1} \circ a = e$; furthermore, $(a \circ b)^{-1} = b^{-1} \circ a^{-1}$.

Extended associative law: $(a \circ (b \circ (c \circ d))) \circ e = a \circ b \circ c \circ d \circ e$, etc. Unnecessary parentheses will be omitted from now on. Also

$$(a \circ b \circ \cdots \circ x \circ y)^{-1} = y^{-1} \circ x^{-1} \circ \cdots \circ b^{-1} \circ a^{-1}.$$

If a is in G, and m is any integer, then a^m is defined as follows:

$$a^0 = e,$$

$$a^1 = a,$$

$$a^2 = a \circ a,$$

$$\vdots$$

$$a^{n+1} = a^n \circ a,$$

$$a^{-m} = (a^{-1})^m.$$

Clearly, these powers all commute, and $a^n \circ a^m = a^{n+m}$. Generally, two elements a and b of G are said to *commute* if $a \circ b = b \circ a$. If all pairs of elements of G commute, G is said to be a *commutative* or *Abelian* group. If all the elements a^n ($n = 0, \pm 1, \pm 2, \ldots$) are distinct, then the element a is of *infinite order*; otherwise, as it is easily seen, there is a smallest positive integer l, the *order* of a, such that $a^l = e$; then $a^m = e$ if and only if l is a divisor of m, and every power of a is equal to one of the elements $\{e, a, a^2, \ldots, a^{l-1}\}$.

A *subgroup* of G is a subset G' of the elements of G which is itself a group under the same law of composition $\circ$ that appears in G. The rotations about the z-axis constitute a subgroup of the group of rotations in 3-space. The distinct powers of an element a constitute a subgroup called the subgroup *generated by* the element a; such a subgroup is a *cyclic* group of finite or infinite order. The *order* of a group is the number of elements in it (finite or infinite). If G' is a subgroup of G, we write $G' < G$. In any case, $G < G$ and $\{e\} < G$. If $G' \neq G$, G' is a *proper* subgroup; if $G' = \{e\}$, G' is the *trivial* subgroup.

MISCELLANEOUS QUESTIONS AND EXERCISES

1.   What is the inverse of the element R_φ in SO(2)? What is the identity element?

2.   Show that SO(2) is commutative, while SO(3) is not.

3.   Show that the group of rotations that leave a cube invariant is of order 24, as claimed in Section 18.1.

4.   Describe the group of rotations under which a right circular cylinder is invariant; same for a regular icosahedron.

5.   Derive the three laws at the beginning of this section from the group axioms.

6.   Determine which of the following are groups:
    (a)   The set of all nonzero complex numbers, under multiplication.
    (b)   The set of all nonzero $n \times n$ matrices under multiplication.
    (c)   The set of all positive rational numbers, under multiplication.
    (d)   The set of all positive irrational numbers, under multiplication.
    (e)   The set of all positive algebraic numbers, under multiplication.
    (f)   The set of all $n \times n$ matrices under addition.
    (g)   The set of all $n \times n$ matrices of the form e^A, under multiplication.
    (h)   The integers $1, 2, \ldots, p - 1$, under multiplication modulo p, p a prime.
    (i)   The integers $1, 2, \ldots, m - 1$, under multiplication modulo m, m composite.
    (j)   The set of all vectors in E^3, under vector addition.

(k)   The set of all nonzero vectors in E^3, under vector multiplication.
(l)   The set of all complex numbers z such that $|z| = 1$, under multiplication.
(m) The set of all $n \times n$ unitary matrices, under addition.
(n)   The set of all $n \times n$ unitary matrices, under multiplication.
(o)   The set of all Möbius transformations

$$z \to z' = \frac{az + b}{cz + d} \qquad (ad - bc \neq 0)$$

in the complex plane.

18.3 Isomorphism

If there is a one-to-one mapping φ of a group G onto a group G' such that

$$\varphi(a \circ b) = \varphi(a) \circ \varphi(b) \qquad (18.3\text{-}1)$$

for all a and b in G, then φ is an *isomorphism*, and the groups are *isomorphic*; in symbols, $G \cong G'$. [In (18.3-1), the first little circle denotes the law of composition in G, the second that in G'.] One says that products are mapped onto products. In this case, G and G' may be regarded as merely two different realizations of the same abstract group.

For example, if G is the set of numbers $\{1, i, -1, -i\}$ under multiplication, and G' is the set of matrices $\{I, A, B, C\}$ under matrix multiplication, where

$$I = \begin{pmatrix} 1 & 0 \\ 0 & 1 \end{pmatrix}, \qquad A = \begin{pmatrix} 0 & 1 \\ -1 & 0 \end{pmatrix}, \qquad B = \begin{pmatrix} -1 & 0 \\ 0 & -1 \end{pmatrix},$$

$$C = \begin{pmatrix} 0 & -1 \\ 1 & 0 \end{pmatrix},$$

then the mapping

$$\varphi: 1 \to I, \qquad i \to A, \qquad -1 \to B, \qquad -i \to C$$

is an isomorphism of G onto G'; the law (18.3-1) is easily verified for each of the 16 possible pairs (a, b) of elements of G. For example, $(-i) = (-1)(i)$; hence $\varphi(-i)$ ought to be $= \varphi(-1)\varphi(i)$, i.e., C ought to $= BA$, which in fact it is. It should be noted that the mapping

$$\varphi_1: 1 \to I, \qquad i \to C, \qquad -1 \to B, \qquad -i \to A$$

is another isomorphism of G onto G'.

If G is the group of all complex numbers z such that $|z| = 1$, under multiplication, then the mapping

$$\varphi: e^{i\theta} \to \begin{pmatrix} \cos\theta & -\sin\theta \\ \sin\theta & \cos\theta \end{pmatrix}$$

is an isomorphism of G onto the 2-dimensional rotation group SO(2).

An isomorphism of a group onto itself is an *automorphism*. An example in the group $\{I, A, B, C\}$ of matrices described above is the mapping

$$I \rightarrow I, \qquad A \rightarrow C, \qquad B \rightarrow B, \qquad C \rightarrow A.$$

An automorphism of SO(2) onto itself is given by

$$\begin{pmatrix} \cos\theta & -\sin\theta \\ \sin\theta & \cos\theta \end{pmatrix} \rightarrow \begin{pmatrix} \cos\theta & \sin\theta \\ -\sin\theta & \cos\theta \end{pmatrix}.$$

Any mapping φ (not necessarily one-to-one or onto) of a group G into a group G' such that (18.3-1) is satisfied is a *homomorphism*. If G is the group GL(n, C) of all $n \times n$ nonsingular complex matrices under multiplication, then the mapping $A \rightarrow \det A$ is a homomorphism of G onto the group of all nonzero complex numbers under multiplication. As a second example, let G be the group M_2 of all rigid motions in a plane, i.e., the group of all transformations of the form

$$T_{\theta, a, b} \begin{cases} x \rightarrow x' = x\cos\theta - y\sin\theta + a, \\ y \rightarrow y' = x\sin\theta + y\cos\theta + b, \end{cases} \tag{18.3-2}$$

where $0 \le \theta < 2\pi$ and where a and b are arbitrary real numbers. Then the mapping

$$T_{\theta, a, b} \rightarrow \begin{pmatrix} \cos\theta & -\sin\theta \\ \sin\theta & \cos\theta \end{pmatrix} \tag{18.3-3}$$

is a homomorphism of G onto SO(2) as can be seen by performing two transformations of the form (18.3-2) in succession. Since SO(2) is a subgroup of G (with $a = b = 0$), the mapping (18.3-3) may be regarded as a homomorphism of G into itself.

18.4 Permutation Groups

A *permutation* is a one-to-one mapping of a set C (usually finite) of objects or symbols onto itself. For example, if C consists of the first seven digits, $C = \{1, 2, \ldots, 7\}$, then a particular permutation is the mapping $\pi: j \rightarrow \pi(j)$, where the function $\pi(j)$ is given by

$$\pi(1) = 7, \qquad \pi(2) = 3, \qquad \pi(3) = 1, \qquad \pi(4) = 4$$
$$\pi(5) = 6, \qquad \pi(6) = 5, \qquad \pi(7) = 2.$$

This permutation is written in condensed notation as

$$\pi = \begin{pmatrix} 1 & 2 & 3 & 4 & 5 & 6 & 7 \\ 7 & 3 & 1 & 4 & 6 & 5 & 2 \end{pmatrix}, \tag{18.4-1}$$

where it is understood that each symbol in the upper row is mapped onto the symbol below it. A *cycle* is a permutation which can be obtained by arranging the symbols in a circle and mapping each symbol onto the one following it (say clockwise) around the circle, for example, as in Figure 18.1;

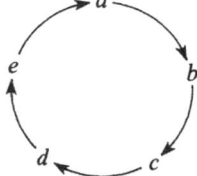

Figure 18.1   A cyclic permutation.

this cycle is written in a still further condensed notation as $(a\ b\ c\ d\ e)$, which is of course the same as $(b\ c\ d\ e\ a)$, etc. Any permutation of the set C can be expressed in terms of cyclic permutations of various subsets of C; for example, the permutation π given by (18.4-1) can be written as

$$\pi = (1723)(4)(56). \tag{18.4-2}$$

The *length* of a cycle is the number of symbols in it. Cycles of length 1 [e.g., (4)] are usually omitted, since they represent the identity mapping, in which nothing is permuted. A cycle of length 2 is a *transposition*; it simply interchanges two of the symbols and leaves the rest unaltered.

Any permutation can be expressed as the resultant of successive transpositions. For example, if

$$\pi_1 = (17), \qquad \pi_2 = (72), \qquad \pi_3 = (23), \qquad \pi_4 = (56),$$

then the permutation (18.4-2) can be written as

$$\pi = \pi_1 \circ \pi_2 \circ \pi_3 \circ \pi_4 = (17)(72)(23)(56), \tag{18.4-3}$$

where it is understood that the transpositions are to be performed in the order reading from right to left. The decomposition of a given permutation into transpositions is not unique, but it will now be proved that for a given permutation the number of transpositions is either always even or always odd.

Namely, let $f(\cdot\cdot\cdot)$ be the function of n real or complex variables defined by

$$f(x_1, \ldots, x_n) = \prod_{1 \le j < k \le n} (x_k - x_j); \tag{18.4-4}$$

let $\pi: j \to \pi(j)$ be a permutation of the integers $1, 2, \ldots, n$, and call

$$f_\pi(x_1, \ldots, x_n) = \prod_{1 \le j < k \le n} (x_{\pi(k)} - x_{\pi(j)}). \tag{18.4-5}$$

If $x_k - x_j$ is any one of the factors in (18.4-4), then either $x_k - x_j$ or $x_j - x_k$ appears precisely once as one of the factors in (18.4-5), so that either $f_\pi \equiv f$ or $f_\pi \equiv -f$; in these cases π is called an *even* or an *odd* permutation, respectively. This property is called the *parity* of the purmutation π. If two permutations have the same parity (i.e., if they are either both even or both odd), then their product or resultant is even; if they have opposite parities, their product is odd. Clearly, the transposition (12) is odd, for then all the factors of (18.4-5) have the same sign as in (18.4-4), except $(x_2 - x_1)$. Next, it is seen that the transposition $(1l) = (2l)(12)(2l)$ is necessarily odd, and finally that the general transposition $(jk) = (1j)(1k)(1j)$ is always odd. Hence, in any decomposition of an even (or odd) permutation into transpositions, the number of factors is always even (or always odd).

There are just as many odd as even permutations of a given set C, because if π_1 is any fixed odd permutation, then the mapping $\pi_1\pi \to \pi$ of even permutations onto odd ones is a one-to-one mapping.

It is clear that the set of all permutations of n symbols (including, of course, the identity permutation in which each symbol is mapped onto itself) is a group of order $n!$ under the usual law of composition of mappings (from right to left); this group is called the *symmetric group* on n symbols, and is denoted by $\mathscr{S}_n$. The subgroup of all $n!/2$ even permutations is called the *alternating group* on n symbols, and is denoted by $\mathscr{A}_n$.

18.5 Homomorphisms; Normal Subgroups

It is recalled that a *homomorphism* is a mapping $\varphi\colon G \to G'$ of a group G into a group G' such that $\varphi(ab) = \varphi(a)\varphi(b)$ for all a and b in G. (If φ is also one-to-one and onto, then it is an isomorphism.) Certain subsets (it will be seen that they are subgroups) of G and G' are defined as follows:

$$G_0 = \{x\colon \varphi(x) = e' \qquad (\text{identity of } G')\}$$
$$= kernel \text{ of the mapping} = \ker(\varphi)$$
$$G'_1 = \{\varphi(x)\colon x \in G\}$$
$$= image \text{ of } G \text{ under } \varphi = \varphi(G).$$

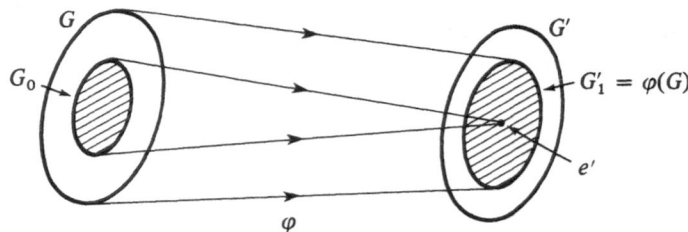

Figure 18.2   A group homomorphism.

The main interest is in the case in which the image $G'_1 = \varphi(G)$ is a simpler group than G (in this case φ cannot be one-to-one) but yet $\varphi(G)$ is not merely the trivial group $\{e'\}$. One may then regard the image as having the main features of G but without some of the fine detail. Insofar as $\varphi(G)$ approximates G, it does so accurately, in that the image of a product of two elements in G is always the product of their images.

Theorem. G_0 and G'_1 are subgroups of G and G' respectively; furthermore, $yxy^{-1} \in G_0$ for any x in G_0 and any y in G.

The proof of this theorem is given in some detail, to serve as a sort of model. Proofs of the other theorems in this Chapter are mostly left to the reader.

(1) PROOF (that $G_0 < G$). It must be proved that the group axioms are satisfied in G_0. First, suppose that x and y are in G_0, i.e., that $\varphi(x) = e'$ and $\varphi(y) = e'$. It must be shown that xy is in G_0, but $\varphi(xy) = \varphi(x)\varphi(y)$ (by the homomorphism property of the mapping φ) $= e'e' = e'$ (by one of the properties of the identity in any group). Therefore, $\varphi(xy) = e'$, but G_0 was defined as the set of *all* elements in G that are mapped onto e'; hence xy is in G_0. Second, the associative axiom is automatically satisfied in G_0 because it is satisfied in all of G [if x, y, and z are in G_0, then $(xy)z = x(yz)$, because x, y, and z are in G, which is a group]. The third axiom will be verified in the alternative form given in Section 18.1: *Identity*: G_0 contains the identity e, for if a is any element of G, then $\varphi(a)\varphi(e) = \varphi(ae) = \varphi(a)$; therefore $\varphi(e) = e'$. *Inverse*: Suppose that a is in G_0 (hence, $\varphi(a) = e'$). Then, $e' = \varphi(e) = \varphi(aa^{-1}) = \varphi(a)\varphi(a^{-1}) = e'\varphi(a^{-1}) = \varphi(a^{-1})$; therefore a^{-1} is in G_0.

(2) PROOF (that $G_1' < G'$). It will be similarly shown that the group axioms are satisfied in G_1'. First, if x' and y' are in G_1', then $x' = \varphi(x)$ and $y' = \varphi(y)$, for some x and y in G, but $x'y' = \varphi(x)\varphi(y) = \varphi(xy)$, and G_1' is defined as the set of all elements of G' which are $= \varphi(z)$ for some z in G; therefore $x'y'$ is in G_1'. Second, multiplication is associative in G_1', because it is so in all G'. Third, it has been shown above that $\varphi(e) = e'$; hence e' is in G_1'. Lastly, if x' is in G_1', then $x' = \varphi(x)$ for some x in G. Now, $x'\varphi(x^{-1}) = \varphi(x)\varphi(x^{-1}) = \varphi(xx^{-1}) = \varphi(e) = e'$, which shows that $\varphi(x^{-1})$ is x'^{-1}; hence x'^{-1} is in G_1'.

(3) PROOF (that yxy^{-1} is in G_0 if x is in G_0, while y is arbitrary in G).

$$\varphi(yxy^{-1}) = \varphi(y)\varphi(x)\varphi(y^{-1}) = \varphi(y)e'\varphi(y^{-1}) = \varphi(y)\varphi(y^{-1}) = \varphi(y)(\varphi(y))^{-1} = e';$$

therefore, yxy^{-1} is in G_0.

Note. All the conclusions hold, even if G' is not a group, but merely a set in which a product $x'y'$ is defined. In particular, it then follows that the subset G_1' is necessarily a group, even though G' may not be. All the arguments are unchanged except the second step in part (2). To prove associativity in G_1', let x', y', and z' be elements in G_1', hence elements of the form $\varphi(x)$, $\varphi(y)$, and $\varphi(z)$. Then $(x'y')z' = (\varphi(x)\varphi(y))\varphi(z) = \varphi(xy)\varphi(z) = \varphi'((xy)z) = \varphi(x(yz))$ because multiplication is associative in G) $= \varphi(x)\varphi(yz) = \varphi(x)(\varphi(y)\varphi(z)) = x'(y'z')$.

Any subgroup G_0 of a group G such that $yxy^{-1} \in G_0$ for any $x \in G_0$, any $y \in G$, is called a *normal subgroup* (or *normal divisor*) or *invariant* or *self-conjugate* subgroup of G; in symbols, $G_0 \lhd G$. Clearly, a homomorphism of the kind we are interested in can exist only if G contains a nontrivial ($\neq \{e\}$) proper ($\neq G$ itself) normal subgroup G_0. If G contains no such, it is called *simple*, because it cannot be mapped homomorphically onto any simpler nontrivial group. It will be shown in Section 18.8 that, conversely, if G contains a normal subgroup, one can construct a homomorphism of which G_0 is the kernel, and that all (in the sense of isomorphism) homomorphic images of G are obtained by this construction.

For any element x, the elements of the form yxy^{-1} ($y \in G$) are the *conjugates* of x. A subgroup is normal if and only if it contains all the conjugates of all its members. In an Abelian group, yxy^{-1} is always equal to x, hence every subgroup is normal.

18.6   Cosets

Let G_0 be a subgroup of G, and y any fixed element of G. The subsets

$$S_y^{(l)} = \{y \circ x : x \in G_0\}, \qquad S_y^{(r)} = \{x \circ y : x \in G_0\}$$

are called *left* and *right* cosets, respectively, of G_0 in G. The element y is called a *representative* of $S_y^{(l)}$ (or of $S_y^{(r)}$). Any member of a coset may be taken as its representative. G_0 itself is both a left and a right coset; its representative may be taken as the identity element e of the group. The association $x \rightarrow y \circ x$, for fixed y, all x in G_0, is a one-to-one mapping of G_0 onto $S_y^{(l)}$ (and similarly for right cosets); hence, each coset contains the same number (finite or infinite) of elements as G_0. Furthermore, it is easily proved that any two left cosets (or any two right cosets) are either identical or disjoint, so that the number of elements in G (if finite) is equal to the number in G_0 times the number of (say left) cosets (including G_0 itself). *Lagrange's theorem* for finite groups follows: *the order of a subgroup divides the order of the group.* It follows, for example, that if the number of elements in G is prime, then G has no subgroups other than $\{e\}$ and G itself. The cosets $S_y^{(l)}$ and $S_y^{(r)}$ are also denoted by yG_0 and G_0y. Note that the cosets, except G_0 itself, are *not* subgroups.

18.7   Factor Groups

Theorem 1.   *A subgroup G_0 in G is normal if and only if every left coset (of G_0 in G) is a right coset, and conversely; the coset containing y is then denoted by S_y. (The proof is elementary and is left as an exercise.)*

Definition.   If S_1 and S_2 are any two subsets (not necessarily subgroups or cosets) of a group, their *product* is defined as the subset

$$S_1 S_2 \stackrel{\text{def}}{=} \{z_1 \circ z_2 : z_1 \in S_1, z_2 \in S_2\}; \qquad (18.7\text{-}1)$$

note that generally $S_1 S_2 \neq S_2 S_1$.

If S_1 and S_2 are cosets (left or right) their product is not generally a coset (it is usually larger than any coset), unless $G_0 \triangleleft G$.

Theorem 2.   *If G_0 is a normal subgroup of G ($G_0 \triangleleft G$), then $S_{y_1} S_{y_2} = S_{y_1 y_2}$ for all y_1 and y_2 in G. Conversely, if G_0 is a subgroup such that the product of any two (say left) cosets is always a (left) coset, then $G_0 \triangleleft G$ (in which*

*case the distinction between right and left cosets disappears). Under these
circumstances, the collection of cosets*

$$\{S_y: y \in G\}, \quad \circ = \text{multiplication defined by (18.7-1)},$$

is a group, called the factor (*or* quotient) group *of G with respect to* G_0
and is denoted by G/G_0. *Furthermore, the mapping* $\varphi_n: G \to G/G_0$ *defined
by* $\varphi_n(y) = S_y$ (*each element of G is mapped onto the coset in which it resides*)
is a homomorphism called the natural homomorphism *of G onto* G/G_0.

According to this theorem, whose proof is also left to the reader, one
can always construct a homomorphic image and a homomorphism corresponding to any normal subgroup G_0. The theorem of the next section shows
that these are essentially the only homomorphisms of the given group G.

18.8 The Law of Homomorphism

Theorem. *Let* $\psi: G \to G'$ *be any homomorphism. Let* G_0 *be its kernel,
and let* $\psi(G)$ *be the image of G in G' under the mapping* ψ. *Let* S_y *denote the
coset of* G_0 *that contains y. Then* $\psi(G)$ *and the factor group* G/G_0 *are isomorphic, and, specifically, the mapping* $\Psi: G/G_0 \to \psi(G)$ *given by*

$$\Psi(S_y) = \psi(y) \tag{18.8-1}$$

*is (1) well-defined (i.e., independent of the particular choice of the repre-
sentative y of* S_y) *and is (2) an isomorphism. Schematically,*

$$\psi(G) \leftarrow G \to G/G_0$$
$$\psi \qquad \varphi_n$$
$$\text{(hom)} \quad \text{(hom)}$$
$$\Psi\text{(iso)}$$

The reader is urged to complete the proof in detail, but let us point out
what has to be proved. To prove that Ψ is well defined, one must show that
if $S_{y_1} = S_{y_2}$, then $\psi(y_1) = \psi(y_2)$. To show that Ψ is one-to-one, we must show
that if $\psi(y_1) = \psi(y_2)$, then $S_{y_1} = S_{y_2}$. To show that Ψ has the homomorphism
property, one must show that

$$\Psi(S_{y_1}S_{y_2}) = \Psi(s_{y_1})\Psi(s_{y_2}).$$

Lastly, it is obvious that Ψ is onto, for any element of $\psi(G)$ is $\psi(y)$ for some
y in G.

18.9 The Structure of Cyclic Groups

As noted in Section 18.6, a group of prime order p has no nontrivial proper
subgroups, hence is necessarily cyclic (hence Abelian), and can be generated
by any of its elements (except the identity) because the order of any element

(that is, the order of the subgroup generated by that element) must be either 1 or p, by Lagrange's theorem. (The only element of order 1 in a group is the identity e, for to say that $a^l = e$, when $l = 1$, is equivalent to saying that $a = e$). If $G = \{e, a, a^2, \ldots, a^{n-1}\}$ is a cyclic group of order n, and m is a divisor of n, then the elements $e, a^m, a^{2m}, \ldots$ constitute a cyclic subgroup of order n/m, and these are the only subgroups of G. In an infinite cyclic group, any element that is $\neq e$ generates an infinite cyclic subgroup; these are the only nontrivial subgroups, and they are all distinct, but isomorphic.

18.10 Translations, Inner Automorphisms

If a is an element of a group G, we denote by T_a the mapping of G onto itself given by $x \to ax$ ($x \in G$); this is a *left translation* in G; a mapping of the form $x \to xa$, for fixed a, is a *right translation*.

EXERCISE

1.   Show that the set $\mathcal{T}$ of all left translations in G is itself a group under the usual law for the composition of mappings, and that this group $\mathcal{T}$ is isomorphic to G.

N.B.1. Any homomorphism of a group G (abstract or otherwise) onto a group of mappings is called a *representation* of G. The isomorphism $G \cong \mathcal{T}$ is called the *regular representation* of G. If a representation is an isomorphism (not merely a homomorphism), it is called *faithful*. The regular representation is faithful.

N.B.2. If G is a finite group, the mapping T_a is a permutation of the elements of G; therefore, $\mathcal{T}$ is some subgroup of $\mathcal{S}_n$, where n is the order of G, i.e., *any finite group is isomorphic to some group of permutations.* (This is *Cayley's theorem.*)

If a is an element of a group G, we denote by A_a the mapping of G onto itself given by $x \to axa^{-1}$.

EXERCISE

2(a).   Show that the set $\mathcal{I}$ of all the mappings A_a in G is itself a group, under the usual law of composition of mappings.
   (b).   If $G = \mathcal{A}_3$, find $\mathcal{I}$.
   (c).   If $G = \mathcal{A}_4$, find $\mathcal{I}$.

N.B.3. $\mathcal{I}$ is in any case some subgroup of $\mathcal{S}_{6/2} = \mathcal{S}_3$ for exercise 2(b) and some subgroup of $\mathcal{S}_{12}$ for exercise 2(c).

N.B.4. "Find $\mathcal{I}$" means to identify $\mathcal{I}$ as being isomorphic to some known group.

N.B.5. Each of the mappings A_a is an automorphism called an *inner automorphism* of G: (1) It is one-to-one and onto, because the equation $axa^{-1} = y$ can always be solved for a unique $x(x = a^{-1}ya)$; (2) it has the homomorphism property that products are mapped into products, because $axwa^{-1} = axa^{-1}awa^{-1}$. $\mathscr{I}$ is called the *inner automorphism group* of G.

Inner automorphisms of $\mathscr{S}_n$ have a special property. Let $\pi \in \mathscr{S}_n$, and let π be written as a product of independent cycles, as in (18.4-2) [two cycles are *independent* if they contain no common symbol; (173) and (24) are independent, but (173) and (34) are not], the longer cycles being written first and the cycles of length 1 being included. Then the lengths of the cycles constitute a *partition* of n, that is, a set of positive integers whose sum is n. The special property referred to is that the image of π under an inner automorphism of $\mathscr{S}_n$ ($\pi \to \sigma\pi\sigma^{-1}$, where σ is a given element of $\mathscr{S}_n$) always corresponds to the same partition of n as π itself, because if $(a, b, \ldots, f)$ is any cycle, then $\sigma(a, b, \ldots, f)\sigma^{-1}$ is the cycle $(\sigma(a), \sigma(b), \ldots, \sigma(f))$.

If a and x are in a group G, then the element axa^{-1} is called *conjugate* to x. The inner automorphism $A_a: x \to axa^{-1}$ (a fixed) maps each group element x onto one of its conjugates. If G_0 is a subgroup of G, then the set

$$G_0' = \{axa^{-1}: x \in G_0\}$$

is also a subgroup, and is often denoted by aG_0a^{-1}; it is said to be *conjugate to* G_0. If $aG_0a^{-1} = G_0$, for all a in G, then $G_0 \triangleleft G$. Hence, normal subgroups are sometimes called "self-conjugate" subgroups. The conjugate elements of $\mathscr{S}_n$ are those that have the same structure when written as products of independent cycles, e.g., (1732)(56)(4) and (4531)(76)(2).

18.11 The Subgroups of $\mathscr{S}_4$

The symmetric group $\mathscr{S}_4$ has 22 subgroups in addition to the trivial subgroup $\{e\}$ and G itself. Arranged in classes of conjugate subgroups, they are:

i. $\{e, (12)\}$, etc—six subgroups of this type.
ii. $\{e, (123), (132)\}$, etc.—four subgroups.
iii. $\{e, (1234), (13)(24), (1432)\}$, etc.—three subgroups.
iv. $\{e, (12)(34)\}$, etc.—three subgroups.
v. $\{e, (12)(34), (13)(24), (14)(23)\} = V_4$—one subgroup.
vi. $\{e, (123), (124), (134), (234), (321), (421), (431), (432), (12)(34), (13)(24), (14)(23))\} = \mathscr{A}_4$—one subgroup.
vii. $\{e, (12), (13), (23), (123), (321)\}$—four subgroups.

Any subgroup in a given line of the table can be obtained from any other subgroup in the same line by an inner automorphism of the whole group $\mathscr{S}_4$. For instance, under the mapping $\pi \to (12)\pi(12)$, the elements of the group $\{e, (134), (431)\}$ go into the elements of the group $\{e, (234), (432)\}$. Therefore, the only normal (i.e., invariant, i.e., self-conjugate) subgroups are V_4 and $\mathscr{A}_4$, each of which occupies a line of the table by itself. However, each of the subgroups in line iv is a normal subgroup of V_4 (which shows, incidentally, that $G_1 \lhd G_2 \lhd G_3$ does not imply $G_1 \lhd G_3$). A complete so-called composition series (see below) of $\mathscr{S}_4$ is the series

$$\{e\} \lhd \{e, (12)(34)\} \lhd V_4 \lhd \mathscr{A}_4 \lhd \mathscr{S}_4. \qquad (18.11\text{-}1)$$

For $n \geq 5$, $\mathscr{A}_n$ is a simple group (it has no nontrivial proper normal subgroups), so the composition series is merely

$$\{e\} \lhd \mathscr{A}_n \lhd \mathscr{S}_n.$$

EXERCISE

1. Show that $\mathscr{A}_5$ is a simple group. Outline of the proof: Assume that $G_0 \lhd \mathscr{A}_5$, but $G_0 \neq \{e\}$; then, it must be proved that $G_0 = \mathscr{A}_5$. Show that G_0 must contain an element π of one of the types

(a) $(a \quad b \quad c)$,
(b) $(a \quad b)(c \quad d)$,
(c) $(a \quad b \quad c \quad d \quad e)$.

Then it contains all elements $\sigma\pi\sigma^{-1}$, where $\sigma \in \mathscr{A}_5$; show that it contains all elements of the type π. Show that if G_0 contains all elements of one of the above types, then it contains elements (hence all elements) of both the other types, e.g., if G_0 contains type (a), then it contains $(1 \ 2 \ 3)(2 \ 3 \ 4) = (21)(34)$. Why does this proof fail for $\mathscr{A}_4$? The simplicity of $\mathscr{A}_5$ is a key step in Galois proof that the quintic equation cannot be solved by radicals.

A *composition series* of a group G is a sequence of subgroups $\{G_i\}$ such that

$$\{e\} = G_0 \lhd G_1 \lhd \cdots \lhd G_k = G, \qquad (18.11\text{-}2)$$

where $G_i \neq G_{i+1}$, and where no refinement is possible, i.e., if $G_i \lhd H \lhd G_{i+1}$, then either $H = G_i$ or $H = G_{i+1}$. A finite group always has a composition series. If G is an infinite group, it may happen (see Exercise 3, below) that every series of the above type has a refinement. The famous *Jordan–Hölder theorem* says that if G has a composition series (18.11-2), and if

$$\{e\} = H_0 \lhd H_1 \lhd \cdots \lhd H_l = G$$

is any other composition series for G, then (1) $l = k$, and (2) although the subgroups $H_1, \ldots, H_{k-1}$ may be different from $G_1, \ldots, G_{k-1}$, at least the factor groups

$$H_1/H_0, H_2/H_1, \ldots, H_k/H_{k-1} \qquad (18.11\text{-}3)$$

are the same (except for order) as

$$G_1/G_0, G_2/G_1, \ldots, G_k/G_{k-1}; \qquad (18.11\text{-}4)$$

that is, the list (18.11-3) can be arranged in such an order that each group in it is isomorphic to the corresponding group in the list (18.11-4).

EXERCISES

2. How does the Jordan–Hölder theorem apply to the composition series (18.11-1) for $\mathscr{S}_4$?

3. Show that the infinite cyclic group has no composition series.

4. Let $\mathscr{S}_\infty$ denote the set of all finite permutations of the positive integers $\{1, 2, \ldots\}$. (Each permutation in $\mathscr{S}_\infty$ maps all but a finite number of these integers onto themselves.) Show that $\mathscr{S}_\infty$ has the composition series $\{e\} \lhd \mathscr{A}_\infty \lhd \mathscr{S}_\infty$.

18.12 Generators and Relations; Free Groups

Let G be a group; a subset $S = \{a, b, \ldots\}$ of the elements of G is said to *generate* G if every element of G can be written as a (finite) product of elements, each of which is either an element in S or the inverse of an element in S. (Equivalently, any g in G is a product of powers of elements in S—the powers of an element were defined in Section 18.2.) To describe a group completely, one must not only say what its generators are, but also give various relations among them. For example, a cyclic group of order n can be described by one generator a and one relation $a^n = e$.

A group without any relations is called a free group and is constructed as follows: Let S be a (finite or infinite) set of letters, $S = \{a, b, \ldots\}$, called *generators*. A *word* is a finite string of symbols $x_1 x_2 \cdots x_k$, where each x_i is either one of the letters in S or a symbol of the form α^{-1}, where α is in S. Two words are *equal* if one of them can be obtained from the other by inserting and deleting symbol pairs of the form $\alpha\alpha^{-1}$ and $\alpha^{-1}\alpha$ (α in S). It is obviously desirable to delete as many such pairs as possible; this may result in the empty word—the word with no letters in it; that word is denoted by e, and it is understood that "e" is *not* one of the letters in S. (For example, $abc^{-1}cb^{-1}a^{-1} = e$.) The *product* of two words is obtained by simply stringing them together: If $w = x_1 x_2 \ldots x_k$ and $u = y_1 y_2 \ldots y_l$, then

$$w \circ u = x_1 \ldots x_k y_1 \ldots y_l, \qquad u \circ w = y_1 \ldots y_l x_1 \ldots x_k.$$

It is an elementary exercise to verify that the set of all words, using a given set S of generators, is a group G under this definition of product. It is called the *free group* generated by S. The identity element is the empty word e, and the inverse of $x_1 x_2 \ldots x_k$ is the word $y_k y_{k-1} \ldots y_1$, where $y_i = \alpha^{-1}$ or α according as x_i is α or α^{-1}.

Relations can be established among the elements of the group (it is then no longer free) by means of equations $w_1 = e$, $w_2 = e$, etc., where w_1, w_2, etc. are certain words. The relations establish a structure in the group.

If among the relations we have $aba^{-1}b^{-1} = e$ (which is equivalent to $ab = ba$) for every pair a, b in S, then the group is Abelian. If these are the *only* relations, then G is called a *free Abelian group*. Free groups and free Abelian groups appear in Section 23.7, in the study of the kinds of multiple connectedness that a manifold can possess. The structure of a free group or of a free Abelian group is determined solely by the number of generators.

Any finite group G is equivalent (i.e., isomorphic) to a group defined by generators and relations: S can be taken as the set of all group elements in G, and the relations taken so as to give all the information provided by the multiplication table of the group; whenever the table says that $ab = c$, then one of the relations is $abc^{-1} = e$.

If a group is defined by finitely many generators and finitely many relations, it is called *finitely presented*; it may still be a group of infinite order, and such groups have provided many intricate and difficult problems for current research. The so-called word problem, though by now not exactly new, illustrates the flavor of the subject. In addition to the defining relations $w_1 = e, \ldots, w_n = e$, there are always other equations of the form $w = e$; for instance, w can be $w_1 w_2$, or it can be $w_0 w_1 w_0^{-1}$, where w_0 is an arbitrary word. The *word problem*, which was first posed in 1912, is this: Given a finitely presented group, find a procedure (i.e., an algorithm) such that if an arbitrary word w in the generators of the group is given, the procedure will allow one to decide, in a finite number of steps, whether the equation $w = e$ is true or false. W. Magnus showed in 1932 how to produce such an algorithm for any group with a single defining relation. In 1955, Novikov gave a (very long) proof that the word problem is in general unsolvable, and by now groups are known, defined by a quite small number of generators and relations, for which it can be proved that no algorithm of the desired kind exists.

18.13 Multiply Periodic Functions and Crystals

A crystal, ideally, is composed of an enormous number of identical unit structures, called cells, arranged in a triply periodic array or lattice in space. If $f(\mathbf{x})$ is the density of mass or charge or a similar quantity within the structure, then $f(\mathbf{x})$ is a triply periodic function. Generally, a function $f(x_1, \ldots, x_n) = f(\mathbf{x})$ of n real variables is called *n-tuply periodic* if there are n linearly independent vectors $\mathbf{v}(1), \ldots, \mathbf{v}(n)$, i.e., vectors for which

$$\det \begin{pmatrix} v_1(1) & \cdots & v_1(n) \\ \vdots & & \vdots \\ v_n(1) & \cdots & v_n(n) \end{pmatrix} \neq 0, \tag{18.13-1}$$

such that $f(\mathbf{x})$ satisfies the equations

$$f(\mathbf{x} + \mathbf{v}(1)) = f(\mathbf{x}),$$
$$f(\mathbf{x} + \mathbf{v}(2)) = f(\mathbf{x}),$$
$$\vdots$$
$$f(\mathbf{x} + \mathbf{v}(n)) = f(\mathbf{x}) \tag{18.13-2}$$

for all $\mathbf{x}$. The vectors $\mathbf{v}(i)$ are called *periods* of $f(\mathbf{x})$; if $\mathbf{w}$ is any vector of the form

$$\mathbf{w} = m_1 \mathbf{v}(1) + \cdots + m_n \mathbf{v}(n), \tag{18.13-3}$$

where the m_i are integers (positive, negative, or zero), then $f(\mathbf{x}) = f(\mathbf{x} + \mathbf{w})$ for all $\mathbf{x}$; hence $\mathbf{w}$ is also a period of $f(\mathbf{x})$. It is furthermore assumed that every period of $f(\mathbf{x})$ is of the form (18.13-3) with integer coefficients; then, the vectors $\mathbf{v}(1), \ldots, \mathbf{v}(n)$ are said to constitute a *fundamental* set of periods. Certain functions may not have a fundamental set, even though they are periodic in the strict sense, for example, constant functions and functions that are periodic in some coordinates and independent of the others. Such functions will be called *degenerate*, and are excluded on the physical ground that each atom occupies a certain volume, and that functions like the potential and the charge density vary from the center of an atom to the outside, so that some variation is unavoidable in any given direction in space. A multiply periodic function is called *nondegenerate* if it has a fundamental set of periods.

The set of points $\mathbf{x}$ in $\mathbb{R}^n$ determined by

$$\mathbf{x} = m_1\mathbf{v}(1) + \cdots + m_n\mathbf{v}(n), \tag{18.13-4}$$

where the m_i are integers, is called the *lattice* of $f(\mathbf{x})$.

If new vectors $\mathbf{v}'(1), \ldots, \mathbf{v}'(n)$ are given by

$$\mathbf{v}'(j) = m_{j1}\mathbf{v}(1) + \cdots + m_{jn}\mathbf{v}(n), \tag{18.13-5}$$

where the m_{jk} are integers such that

$$\det\begin{pmatrix} m_{11} & \cdots & m_{1n} \\ \vdots & & \vdots \\ m_{n1} & \cdots & m_{nn} \end{pmatrix} = \pm 1, \tag{18.13-6}$$

then $\mathbf{v}'(1), \ldots, \mathbf{v}'(n)$ also constitute a fundamental set of periods, because when (18.13-5) is solved by the use of determinants, the $\mathbf{v}(j)$ are seen to be linear combinations of the $\mathbf{v}'(j)$ with integer coefficients. Both fundamental sets generate the same lattice.

18.14 The Space and Point Groups

If, for any fixed $\mathbf{w}$, $T_\mathbf{w}$ denotes the translation

$$T_\mathbf{w}: \mathbf{x} \to \mathbf{x} + \mathbf{w} \qquad \text{(for all } \mathbf{x}\text{)} \tag{18.14-1}$$

of $\mathbb{R}^n$ onto itself, then a multiply periodic function $f(\mathbf{x})$ is invariant under all transformations of the Abelian group

$$\mathcal{T} = \{T_\mathbf{w}: \mathbf{w} \text{ is a period of } f(\mathbf{x})\}, \tag{18.14-2}$$

which is called the *translation group* of $f(\mathbf{x})$. The function $f(\mathbf{x})$ may, of course, also be invariant under other transformations, such as certain rotations about certain axes, reflections in certain planes, and the like. The group G_s of all (homogeneous and inhomogeneous) linear transformations of $\mathbb{R}^n$ under which $f(\mathbf{x})$ is invariant is called the *space group of* $f(\mathbf{x})$ or, if $f(\mathbf{x})$ represents a crystal, the *space group of the crystal*.

An element of G_s is a transformation of the form

$$\mathbf{x} \to \mathbf{x}' = M\mathbf{x} + \boldsymbol{\xi}, \tag{18.14-3}$$

where M is a nonsingular matrix and ξ is a vector; the transformation is denoted by (ξ, M). The law of composition in the group G_s is found by applying two such transformations in succession:

$$\mathbf{x}'' = M_1\mathbf{x}' + \xi_1 = M_1(M_2\mathbf{x} + \xi_2) + \xi_1.$$

It is seen that

$$(\xi_1, M_1) \circ (\xi_2, M_2) = (M_1\xi_2 + \xi_1, M_1M_2). \qquad (18.14\text{-}4)$$

A transformation (18.14-3) under which $f(\mathbf{x})$ is invariant is called a *symmetry operation* of $f(\mathbf{x})$.

The transformation (18.14-3) generally combines translation, rotation, dilation, and shear. However, it is easy to see that if $f(\mathbf{x})$ is continuous and nondegenerate (hence, in particular if it represents a real crystal), dilation and shear are ruled out. To show the general nature of the argument, we discuss shear in two dimensions: The square lattice of points with integer coordinates in the x, y plane is invariant under the translation group $\mathcal{T}$ consisting of translations $x \to x + k, y \to y + l$ (k, l integers) and also under various transformations involving shear, such as the transformation

$$S_n: x \to x + ny, y \to y \qquad (n \text{ integer}). \qquad (18.14\text{-}5)$$

If $f(x, y)$ is the function that is equal to 1 at the lattice points (x, y integers) and equal to 0 elsewhere, then f is also invariant under (18.14-5). However, a *continuous* nondegenerate doubly periodic function $f(x, y)$ cannot be invariant under (18.14-5). If it were, then the identity

$$f(x + ny - k, y) \equiv f(x, y)$$

would hold, for all n and k; for y irrational, the numbers $ny - k$ are dense on $\mathbb{R}$, so by continuity $f(x, y)$ would have to be independent of x for all irrational y, hence (by continuity again) for all y; hence $f(x, y)$ would be degenerate. Therefore shear must be excluded. By arguments of this sort, it is concluded that the matrix M in (18.14-3) must be an orthogonal matrix.

The set of all orthogonal matrices M such that (ξ, M) is in G_s for some ξ is also a group; it is called the *point group* of $f(\mathbf{x})$ and is denoted by G_p. Clearly the mapping

$$\varphi: G_s \to G_p: (\xi, M) \to M \qquad (18.4\text{-}6)$$

is a homomorphism, whose kernel is $\mathcal{T}$; hence, by the homomorphism law for groups, $\mathcal{T}$ is a normal subgroup of G_s, and G_p is isomorphic to the factor group $G_s/\mathcal{T}$.

EXERCISE

1. Using (18.14-4), find the formula for $(\xi, M)^{-1}$. Then verify directly that $\mathcal{T}$ is a normal subgroup of G_s by showing that, if (ξ, I) is any pure translation (I being the unit matrix), then any group element of the form $(\eta, M)(\xi, I)(\eta, M)^{-1}$ is also a pure translation.

In crystallography, much information about a crystal structure can be given by specifying G_p and $\mathcal{T}$ (the latter by describing the lattice it generates). However, that does not generally give as much information as is given by the space group G_s. In particular, G_s may or may not contain G_p as a subgroup, because G_s may contain (ξ, M) for some $\xi \neq 0$ but not $(0, M)$.

Note. $\mathcal{T}$, as an abstract group, is isomorphic to the free Abelian group on n generators, hence gives no information. However, the lattice generated by $\mathcal{T}$, when a fundamental set of periods is given, does give information about $f(\mathbf{x})$. The space group under which the lattice is transformed into itself contains the space group of $f(\mathbf{x})$ as a subgroup.

For $n = 3$, a detailed description of the possible symmetry operations and of the space and point groups, is contained in the *International Tables for X-ray Crystallography*, Henry and Lonsdale (1965). The symmetry operations are: pure translation, pure rotation, reflection in a plane, reflection together with a rotation about an axis perpendicular to the plane of reflection, reflection together with a translation parallel to the plane, and a rotation with a translation parallel to the axis of rotation. Whenever a rotation occurs, the possible angles of rotation are $\pm 2\pi/n$, where $n = 1, 2, 3, 4$, or 6, as shown for the case of pure rotations by Exercises 2 and 3 below. There are 32 point groups, 14 types of lattice, and 230 space groups.

Exercises

2. (The purpose of this exercise is to show that the only possible pure rotational symmetries of a 2-dimensional crystal are ones with an n-fold axis, where $n = 1, 2, 3, 4$, or 6.) Consider a nondegenerate doubly periodic function $f(x, y)$, and write it as $f(z)$, a (nonanalytic) real function of the complex variable $z = x + iy$. Let α and β be a fundamental pair of period; then $\mathrm{Re}(\alpha/\beta) \neq 0$, and $f(z + n\alpha + m\beta) \equiv f(z)$, when n and m are integers. By suitable scaling and suitable orientation of the x and y axes, take $\beta = 1$ for simplicity. Assume that $f(z)$ is also invariant under a rotation $z \rightarrow e^{i\theta}z$. From the equations

$$f(z) = f(ze^{i\theta}),$$

$$f(z + 1) = f(ze^{i\theta} + e^{i\theta}),$$

$$f(z + \alpha) = f(ze^{i\theta} + \alpha e^{i\theta}),$$

conclude that $e^{i\theta}$ and $\alpha e^{i\theta}$ are also periods of $f(z)$. Show from that that α satisfies the equation

$$r\alpha^2 + (s - p)\alpha - q = 0,$$

$$-r\alpha^2 + (s - p)\alpha - q = 0,$$

where p, q, r, and s are integers such that

$$ps - rq = 1,$$

and hence that $e^{i\theta}$ is given by

$$e^{i\theta} = \frac{l \pm \sqrt{l^2 - 4}}{2}, \qquad l = p + s.$$

For θ real, l must be in $[-2, 2]$. Conclude that the possible values of θ are 0, $\pm \pi/3$, $\pm \pi/2$, $\pm 2\pi/3$, $\pm \pi$.

3. Extend the conclusion of Exercise 2 to the 3-dimensional case, as follows: Assume that the function $f(x, y, z) = f(\mathbf{x})$ is triply periodic and that $\{\mathbf{u}, \mathbf{v}, \mathbf{w}\}$ is a fundamental set of periods. Suppose, furthermore, that $f(\mathbf{x})$ is invariant under a rotation through an angle θ about some axis in space. By choosing the origin to lie on the axis, the rotation can be written as $\mathbf{x} \rightarrow R\mathbf{x}$, where R is a 3×3 orthogonal matrix of determinant $= 1$. Show that the vectors

$$\mathbf{u}' = R\mathbf{u} - \mathbf{u},$$

$$\mathbf{v}' = R\mathbf{v} - \mathbf{v},$$

$$\mathbf{w}' = R\mathbf{w} - \mathbf{w},$$

are all periods of $f(\mathbf{x})$, are all perpendicular to the axis of rotation, and are not collinear. It follows that $f(\mathbf{x})$ is doubly periodic in any plane perpendicular to the axis of rotation, hence Exercise 2 applies.

In some books, the restrictions to 2-, 3-, 4-, and 6-fold rotation axes is derived from a somewhat mysterious "principle of rational indices," which is said to be of empirical origin. We have seen, however, that the restriction follows directly from the existence of a triply periodic structure; hence, the "principle of rational indices" is unnecessary.

18.15 Direct and Semidirect Products of Groups; Symmorphic Space Groups

If G_0 is a normal subgroup of G, one cannot in general think of G as a product of G_0 and G/G_0; in fact, G does not in general even contain a subgroup isomorphic to G/G_0. Below, two exceptions to that general rule are discussed.

In the simplest case, H and K are subgroups of G such that every g in G can be written uniquely as hk, with h in H and k in K, and such that every h in H commutes with every k in K. Then, G is said to be the *direct product* of H and K. In symbols, $G = H \times K$ or $K \times H$. The identity e is the only element in $H \cap K$ (for if any other a were in $H \cap K$, it would have two representations in the form hk, namely ae and ea). Furthermore, H and K are normal subgroups of G, for if $a = hk_1$ is any element of G and k_2 is any element of K, then $ak_2a^{-1} = hk_1k_2k_1^{-1}h^{-1}$ but this is $= k_1k_2k_1^{-1}$, because h commutes with k_1, k_2, and k_1^{-1}; hence, ak_2a^{-1} is in K, and $K \lhd G$; similarly, $H \lhd G$. Furthermore, the factor group G/H is isomorphic to K, and G/K is isomorphic to H, because any element of G/H is a coset $aH = \{ah: h \in H\}$ and can be written uniquely as a coset kH (where $a = kh$, $k \in K$, and $h \in H$); the mapping $k \rightarrow kH$ is clearly an isomorphism of K onto G/H, because $k_1Hk_2H = k_1k_2H$.

Another point of view is to assume that H_0 and K_0 are arbitrary given groups and to construct a group G, called their *direct product*, whose elements are pairs (h, k), where $h \in H_0$ and $k \in K_0$, with the law of composition

$$(h_1, k_1) \circ (h_2, k_2) \overset{\text{def}}{=} (h_1h_2, k_1k_2). \tag{18.15-1}$$

The identity element of G is the pair (e, e'), where e and e' are the identity elements of H_0 and of K_0, respectively; furthermore, $(h, k)^{-1} = (h^{-1}, k^{-1})$. Let H and K be the subsets of G defined as

$$H = \{(h, e'): h \in H_0\} \quad \text{and} \quad K = \{(e, k): k \in K_0\}.$$

It is easy to verify that G, H, and K are groups, that H and K are normal subgroups of G, that $H \cong H_0$ and $K \cong K_0$, and lastly that $G = H \times K$.

EXERCISE

1. Verify the statements made in the last sentence.

Because of the isomorphisms $H \cong H_0$ and $K \cong K_0$, G is also called the direct product of H_0 and K_0; in fact, it is usual to *identify* H_0 and K_0 with the subgroups H and K, and to drop the subscript 0 altogether.

EXERCISE

2. Assume that the identity e is the only element common to two subgroups H and K of a group G. Show that h in H commutes with every k in K if and only if $H \lhd G$ and $K \lhd G$.

In the next simplest case, the semidirect product, it is still assumed that every g in G can be uniquely expressed as hk, with $h \in H$ and $k \in K$, but it is assumed only that $H \lhd G$, while K is not necessarily normal. Then G is a so-called *semidirect product* of H and K. Any coset of H in G (i.e., any element of the factor group G/H) has a unique representation $kH = Hk$, with k in K; furthermore, $Hk_1 Hk_2 = Hk_1 k_2$, and hence the factor group G/H is isomorphic with K. If g_1 and g_2 in G are expressed uniquely as $g_1 = h_1 k_1$ and $g_2 = h_2 k_2$, then the unique expression of $g_1 g_2 = g_3$ is $h_3 k_3$, where

$$h_3 = h_1 k_1 h_2 k_1^{-1}, \qquad k_3 = k_1 k_2.$$

(Note that $k_1 h_2 k_1^{-1}$ is in H, because H is normal, but is not necessarily $= h_2$ unless K is also normal.)

The group G of rigid motions in the plane (or in n-space) provides an example of a semidirect product. A rigid motion is a transformation

$$\mathbf{x} \to \mathbf{x}' = M\mathbf{x} + \boldsymbol{\xi}, \tag{18.15-2}$$

where M is a 2×2 (or $n \times n$) real orthogonal matrix with determinant $= 1$, and $\boldsymbol{\xi}$ is an arbitrary vector. G is generated by the group $\mathscr{T}$ of translations

$$\mathbf{x} \to \mathbf{x}' = \mathbf{x} + \boldsymbol{\xi}$$

and the group $\mathscr{R}$ of rotations about the origin

$$\mathbf{x} \to \mathbf{x}' = M\mathbf{x}.$$

(The law of composition in $\mathscr{T}$ is vector addition $\boldsymbol{\xi}_1 + \boldsymbol{\xi}_2$, and that in $\mathscr{R}$ is matrix multiplication $M_1 M_2$. Here, $\mathscr{T}$ stands for the group of *all* translations, not merely those of a lattice.) The combined transformation (18.15-2) is denoted by (ξ, M), as in the preceding section, where it was shown that the law of composition in G is given by

$$(\xi_1, M_1) \circ (\xi_2, M_2) = (\xi_1 + M_1\xi_2, M_1M_2). \tag{18.15-3}$$

This would make G into the direct product $\mathcal{T} \times \mathcal{R}$ if the right member of the above equation were replaced by $(\xi_1 + \xi_2, M_1M_2)$. The significance of the term $M_1\xi_2$ for the two groups is as follows: First, for fixed M, the mapping

$$\xi \to M\xi, \quad \text{for all } \xi$$

of $\mathcal{T}$ onto itself is an automorphism of $\mathcal{T}$, because it is one-to-one and onto (M has an inverse), and it maps $\xi + \eta$ onto $M\xi + M\eta$; this automorphism is called $\tau(M)$. Second, as M varies in all of $\mathcal{R}$, these automorphisms form a group $\mathcal{A}$, a subgroup of the group of all automorphisms of $\mathcal{T}$. Third, the mapping

$$M \to \tau(M), \quad \text{for all } M \text{ in } \mathcal{R},$$

is a homomorphism of $\mathcal{R}$ onto $\mathcal{A}$, because if the automorphism $\tau(M)$: $\xi \to \xi' = M\xi$ is first performed on an element ξ of $\mathcal{T}$, followed by $\tau(N)$: $\xi' \to \xi'' = N\xi'$, the result is $\xi \to NM\xi$; that is, $\tau(N)\tau(M) = \tau(NM)$.

Definition. Let H and K be any two groups (the law of composition will be written multiplicatively in both), and let there be given a homomorphism $k \to \tau(k)$ of K into the automorphism group of H [for fixed k, $\tau(k)$ maps h onto $\tau(k)h$, for all h in H]. Then, the set of all pairs (h, k), with h in H and k in K, and with the law of composition of such pairs given by

$$(h_1, k_1) \circ (h_2, k_2) = (h_1\tau(k_1)h_2, k_1k_2) \tag{18.15-4}$$

is a group, called the *semidirect product* of H and K (or of H by K) and is denoted by

$$H \times_\tau K.$$

The main properties of the semidirect product are summarized in the following exercises. The reader is urged to do these exercises, because they reveal that the definition given above, which may have seemed rather arbitrary, contains just the right ingredients to give the semi-direct product its various desirable properties.

EXERCISES

3. Show that the identity elements of $G = H \times_\tau K$ is (e, e'), where e and e' are the identity elements of H and K.

4. Show that the element $(\tau(k^{-1})h^{-1}, k^{-1})$ is the inverse (i.e., both right and left inverse) of (h, k).

5. Show that the associative law holds in G. *Warning*: $\tau(k)[h_1h_2]$ is not the same as $[\tau(k)h_1]h_2$, because $\tau(k)$ is a mapping, not a group element. Exercises 3, 4, and 5 show that G is a group, as claimed in the definition of semidirect product.

6.   Now identify H and K with the subgroups $\{(h, e')$: all h in $H\}$ and $\{(e, k)$: all k in $K\}$, respectively, and show that H is a normal subgroup of G.

7.   Construct the factor group G/H, and show that it is isomorphic to K.

8.   Conversely, suppose that a group G contains subgroups H and K, of which H is normal, such that $H \cap K = \{e\}$, and such that the factor group G/H is isomorphic to K; show that G is the semidirect product $H \times_\tau K$, where, for any k in K, $\tau(k)$ is the mapping $h \rightarrow khk^{-1}$, for all h in H.

9.   Show that the semidirect product is equal to the direct product $H \times K$ if and only if $\tau(k) \equiv I$ [that is, the homomorphism $k \rightarrow \tau(k)$ of K into the automorphism group of H maps all of K onto the identity element (identity mapping of H onto itself)].

10.   Show that K, as well as H, is a normal subgroup of G if and only if $\tau(k) \equiv I$, i.e., if and only if the product is direct.

Exercise 8 shows that the automorphisms $\tau(k)$ that appear in the semidirect product of two given groups H and K become inner automorphisms of the group $H \times_\tau K$ that is being defined. This fact is somewhat obscured, in the case of the rigid motion group G, by the use of the additive notation for the translation group $\mathcal{T}$. If, instead, the translation $\mathbf{x} \rightarrow \mathbf{x}' = \mathbf{x} + \boldsymbol{\xi}$ is denoted by T_ξ and the multiplicative notation is used, so that $\{\boldsymbol{\xi}, M\}$ is simply the combined operation $T_\xi M$, then $T_{M\xi} = M T_\xi M^{-1}$, so that the automorphism $\tau(M)$: $\boldsymbol{\xi} \rightarrow M\boldsymbol{\xi}$ in $\mathcal{T}$ takes the form

$$\tau(M)\colon T_\xi \rightarrow M T_\xi M^{-1},$$

hence is an inner automorphism in G.

According to the preceding section, the translation group $\mathcal{T}$ of a crystal structure, given by (18.14-2), is a normal subgroup of the space group G_s, and the point group G_p is isomorphic to the factor group $G_s/\mathcal{T}$ [it is recalled that the point group is the group of all rotations and reflections $\mathbf{x} \rightarrow M\mathbf{x}$ such that $(\boldsymbol{\xi}, M)$ is in G_s, for some $\boldsymbol{\xi}$]. G_s may or may not contain a subgroup, say G_p', isomorphic to G_p; if it does, then $\mathcal{T}$ and G_p' can have only the element e in common, because all the other elements of $\mathcal{T}$ are of infinite order (if $T \in \mathcal{T}$, then $T^m \neq I$ for all $m \neq 0$), while all elements of G_p are of finite order. Therefore, according to Exercises 5 and 6, above, G_s contains such a subgroup if and only if it is a semidirect product $\mathcal{T} \times_\tau G_p$. In this case, the space group is called *symmorphic* by the crystallographers.

When an x-ray crystallographer starts to analyze a set of x-ray reflection data to determine a crystal structure, he often knows the point group in advance, from measurement of the angles between crystal faces and cleavage planes, and from other macroscopic properties of the crystals. However, he cannot assume that the space group contains a copy of the point group, i.e., that the space group is symmorphic.

A simple 2-dimensional example of a nonsymmorphic space group is that of the function $f(x, y)$, which is equal to 1 in the shaded triangles in Figure 18.3, and 0 otherwise. In complex notation, the space group is generated by the translations $z \rightarrow z + \alpha$, $z \rightarrow z + i\beta$ and the so-called glide-reflection $z \rightarrow \bar{z} + \frac{1}{2}\alpha$; hence the point groups contains the reflection $z \rightarrow \bar{z}$, while the space group does not.

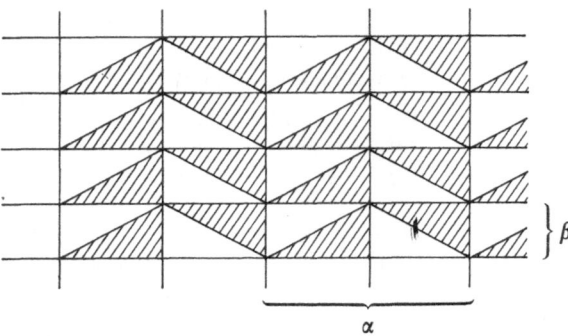

Figure 18.3

The semidirect product $H \times_\tau K$ is an instance of a so-called *extension* of the group H by the group K. For a fuller discussion of group extensions, see Chapter 12 of Kurosh, *Theory of Groups* (1956) or §50 of Rédei, *Algebra* (1959) (in German).

CHAPTER 19

Continuous Groups

General linear, special linear, orthogonal, and unitary groups; rotation and
Lorentz groups; Euler's theorem, the four components of the full Lorentz
group, the Thomas presession; group manifolds; intrinsic coordinates;
double connectivity of the rotation group; homomorphism of SU(2) onto
SO(3) and of SL(2) onto $\mathscr{L}_p$; simplicity of the rotation and Lorentz groups.

Prerequisites: Chapter 18 and a little algebra.

Groups of matrices (or of the corresponding linear transformations) in
which the group elements depend continuously on certain parameters, like
the Euler angles in the case of the rotation group, are called *continuous* (the
formal definition will be given in Chapter 27). Here, properties of some
familiar continuous groups are described.

The group of all nonsingular (generally complex) $n \times n$ matrices is called
the *general linear group* and is denoted by GL(n, $\mathbb{C}$). The subgroup of real
matrices is denoted by GL(n, $\mathbb{R}$). The subgroups of GL(n, $\mathbb{C}$) and GL(n, $\mathbb{R}$)
consisting of matrices with determinant $= 1$ are called *special* (or *unimodular*)
linear groups and are denoted by SL(n, $\mathbb{C}$) and SL(n, $\mathbb{R}$). The notation for
other subgroups of GL(n, $\mathbb{C}$) will be given in the course of the discussion.

19.1 Orthogonal and Rotation Groups

In matrix-vector notation, the plane rotation (18.1-1) is

$$\mathbf{x} \to \mathbf{x}' = R\mathbf{x}, \tag{19.1-1}$$

where

$$\mathbf{x} = \begin{pmatrix} x \\ y \end{pmatrix}, \qquad \mathbf{x}' = \begin{pmatrix} x' \\ y' \end{pmatrix}, \qquad R = R_\varphi = \begin{pmatrix} \cos\varphi & -\sin\varphi \\ \sin\varphi & \cos\varphi \end{pmatrix}. \tag{19.1-2}$$

Under the transformation (19.1-1), the length of any vector and the
angle between any two vectors are preserved, so that if $\mathbf{x}' = R\mathbf{x}$ and $\mathbf{w}' = R\mathbf{w}$,
then $\mathbf{x}' \cdot \mathbf{w}' = \mathbf{x} \cdot \mathbf{w}$ for any two vectors $\mathbf{x}$ and $\mathbf{w}$.

We now look for transformations in n dimensions having this same property; i.e., we call

$$\mathbf{x} = \begin{pmatrix} x_1 \\ \vdots \\ x_n \end{pmatrix} \quad \text{etc.} \quad \text{and} \quad R = \begin{pmatrix} R_{11} & \cdots & R_{1n} \\ \vdots & & \vdots \\ R_{n1} & \cdots & R_{nn} \end{pmatrix},$$

and we look for matrices R such that $(R\mathbf{x}) \cdot (R\mathbf{y}) = \mathbf{x} \cdot \mathbf{y}$ for all vectors $\mathbf{x}$ and $\mathbf{y}$. If $\xi(j)$ denotes the vector whose jth component is $= 1$ and whose other components are $= 0$, then, in particular, R must be such that

$$R\xi(j) \cdot R\xi(k) = \xi(j) \cdot \xi(k) = \begin{cases} 1 & \text{if } j = k \\ 0 & \text{if } j \neq k \end{cases}.$$

Since the vector

$$R\xi(j) = \begin{pmatrix} R_{1j} \\ \vdots \\ R_{nj} \end{pmatrix}$$

is the jth column of R, it follows that the columns of R are pairwise orthogonal unit vectors. Any such matrix is called *orthogonal*. Conversely, if R has that property, then $R\mathbf{x} \cdot R\mathbf{y} = \mathbf{x} \cdot \mathbf{y}$ for all $\mathbf{x}$, $\mathbf{y}$. If R^T denotes the transpose of R, then the rows of R^T are the columns of R, so that the law of matrix multiplication gives

$$R^T R = \begin{pmatrix} 1 & 0 & \cdots & 0 \\ 0 & 1 & \cdots & 0 \\ \vdots & & \ddots & \vdots \\ 0 & 0 & \cdots & 1 \end{pmatrix} = I, \tag{19.1-3}$$

which is another characterization of an orthogonal matrix. Since $R^T = R^{-1}$ is the matrix of the inverse transformation, which also preserves dot products, it follows that R^T is also an orthogonal matrix; hence, the columns of R^T, that is, the rows of R, are another set of n pairwise orthogonal unit vectors. Since $\det R^T = \det R$, equation (19.1-3) shows that $\det R = \pm 1$. We now define

$$O(n) = \{R : R = n \times n \text{ real orthogonal matrix}\}$$

($\circ$ = matrix multiplication) as the *orthogonal goup* in n dimensions. Then, the subgroup

$$SO(n) = \{R \in O(n) : \det R = 1\}$$

is the *special* (or *unimodular*) orthogonal group in n dimensions. We can think of these groups either as groups of matrices or as groups of the corresponding transformations $\mathbf{x} \to R\mathbf{x}$ of n-space.

19.2 The Rotation Group SO(3);
Euler's Theorem

We now show that if R is a 3×3 orthogonal (real) matrix with det $R = 1$, then the transformation $\mathbf{x} \to R\mathbf{x}$ can be obtained by first choosing a fixed direction in space through the origin and then rotating the coordinate system through a suitable angle about this direction as an axis. This is *Euler's theorem.* Let λ_i and $\mathbf{v}_i$ $(i = 1, 2, 3)$ be the eigenvalues and eigenvectors of R (they may be complex, even though R is real); they satisfy the equations

$$R\mathbf{v}_i = \lambda_i \mathbf{v}_i \qquad (i = 1, 2, 3). \tag{19.2-1}$$

Since R is also a unitary matrix, we have $\|R\mathbf{v}_i\| = \|\mathbf{v}_i\|$, where, for any (generally complex) vector $\mathbf{v}$, $\|\mathbf{v}\|$ denotes $(|v_x|^2 + |v_y|^2 + |v_z|^2)^{1/2}$; therefore,

$$|\lambda_i| = 1 \qquad (i = 1, 2, 3). \tag{19.2-2}$$

The λ_i are the roots of the real cubic equation

$$\det(\lambda I - R) = 0, \tag{19.2-3}$$

and the product of the roots is

$$\lambda_1 \lambda_2 \lambda_3 = \det R = 1. \tag{19.2-4}$$

At least one of the roots is real; if the other two (say λ_2 and λ_3) are complex, then $\lambda_3 = \bar{\lambda}_2$ and, by (19.2-2), $\lambda_2 \lambda_3 = 1$; hence $\lambda_1 = 1$. If all three roots are real, they can be 1, 1, 1 or 1, -1, -1. In any case there is always one root, say λ_1, equal to $+1$; hence

$$R\mathbf{v}_1 = \mathbf{v}_1,$$

which shows that the straight line through the origin in the direction of $\mathbf{v}_1$ ($\mathbf{v}_1$ can be taken as real) is invariant under the transformation $\mathbf{x} \to R\mathbf{x}$; evidently this line is the axis of rotation.

[**Reminder.** If M is any normal matrix, i.e., a matrix that commutes with its Hermitian conjugate, i.e., if $MM^* = M^*M$ (in particular, Hermitian matrices and unitary matrices are normal), then the eigenvectors of M can be taken to be a complete orthonormal (with respect to the Hermitian inner product) system of vectors in the n-dimensional space. If the eigenvalues are all distinct, the eigenvectors are automatically orthogonal; if λ is an r-fold eigenvalue, the corresponding eigenspace is r-dimensional, and one can choose an orthonormal system in it; if that is done for each eigenspace, a complete orthonormal system results.]

Let $\lambda_1 = 1$, $\lambda_2 = e^{i\theta}$, $\lambda_3 = e^{-i\theta}$, and suppose that $\mathbf{v}_1$, $\mathbf{v}_2$, $\mathbf{v}_3$ form an orthonormal set (they are the eigenvectors). Call

$$\mathbf{u}_1 = \mathbf{v}_1,$$

$$\mathbf{u}_2 = \frac{1}{\sqrt{2}}(\mathbf{v}_1 + \mathbf{v}_2),$$

$$\mathbf{u}_3 = \frac{i}{\sqrt{2}}(\mathbf{v}_1 - \mathbf{v}_2);$$

(19.2-5)

these also form an orthonormal set (they can all be taken as real, for $\mathbf{v}_2$ and $\mathbf{v}_3$ can be taken as complex conjugates), and

$$R\mathbf{u}_1 = \mathbf{u}_1,$$

$$R\mathbf{u}_2 = \cos\theta\mathbf{u}_2 + \sin\theta\mathbf{u}_3,$$

$$R\mathbf{u}_3 = -\sin\theta\mathbf{u}_2 + \cos\theta\mathbf{u}_3.$$

(19.2-6)

It is seen that the transformation of R is a rotation in planes perpendicular to $\mathbf{u}_1$.

The practical calculation of the angle and axis of rotation, when the matrix R is given, proceeds as follows: Since the sum of the eigenvalues of a matrix is equal to its trace, the angle θ is given by

$$1 + e^{i\theta} + e^{-i\theta} = R_{11} + R_{22} + R_{33},$$

or

$$\cos\theta = \tfrac{1}{2}(R_{11} + R_{22} + R_{33} - 1).$$

(19.2-7)

Next, the axis of rotation is in the direction of the eigenvector $\mathbf{v}$ (called $\mathbf{v}_1$ above) that corresponds to the eigenvalue $\lambda = 1$; hence $R\mathbf{v} = \mathbf{v}$. But, since R is an orthogonal matrix, $R^T R = I$; hence $\mathbf{v} = R^T\mathbf{v}$. Therefore, $(R - R^T)\mathbf{v} = 0$, so that the components v_1, v_2, and v_3 of $\mathbf{v}$ are in the ratio

$$v_1 : v_2 : v_3 = R_{23} - R_{32} : R_{31} - R_{13} : R_{12} - R_{21}.$$

(19.2-8)

EXERCISE

Make a similar analysis of the group SO(n) for arbitrary n.

19.3 Unitary Groups

Orthogonal groups can be generalized to the complex case in two ways. First, a complex orthogonal matrix is any complex matrix M which satisfies $M^T M = I$, just as for the real case. The resulting groups do not seem to be of much importance.

Second, a matrix U such that $U^*U = I$ (then $UU^* = I$) is called a *unitary* matrix. Under a unitary transformation $\mathbf{x} \to U\mathbf{x}$ of the n-dimensional complex space C^n, the Hermitian scalar product

$$\mathbf{x}^*\mathbf{y} = \sum_{j=1}^{n} \bar{x}_j y_j$$

of any two vectors $\mathbf{x}$ and $\mathbf{y}$ is an invariant. Conversely, if this form is invariant, for all $\mathbf{x}$ and $\mathbf{y}$, then U is unitary. The *unitary group* $U(n)$ is the group of all $n \times n$ unitary matrices (or unitary transformations in C^n).

Since $\det U^*$ is the complex conjugate of $\det U$, the equation $U^*U = I$ shows that $|\det U| = 1$, i.e., $\det U$ is a number on the unit circle in the complex plane. The subgroup of $U(n)$ consisting of unitary matrices such that $\det U = 1$ is called the *special* (or *unimodular*) unitary group, and is denoted by $SU(n)$.

19.4 The Lorenz Groups

According to the theory of special relativity, if x, y, z, t and x', y', z', t' are Cartesian coordinates in two inertial frames of reverence whose axes are parallel, but are such that the second frame is moving relative to the first with speed V in the $+x$ direction, and if the origins are coincident at time $t = t' = 0$, then

$$x' = \frac{x - Vt}{\sqrt{1 - v^2/c^2}}, \qquad y' = y, \qquad z' = z, \qquad t' = \frac{t - (V/c^2)x}{\sqrt{1 - v^2/c^2}}. \quad (19.4\text{-}1)$$

Other inertial frames can be obtained by relative velocities in other directions, by rotations in space, by displacements of the origin of space-time, by spatial reflections, and by time reversal. When reflections and time reversal are included, we have the full Lorentz group, otherwise the proper Lorentz group. Displacements of the origin will not be discussed here, so that the transformations considered are homogeneous, i.e., the equations contain no constant terms.

If the notation (customary in relativity theory)

$$x^1 = x, \qquad x^2 = y, \qquad x^3 = z, \qquad x^4 = ct$$

(sometimes, ct is called x^0), is introduced, and φ is defined by the equation

$$\sinh \varphi = -\frac{V/c}{\sqrt{1 - v^2/c^2}}, \quad (19.4\text{-}2)$$

then (19.4-1) is written as

$$x'^{\mu} = \sum_{v=1}^{4} p_v^{\mu} x^v, \quad (19.4\text{-}3)$$

where the coefficients p_ν^μ, when written in matrix format, are

$$[p_\nu^\mu] = P(\varphi) = \begin{pmatrix} \cosh\varphi & 0 & 0 & \sinh\varphi \\ 0 & 1 & 0 & 0 \\ 0 & 0 & 1 & 0 \\ \sinh\varphi & 0 & 0 & \cosh\varphi \end{pmatrix}. \qquad (19.4\text{-}4)$$

In the remainder of this section, the *summation convention* will be used, according to which any term containing a repeated index, say ν as in (19.4-3), is understood to be summed for $\nu = 1, 2, 3, 4$, so that, with this convention, (19.4-3) is written simply as $x'^\mu = p_\nu^\mu x^\nu$. In relativity theory, Greek indices usually go from 1 to 4, and Latin indices from 1 to 3.

The set $\{P(\varphi)\}$ of matrices (or transformations) of this kind, obtained by letting φ take on all real values, is a group that will be denoted by $\mathscr{L}_x$—it is a subgroup of the Lorentz group. Note that

$$P(\varphi_1)P(\varphi_2) = P(\varphi_1 + \varphi_2); \qquad (19.4\text{-}5)$$

from this equation, the *law of composition of* (collinear) *velocities* can be obtained, that is, the formula for the velocity with which a third frame moves relative to the first in terms of the relative velocities of the second with respect to the first and of the third with respect to the second; the derivation is left as an exercise.

If the second frame is obtained by merely rotating the first one in space, then the transformation is given by a matrix of the form

$$R = \begin{pmatrix} & & & 0 \\ & (R') & & 0 \\ & & & 0 \\ 0 & 0 & 0 & 1 \end{pmatrix}, \qquad (19.4\text{-}6)$$

where R' denotes a 3×3 proper rotation matrix—an element of SO(3). The set of all such matrices (or transformations) is the *rotation subgroup* of the Lorentz group, and will be denoted by $\mathscr{R}$.

The group generated by the elements of $\mathscr{L}_x$ together with those of $\mathscr{R}$, i.e., the group consisting of all finite products $Q_1 Q_2 \ldots Q_j$, where each Q_i is either of the form (19.4-4) or of the form (19.4-6), is called the *proper* (or *restricted*) *Lorentz group* and is denoted by $\mathscr{L}_p$. This group is *connected*, in the following sense:

Lemma. *If Q_0 is any element of $\mathscr{L}_p$, there is a one-parameter family $Q(\lambda)$ of elements of $\mathscr{L}_p$ such that the matrix elements all depend continuously on λ, for $0 \le \lambda \le \lambda_0$, and such that $Q(0) = I$, $Q(\lambda_0) = Q_0$.*

PROOF.   First, any element $P(\varphi)$ of $\mathscr{L}_x$ can be connected to the identity by letting φ vary continuously from zero to its final value; second, any element R of $\mathscr{R}$ can be connected to the identity by letting the angle of rotation vary continuously from zero to its final value; therefore, if $Q_0 = Q_1 Q_2 \ldots Q_j$, where each Q_i is in $\mathscr{L}_x$ or $\mathscr{R}$, then the interval $[0, \lambda_0]$ can be taken as $[0, j]$, and $Q(\lambda)$ can be chosen so that $Q(0) = I$, $Q(1) = Q_1, Q(2) = Q_1 Q_2$, and so on, and finally $Q(j) = Q_1 Q_2 \ldots Q_j$, with continuous variations in between.

Inspection shows that the *fundamental (quadratic) form* $(x^1)^2 + (x^2)^2 + (x^3)^2 - (x^4)^2$ is an invariant under all transformations of $\mathscr{L}_x$ and of $\mathscr{R}$, hence also under all transformation of $\mathscr{L}_p$. This form is written as $g_{\mu\nu} x^\mu x^\nu$, where

$$[g_{\mu\nu}] = \begin{pmatrix} 1 & 0 & 0 & 0 \\ 0 & 1 & 0 & 0 \\ 0 & 0 & 1 & 0 \\ 0 & 0 & 0 & -1 \end{pmatrix} = G, \qquad (19.4\text{-}7)$$

and where the summation convention applies to both μ and ν. If the 4-vector x^μ in $g_{\mu\nu} x^\mu x^\nu$ is replaced first by $x^\mu + y^\mu$, then by $x^\mu - y^\mu$, and the results subtracted, it is seen that the invariance of the quadratic form $g_{\mu\nu} x^\mu x^\nu$ is equivalent to the invariance of the symmetric bilinear form $g_{\mu\nu} x^\mu y^\nu = x^1 y^1 + x^2 y^2 + x^3 y^3 - x^4 y^4$.

The *full (homogeneous) Lorentz group* is now defined as the group of all homogeneous linear transformations of $x^1, \ldots, x^4$ under which $g_{\mu\nu} x^\mu x^\nu$ is invariant for all 4-vectors x^μ; it is denoted by $\mathscr{L}_f$. If the transformation $x^\mu \to x'^\mu = q^\mu_\nu x^\nu$ is any element of $\mathscr{L}_f$, the invariance of the bilinear form shows that $g_{\mu\nu} q^\mu_\kappa q^\nu_\lambda = g_{\kappa\lambda}$ or, in matrix notation, that

$$Q^T G Q = G; \qquad (19.4\text{-}8)$$

in other words, the columns of Q are pseudo-orthogonal pseudo-unit vectors, in the sense that

$$(q^1_\nu)^2 + (q^2_\nu)^2 + (q^3_\nu)^2 - (q^4_\nu)^2 = \begin{cases} +1, & \text{for } \nu = 1, 2, 3, \\ -1, & \text{for } \nu = 4, \end{cases} \qquad (19.4\text{-}9)$$

and

$$q^1_\nu q^1_\lambda + q^2_\nu q^2_\lambda + q^3_\nu q^3_\lambda - q^4_\nu q^4_\lambda = 0, \qquad \text{for } \nu \neq \lambda. \qquad (19.4\text{-}10)$$

From this it follows that the inverse of Q is its transpose with signs changed according to the pattern

$$\begin{pmatrix} + & + & + & - \\ + & + & + & - \\ + & + & + & - \\ - & - & - & + \end{pmatrix}.$$

Since Q^{-1} is also a Lorentz transformation, it follows that the *rows* of Q are also pseudo-orthogonal pseudo-unit vectors.

Equation (19.4-8) shows that the determinant of Q is ± 1. If $\{x^1, \ldots, x^4\}$ is taken as $\{0, 0, 0, 1\}$, the invariance of the fundamental form shows that

$$\sum_{j=1}^{3} (x'^j)^2 - (x'^4)^2 = \sum_{j=1}^{3} (q_4^j)^2 - (q_4^4)^2 = -1;$$

therefore, q_4^4 is either ≥ 1 or ≤ -1.

Theorem. *The proper Lorentz group $\mathscr{L}_p$, defined above as the group generated by $\mathscr{L}_x$ and $\mathscr{R}$, consists of all those transformations Q of $\mathscr{L}_f$ for which* $\det Q = +1$ *and* $q_4^4 \geq +1$.

PROOF. It is shown first that the connectedness of the group $\mathscr{L}_p$ implies that $\det Q = +1$ and that $q_4^4 \geq 1$ for any Q in $\mathscr{L}_p$: Let Q be connected to the identity I, as in the lemma; since $Q(0) = I$, we have $\det Q(0) = 1$ and $q(0)_4^4 = 1$; $\det Q(\lambda)$ and $q(\lambda)_4^4$ are continuous, and hence cannot jump to negative values as λ varies. (That $\det Q$ is equal to 1 can also be seen directly from the decomposition $Q = Q_1 Q_2 \ldots Q_j$, where each Q_i is either in $\mathscr{L}_x$ or in $\mathscr{R}$). Conversely, let Q be any transformation in $\mathscr{L}_f$ such that $\det Q = 1$ and $q_4^4 \geq 1$. It will be shown that Q can be expressed as $R_1 P R_2$. where R_1 and R_2 are in $\mathscr{R}$ and P is in $\mathscr{L}_x$; hence, Q is in $\mathscr{L}_p$. (This shows, furthermore, that three factors always suffice in the decomposition $Q_1 Q_2 \ldots Q_j$.) First, let R_3 and R_4 be rotations that take the 3-vectors (q_4^1, q_4^2, q_4^3) and (q_1^4, q_2^4, q_3^4), respectively, into the direction of the positive x^1 axis. Then,

$$R_3 Q R_4 = \begin{pmatrix} & & & q_4'^1 \\ & (X') & & 0 \\ & & & 0 \\ q_1'^4 & 0 & 0 & q_4'^4 \end{pmatrix} = Q', \qquad (19.4\text{-}11)$$

where X' is some 3×3 matrix and where $q_4'^1$ and $q_1'^4$ are ≥ 0, and $q_4'^4 = q_4^4 \geq 1$. Since $(q_4'^1)^2 - (q_4'^4)^2$ and $(q_1'^4)^2 - (q_4'^4)^2$ both $= -1$, a parameter φ can be chosen so that

$$q_4'^1 = q_1'^4 = \sinh \varphi, \qquad q_4'^4 = \cosh \varphi.$$

Therefore, if Q' is multiplied by $P(-\varphi)$, the last row and the last column of the product are the same as those of $P(-\varphi)P(\varphi) = I$; hence

$$P(-\varphi)Q' = \begin{pmatrix} & & & 0 \\ & (X'') & & 0 \\ & & & 0 \\ 0 & 0 & 0 & 1 \end{pmatrix} = Q''.$$

This matrix, like Q and Q', is in $\mathscr{L}_p$, and hence leaves the fundamental quadratic form invariant. Therefore, X'' leaves $(x^1)^2 + (x^2)^2 + (x^3)^2$ invariant, hence is in O(3), but $\det X'' = +1$, hence X'' is in SO(3); i.e., Q'' is an element of $\mathscr{R}$, say R_5. That is,

$$P(-\varphi)R_3 Q R_4 = R_5,$$

and it follows that Q is of the required form

$$Q = R_1 P(\varphi) R_2,$$

where $R_1 = R_3^{-1}$ and $R_2 = R_5 R_4^{-1}$.

Comment. Elements of the form $L = R^{-1}PR$, where R is in $\mathcal{R}$ and P is in $\mathcal{L}_x$ are called *pure Lorentz transformations* or *boosts*: In this case the primed and unprimed axes are still parallel, just as for $P(\varphi)$, but the relative velocity can be in any direction. L is a symmetric matrix, but if L_1 and L_2 are two pure Lorentz transformations, it is readily seen that L_1L_2 is not symmetric unless the relative velocities are parallel; hence, the pure Lorentz transformations do not form a subgroup of $\mathcal{L}_p$. Furthermore, if L_1L_2 is not symmetric, a third pure Lorentz transformation L_3 can be found such that $L_1L_2L_3$ is a pure rotation (an element of $\mathcal{R}$), not $=I$. This leads to the pseudoparadoxical result that a body can be rotated by subjecting it to three successive stages of purely linear acceleration, in three different directions, in such a way that the body is brought back to rest by the combined action of the three stages. This phenomenon, expressed a little differently, was discovered by L. H. Thomas in connection with relativistic corrections to atomic energy levels, in the early days of quantum theory. [An electron, orbiting a nucleus, undergoes a (continuous) sequence of Lorentz translations, somewhat as described, and an effect similar to that of electron spin results.]

The group $\mathcal{L}_t$ generated by the elements of $\mathcal{L}_p$ together with the transformation of time reversal, given by the matrix

$$T = \begin{pmatrix} 1 & 0 & 0 & 0 \\ 0 & 1 & 0 & 0 \\ 0 & 0 & 1 & 0 \\ 0 & 0 & 0 & -1 \end{pmatrix},$$

is a subgroup of $\mathcal{L}_f$ because T clearly preserves the fundamental form. Similarly, the group $\mathcal{L}_s$ generated by the elements of $\mathcal{L}_p$ together with spatial inversion, given by the matrix

$$S = \begin{pmatrix} -1 & 0 & 0 & 0 \\ 0 & -1 & 0 & 0 \\ 0 & 0 & -1 & 0 \\ 0 & 0 & 0 & 1 \end{pmatrix},$$

is a subgroup of $\mathcal{L}_f$. Lastly, $\mathcal{L}_f$ itself is generated by the elements of $\mathcal{L}_p$ together with T and S.

Exercises

Prove or disprove the following statements:

1. Every proper Lorentz transformation with a symmetric matrix is a pure Lorentz transformation.

2. If P_1 and P_2 are two pure Lorentz transformations, then $P_1P_2P_1^{-1}P_2^{-1}$ is a pure rotation.

3. The rotation subgroup is a normal subgroup of $\mathcal{L}_p$.

The Lorentz groups can be generalized in an obvious way to groups of transformations that preserve the value of a fundamental form

$$(x^1)^2 + \cdots + (x^r)^2 - (x^{r+1})^2 - \cdots - (x^{r+l})^2;$$

the possibility of spatial reflections appears when r is odd, and of time reversal when l is odd.

19.5 Group Manifolds

If each element

$$R = \begin{pmatrix} R_{11} & R_{12} & R_{13} \\ R_{21} & R_{22} & R_{23} \\ R_{31} & R_{32} & R_{33} \end{pmatrix}$$

of the orthogonal group O(3) is represented by a point in a 9-dimensional (real) space V^9, with coordinates $R_{11}, R_{12}, \ldots, R_{33}$ (taken, let us say, in the order in which they appear in the matrix), then the group is represented by the set of points in V^9 that satisfy the six algebraic equations

$$R_{1j}^2 + R_{2j}^2 + R_{3j}^2 = 1, \qquad j = 1, 2, 3, \tag{19.5-1}$$

$$R_{1j}R_{1k} + R_{2j}R_{2k} + R_{3j}R_{3k} = 0, \qquad (j \quad k) = (1 \quad 2), (1 \quad 3), (2 \quad 3), \tag{19.5-2}$$

which assert that the columns of R are pairwise orthogonal unit vectors, as required by the discussion in Section 19.2. The resulting 3-dimensional algebraic surface in V^9 is called the *manifold* of the group O(3). The exact size, shape, and curvature of this surface are of no interest, but its overall topological properties are closely connected with the structure of the group.

With each continuous group of linear transformations is similarly associated an algebraic surface in some V^m. However, care must be used in drawing conclusions from the number of equations. For SO(3), in addition to the six equations (19.5-1,2) above, there is a seventh algebraic equation

$$\det R = 1. \tag{19.5-3}$$

As will be seen, the surface determined by the first six equations has two separate parts or components, like the two branches of a hyperbola, and equation (19.5-3) merely eliminates one of the components, but does not reduce the dimensionality of the surface. For the full linear group GL(n, ℝ), there is only an algebraic *inequality* $\det M \neq 0$; for GL(n, ℂ) there is the algebraic inequality $(\mathrm{Re}\ \det M)^2 + (\mathrm{Im}\ \det M)^2 \neq 0$.

EXERCISE

Find an algebraic equation $F(x, y, z) = 0$ which determines the surface of the torus in V^3 just as the equation $x^2 + y^2 + z^2 - 1 = 0$ determines the surface of a sphere. Find a group of which the torus is the group manifold.

19.6 Intrinsic Coordinates in the Manifold of the Rotation Group

Although any manifold can be regarded, in principle, as an n-dimensional surface embedded in a space V^m of higher dimension, as in the foregoing examples, the embedding is not always easy to find or describe, and the manifold is more appropriately described by intrinsic coordinates, like the polar angles θ, φ on the surface of a sphere, or more generally by two or more overlapping systems of intrinsic coordinate systems.

Each transformation of the group SO(3) can be obtained by first choosing a fixed axis and then performing a rotation about that axis (Section 19.2). If R is the matrix of a rotation through the angle $\theta \geq 0$ (seen as clockwise when looking along the positive direction of $\mathbf{k}$) about an axis having the direction of the unit vector $\mathbf{k}$, then the numbers θk_x, θk_y, θk_z $(=\theta_x, \theta_y, \theta_z)$ may be taken as intrinsic coordinates in SO(3), and we write $R = R(\boldsymbol{\theta})$, where $\boldsymbol{\theta}$ is the vector $\mathbf{k}\theta$. To make the coordinate system unique, $\boldsymbol{\theta}$ must be restricted to the spherical ball $K = \{\boldsymbol{\theta}: \|\boldsymbol{\theta}\| \leq \pi\}$ in the coordinate space $\mathbb{R}^3$, in which θ_x, θ_y, and θ_z are taken as Cartesian, and we must observe that opposite ends of a diameter in K correspond to the same element of SO(3)— otherwise, each point of K corresponds to a unique element of SO(3), and conversely.

The matrix $R = R(\boldsymbol{\theta})$ is given explicitly in terms of the intrinsic coordinates by the equation

$$R(\boldsymbol{\theta}) = \exp\begin{pmatrix} 0 & -\theta_z & \theta_y \\ \theta_z & 0 & -\theta_x \\ -\theta_y & \theta_x & 0 \end{pmatrix}. \tag{19.6-1}$$

To see this, note first that since R is a nonsingular normal matrix, its logarithm Λ is well defined (though multivalued), so that

$$R = e^\Lambda, \qquad R^{-1} = e^{-\Lambda} = R^T = e^{\Lambda^T};$$

therefore, Λ is antisymmetric and hence must be of the form

$$\Lambda = \begin{pmatrix} 0 & a & b \\ -a & 0 & c \\ -b & -c & 0 \end{pmatrix}.$$

To see that a, b, and c have been correctly interpreted in (19.6-1), note that, with this interpretation, (1) the eigenvalues of Λ are 0, $\pm i\theta$, where $\theta = \sqrt{\theta_x^2 + \theta_y^2 + \theta_z^2}$, so that the eigenvalues of $\exp \Lambda$ are 1 and $e^{\pm i\theta}$, (2) the first eigenvector of Λ (hence also of $\exp \Lambda$) is proportional to $\boldsymbol{\theta}$, (3) since Λ is a normal matrix, its eigenvectors can be taken as an orthonormal set, and (4) it now follows, just as in Section 19.2, that R represents a rotation through an angle θ about an axis in the direction of $\boldsymbol{\theta}$. It is left as an exercise to verify

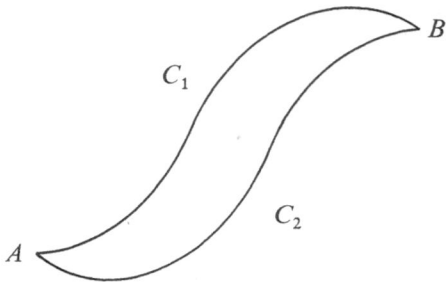

that (19.6-1) defines $R(\theta)$ rather than $R(-\theta)$, if the coordinate system is right-handed.

By means of these intrinsic coordinates $\theta_x, \theta_y, \theta_z$, the properties of the surface $\mathscr{S}$ in V^9 determined by the algebraic equations (19.5-1,2,3) can be found. Each point of the ball K corresponds to a unique point of the surface $\mathscr{S}$, except that opposite ends of a diameter in K correspond to the same point of $\mathscr{S}$, and the nine coordinates of a point in $\mathscr{S}$, are continuous functions of the intrinsic coordinates in K, according to equation (19.6-1). Therefore, $\mathscr{S}$ is a connected surface, because any point of the ball K can be connected to any other point of K by a curve (in fact by a straight line segment) lying in K. However, $\mathscr{S}$ is not *simply* connected.

A connected surface is called *simply connected* if, given any two points A and B and any two curves C_1 and C_2 going from A to B in the surface, C_1 can be deformed into C_2 by a continuous deformation without leaving the surface. The plane and the sphere are simply connected, while the torus, the surface of a cylinder, and the annulus $R_1^2 < x^2 + y^2 < R_2^2$ are not. Among solids, the ball, the cube, and the spherical shell $R_1^2 < x^2 + y^2 + z^2 < R_2^2$ are simply connected, while the solid torus and the pretzel are not.

To show that the manifold $\mathscr{S}$ of SO(3) is not simply connected, let A be the point of $\mathscr{S}$ corresponding to the center of the ball K, and let B be the point of $\mathscr{S}$ that corresponds to the two ends of a diameter of K. Then the two radii which make up this diameter correspond, in $\mathscr{S}$, to two curves or paths going from A to B, and it is evident that neither can be deformed into the other. However, any other path from A to B can be continuously deformed, in $\mathscr{S}$, into one of these two.

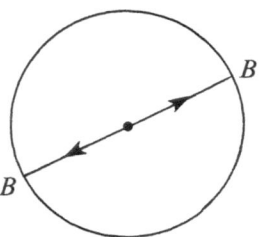

19.7 The Homomorphism of SU(2) onto SO(3)

Let x, y, z be Cartesian coordinates in a Euclidean space E^3, and call A the matrix

$$A = \begin{pmatrix} z & x - iy \\ x + iy & -z \end{pmatrix}, \tag{19.7-1}$$

which is evidently Hermitian and of trace zero. (The trace of a matrix is the sum of its eigenvalues and is equal to the sum of its diagonal elements.) Let U be any 2×2 unitary matrix of determinant $= 1$ [i.e., $U \in SU(2)$], and call

$$A' = UAU^*. \tag{19.7-2}$$

Since the eigenvalues of A' are the same as those of A, the trace of A' is also zero; A' is also Hermitian ($A'^* = A'$), so it can be written as

$$A' = \begin{pmatrix} z' & x' - iy' \\ x' + iy' & -z' \end{pmatrix}, \tag{19.7-3}$$

where x', y' and z' are real. Furthermore, $\det A' = \det A$, because of (19.7-2), so

$$x^2 + y^2 + z^2 = x'^2 + y'^2 + z'^2. \tag{19.7-4}$$

The relation between x, y, z and x', y', z' is obviously linear for given U; hence, if we define a 3×3 real matrix $R = R(U)$ by

$$\begin{pmatrix} x' \\ y' \\ z' \end{pmatrix} = R \begin{pmatrix} x \\ y \\ z \end{pmatrix},$$

then R is an orthogonal matrix.

A General Comment on Orthogonal, Unitary, and Lorentz Transformations
Suppose that a homogeneous linear transformation $\mathbf{x} \to \mathbf{x}'$ in real n-space is such that the quadratic form $\mathbf{x} \cdot \mathbf{x} = x_1^2 + \cdots + x_n^2$ is invariant; i.e., $\mathbf{x} \cdot \mathbf{x} = \mathbf{x}' \cdot \mathbf{x}'$, for all $\mathbf{x}$. Then the bilinear form $\mathbf{x} \cdot \mathbf{y}$ is also invariant, for any $\mathbf{x}$, any $\mathbf{y}$, because

$$\mathbf{x} \cdot \mathbf{y} = \tfrac{1}{4}(\mathbf{x} + \mathbf{y}) \cdot (\mathbf{x} + \mathbf{y}) - \tfrac{1}{4}(\mathbf{x} - \mathbf{y}) \cdot (\mathbf{x} - \mathbf{y}).$$

That is, if all lengths are preserved, then all angles are preserved, too. Similarly, if $(\cdot, \cdot)$ denotes the Hermitian symmetric (i.e., sesquilinear) inner product in complex n-space, then $(\mathbf{x}, \mathbf{y})$ is invariant, for all $\mathbf{x}$, all $\mathbf{y}$, if and only if $(\mathbf{x}, \mathbf{x})$ is invariant, for all $\mathbf{x}$. Therefore, orthogonal and unitary groups can be characterized by invariance of either the quadratic, or the corresponding bilinear form.

EXERCISE

Formulate the analogous result for the Lorentz group.

Furthermore, det $R = 1$, by continuity, since the only possible values of det R are ± 1, since det R depends continuously on U, and since SU(2) is a connected group. [If U is the 2×2 unit matrix, then $R(U)$ is the 3×3 unit matrix and has determinant $= 1$.] $R(U_1)R(U_2)$ is equal to $R(U_1 U_2)$, because $R(U_1)R(U_2)$ is the result of transforming A first into $U_2 A U_2^*$, then into $U_1 U_2 A U_2^* U_1^*$. Therefore the mapping

$$\Phi: U \to R(U)$$

is a homomorphism. We state without proof that it is onto.

The kernel is easily seen to be $\{I, -I\}$, so with each $R \in \mathrm{SO}(3)$ there are associated two elements U and $-U$ of SU(2). It will be proved in Section 19.9 that SO(3) is simple, so no further nontrivial homomorphism of SO(3) is possible.

19.8 The Homomorphism of SL(2, $\mathbb{C}$) onto the Proper Lorentz Group $\mathscr{L}_p$

The homomorphism found in the preceding section is very easily extended. Let x, y, z, t be coordinates in space-time (in units such that $c = 1$). Then the matrix

$$A = \begin{pmatrix} t + z & x - iy \\ x + iy & t - z \end{pmatrix}$$

is Hermitian, and det $A = t^2 - x^2 - y^2 - z^2$. If P is an arbitrary 2×2 (complex) matrix with det $P = 1$, then the matrix $A' = PAP^*$ is also Hermitian, and hence can be written as

$$A' = \begin{pmatrix} t' + z' & x' - iy' \\ x' + iy' & t' - z' \end{pmatrix}.$$

Now det $A' = $ det A; hence

$$t^2 - x^2 - y^2 - z^2 = t'^2 - x'^2 - y'^2 - z'^2.$$

That is, the matrix P of SL(2, C) induces a Lorentz transformation T. Clearly the mapping $P \to T$ has the homomorphism property that if $P_1 \to T_1$ and $P_2 \to T_2$, then $P_1 P_2 \to T_1 T_2$. It is in fact a 2-to-1 homomorphism of SL(2, C) onto $\mathscr{L}_p$.

19.9 Simplicity of the Rotation and Lorentz Groups

Theorem. *The rotation group* SO(3) *is simple*

PROOF.   Suppose that G_0 is a nontrivial normal subgroup; $G_0 \lhd \mathrm{SO}(3)$. It must be proved that $G_0 = \mathrm{SO}(3)$. Let $R_0 \in G_0$, $R_0 \neq I$. It was proved in Section 6.4 that, for any R in SO(3), the transformation $\mathbf{x} \to R\mathbf{x}$ can be described as a rotation through some angle θ about a fixed direction. Let $\boldsymbol{\theta}$ be a vector in that direction and having length $\|\boldsymbol{\theta}\| = \theta$, and call $R = R(\boldsymbol{\theta})$, as in Section 19.6. Then $R_0 = R(\boldsymbol{\theta}_0)$ for some $\boldsymbol{\theta}_0 \neq 0$. Now let $\boldsymbol{\theta}_1$ be any other vector having the same length as $\boldsymbol{\theta}_0$, $\|\boldsymbol{\theta}_1\| = \|\boldsymbol{\theta}_0\|$, and let R_1 be a rotation that carries $\boldsymbol{\theta}_1$ into $\boldsymbol{\theta}_0$. Then $R_1^{-1}R(\boldsymbol{\theta}_0)R_1$ is in the subgroup G_0, because G_0 is normal; but $R_1^{-1}R(\boldsymbol{\theta}_0)R_1 = R(\boldsymbol{\theta}_1)$, and hence the subgroup G_0 contains every $R(\boldsymbol{\theta})$ for $\|\boldsymbol{\theta}\| = \|\boldsymbol{\theta}_0\|$. Next, G_0 also contains every element $R(\boldsymbol{\theta})R(\boldsymbol{\theta}_0)$, for $\|\boldsymbol{\theta}\| = \|\boldsymbol{\theta}_0\|$; this is $R(\boldsymbol{\theta}')$ for some $\boldsymbol{\theta}' = \boldsymbol{\theta}'(\boldsymbol{\theta}, \boldsymbol{\theta}_0)$. Clearly $\|\boldsymbol{\theta}'\|$ is a continuous function of the components of $\boldsymbol{\theta}$. [Recall the explicit formula for $R(\boldsymbol{\theta})$ in terms of θ_x, θ_y, θ_z given in Section 19.6.] For $\boldsymbol{\theta} = -\boldsymbol{\theta}_0$ and $+\boldsymbol{\theta}_0$, $\|\theta'\| = 0$ and $2\|\theta_0\|$, respectively. Therefore, for any θ in $[0, 2\|\boldsymbol{\theta}_0\|]$, G_0 contains at least one element $R(\boldsymbol{\theta})$ such that $\|\boldsymbol{\theta}\| = \theta$. By the first argument, again, G_0 therefore contains every $R(\boldsymbol{\theta})$ such that $0 \leq \|\boldsymbol{\theta}\| \leq 2\|\boldsymbol{\theta}_0\|$; but for any such $R(\boldsymbol{\theta})$, G_0 also contains $R(\boldsymbol{\theta})R(\boldsymbol{\theta}) = R(2\boldsymbol{\theta})$, $R(3\boldsymbol{\theta})$, etc.; hence G_0 contains every $R(\boldsymbol{\theta})$ for $0 \leq \|\boldsymbol{\theta}\| \leq \pi$, that is, $G_0 = SO(3)$.

The proper Lorentz group is also simple, although the proof is more complicated.

CHAPTER 20

Group Representations I: Rotations and Spherical Harmonics

Representation; faithful representation; dimension; reducibility; irreducible representation; tensor transformation laws; representations of SO(2) and SO(3); infinitesimal operators; raising and lowering operators; effective and transitive action of a group; homogeneous space; regular representation; tesseral harmonics; Legendre polynomials and associated Legendre functions; recurrence relations and the differential equation for P_l^m; Rodrigues' formula for P_l^m; orthonormality and completeness of the tesseral harmonics; addition theorem.

Prerequisites: Chapters 18 and 19; elementary theory of matrices; familiarity with the special functions of mathematical physics.

In this chapter and the next two, three apparently remote subjects are shown to have important interconnections: group representations, the classical special functions, and quantum mechanics.

Group representations are intimately connected with various special functions of mathematical physics. In a sense, the primary role of those functions is to exhibit symmetry relations. For instance, the Legendre functions appear (via spherical harmonics) in problems with spherical symmetry in such diverse fields as electrostatics, acoustics, heat flow, neutron transport, and the quantum mechanics of hydrogen-like atoms. Bessel functions (of integer or half-odd-integer order) appear mostly in problems of wave motion, but a closer examination shows that they are associated more with certain symmetries than with the mechanism of the waves. In fact, wave motion in a nonconstant (even spherically symmetric) potential generally involves other functions, for example Laguerre functions in hydrogen-like atoms, while Bessel functions appear when the system is invariant under the full rigid-motion groups, not merely the rotation subgroups—see next chapter. It will be seen in Section 20.5 that the trigonometric functions appear, in the form $e^{im\varphi}$, in the representations of the two-dimensional rotation group, and hence are associated with symmetry about an axis.

20.1 Finite-Dimensional Representations of a Group

Let G be any group, and X an n-dimensional (often complex) vector space. A homomorphism $\rho: g \to \rho(g)$ of G onto a group of linear transformations in X is called an *n-dimensional representation of G (on X)*. The linear transformations are normally described by matrices, and the matrix that describes the transformation $\rho(g)$ is also denoted by $\rho(g)$. Hence, we may also regard a representation of G as a homomorphism of G onto a group of matrices; then the identity in G, the inverse of an element in G, and the law of compositions in G are represented by the unit matrix, the matrix inverse, and the matrix product. If the homomorphism is an isomorphism, then the representation ρ is called *faithful*. If G itself is a group of linear transformations in a vector space Y, then the identity mapping $g \to g$ of G onto itself is a faithful representation. Even in this case, however, it is worthwhile, for many reasons, to consider other representations also.

Note. The usage here differs slightly from that of Section 18.10, where a representation of G was a homomorphism of G onto any group of (not necessarily linear) transformations.

If X contains a proper subspace X_1 that is invariant under all the transformations $\rho(g)$, $g \in G$, then the representation is *reducible*, and the restriction of the transformations $\rho(g)$ to X_1 gives a representation of smaller dimension, called a *subrepresentation*. If there is no such proper subspace X_1, ρ is called *irreducible*.

If G is a continuous group (a Lie group), it is also required that the matrix elements of $\rho(g)$ depend continuously on g. The only continuous groups to be considered here are matrix groups, like the unitary, orthogonal, and Lorentz groups. In this case the requirement is that the matrix elements of $\rho(g)$ be continuous functions of the matrix elements of g, as is immediately seen to be the case in all the examples below.

20.2 Vector and Tensor Transformation Laws

We shall now describe some naturally occurring group representations in physics. According to Chapter 19, the transformations in 3-dimensional space that result from rotations of the Cartesian axes about the origin constitute a group called SO(3). Various representations of this group are given by the transformation laws for the components of tensors. If the Cartesian coordinates are transformed by $\mathbf{x} \to \mathbf{x}'$, where

$$x'_i = \sum_{k=1}^{3} g_{ik} x_k, \qquad (20.2\text{-}1)$$

and where (g_{ik}) is a rotation matrix [an element of SO(3)], then the components T_{ij} of a second rank tensor transform according to the law

$$T'_{ij} = \sum_{(k,l)} g_{ik} g_{jl} T_{kl}. \tag{20.2-2}$$

If the nine quantities T_{ij} are called $X_1, \ldots, X_9$ and are regarded as the coordinates of a point $\mathbf{X}$ in $\mathbb{R}^9$, then each transformation $\mathbf{x} \to \mathbf{x}'$ induces a transformation $\mathbf{X} \to \mathbf{X}'$, and those transformations give a representation of SO(3) on $\mathbb{R}^9$.

Transformation of the components of a third rank tensor give a representation on $\mathbb{R}^{27}$, and so on.

More special representations can appear when certain symmetry relations exist among the tensor components. Suppose that the T_{ij} are the components of the strain-rate tensor at some point in a fluid:

$$T_{ij} = \frac{\partial v_i}{\partial x_j},$$

where $\mathbf{v} = \mathbf{v}(\mathbf{x})$ is the velocity vector field. If the flow is irrotational, then

$$T_{ij} = T_{ji} \qquad (i, j = 1, 2, 3). \tag{20.2-3}$$

It is readily verified that these equations are invariant under rotations; that is, if $T_{ij} = T_{ji}$ for all i and j, then $T'_{ij} = T'_{ji}$ (see Exercise 2 below). Hence, in this case, only six of the T_{ij} are independent, say the quantities Y_l defined as

$$Y_1 = T_{12}, \qquad Y_2 = T_{23}, \qquad Y_3 = T_{31},$$
$$Y_4 = T_{11}, \qquad Y_5 = T_{22}, \qquad Y_6 = T_{33}.$$

Then the rotation $\mathbf{x} \to \mathbf{x}'$ induces a linear transformation $\mathbf{Y} \to \mathbf{Y}'$, and a 6-dimensional representation of SO(3) results.

If, furthermore, the fluid is incompressible, there is an additional relation

$$T_{11} + T_{22} + T_{33} = 0, \tag{20.2-4}$$

which is also invariant under rotations (see Exercise 3 below). In this case Y_6 can be dropped (i.e., it can always be computed as $-Y_4 - Y_5$), and the transformations of $Y_1, \ldots, Y_5$ give a 5-dimensional representation of SO(3).

The relations (20.2-3) and (20.2-4) determine subspaces of $\mathbb{R}^9$ that are invariant under the transformations (20.2-2); the invariance of those subspaces permits the 9-dimensional representation to be reduced to the 6-dimensional and 5-dimensional ones. [An 8-dimensional representation can be obtained by using (20.2-4) alone.] It will appear below that the 5-dimensional subspace cannot be further decomposed into smaller invariant subspaces; hence the 5-dimensional representation is irreducible. It will appear that for each odd integer m there is an irreducible m-dimensional representation of SO(3); when $m > 1$, that representation is faithful.

The original transformation law (20.2-1) for the components of a vector $\mathbf{x}$ provides of course a 3-dimensional representation. (If the rotations are regarded as constituting an abstract group, then that representation is no more trivial than the others.) Furthermore, for the sake of completeness (in a sense to be made precise later), we include the 1-dimensional representation given by the transformation law for scalars, which says simply that they are not transformed at all. Each scalar is a real number x, and each group element (rotation) is mapped onto the identity transformation $x \to x$ in $\mathbb{R}$. [It should not be concluded that a 1-dimensional representation of a group always consists only of the identity transformation. A 1-dimensional representation of $GL(n, \mathbb{R})$ or $GL(n, \mathbb{C})$ is given by mapping the group element M—an $n \times n$ matrix—onto the transformation $x \to (\det M)x$ in $\mathbb{R}$ or $\mathbb{C}$.]

Representations of the Lorentz groups (see Section 19.4) are given similarly by the transformation laws of scalars, vectors (i.e., 4-vectors), and tensors, under Lorentz transformations in space-time; they are of dimension $1, 4, 16, \ldots$.

A 6-dimensional representation of the (restricted) Lorentz group $\mathscr{L}_p$ is given by the transformation law for the components of the electric and magnetic fields $\mathbf{E}$ and $\mathbf{H}$ in free space. (The interpretation of this transformation law in terms of tensors in explained in Exercise 5 below.) Under the particular Lorentz transformation (19.4-1), the electric and magnetic field components transform according to the law

$$E'_x = E_x, \qquad E'_y = \gamma\left(E_y - \frac{v}{c} H_z\right), \qquad E'_z = \gamma\left(E_z + \frac{v}{c} H_y\right),$$

$$H'_x = H_x, \qquad H'_y = \gamma\left(H_y + \frac{v}{c} E_z\right), \qquad H'_z = \gamma\left(H_z - \frac{v}{c} E_y\right),$$

(20.2-5)

where

$$\gamma = \left(1 - \frac{v^2}{c^2}\right)^{-1/2}.$$

(See any good book on electromagnetic theory.) Under a rotation, the components of $\mathbf{E}$ and those of $\mathbf{H}$ transform independently according to the usual vector law (20.2-1). According to Section 19.4 any element of $\mathscr{L}_p$ can be written in the form $R_1 P R_2$, where R_1 and R_2 are rotations in space and P is a transformation of the form (19.4-1), hence the general transformation for $\mathbf{E}$ and $\mathbf{H}$ can be obtained by combining (20.2-1) and (20.2-5).

One of the aims of the theory of group representations is to find all possible transformation laws for physical quantities, that is, all possible representations of physical symmetry groups. As seen in the above examples, there are two main procedures; the *building up* of representations from simpler ones by means of tensors, and the *breaking down* of representations into subrepresentations. Still another procedure, based on the action of a group on spaces of functions, is described below, starting in Section 20.5.

EXERCISES

1. Show that the representation of rotations (20.2-1) by the transformations (20.2-2) is a homomorphism, as required. That is, show that if $\rho(g)$ is the transformation induced in $\mathbb{R}^9$ by the rotation g, then $\rho(g)\rho(g') = \rho(gg')$.

2.   Show that the symmetry or antisymmetry of a second rank tensor is preserved under rotations, when the law (20.2-2) is used; that is, show that if $T_{ij} = T_{ji}$ (or $T_{ij} = -T_{ji}$) for all i, j, then $T'_{ij} = T'_{ji}$ (or $T'_{ij} = -T'_{ji}$) for all i, j.

3.   Show that the trace of a second-rank tensor is preserved under rotation; that is, show that

$$T'_{11} + T'_{22} + T'_{33} = T_{11} + T_{22} + T_{33}.$$

4.   Consider a general linear transformation of the form (20.2-1) in $\mathbb{R}^n$, where (g_{ij}) is an $n \times n$ nonsingular matrix, and show that the transformation is orthogonal $(g^T g = g g^T = I)$ if and only if the trace of every second rank tensor is invariant under (20.2-2).

5.   Show that the transformation law (20.2-5) for the electromagnetic field under the Lorentz transformation (19.4-1) can be obtained by transforming the antisymmetric second rank tensor $T^{\mu\nu}$ defined by

$$T^{\mu\nu} = \begin{pmatrix} 0 & H_z & -H_y & -E_x \\ -H_z & 0 & H_x & -E_y \\ H_y & -H_x & 0 & -E_z \\ E_x & E_y & E_z & 0 \end{pmatrix}$$

according to the law

$$T'^{\mu\nu} = \sum_{(\sigma,\tau)}^{4} q_\sigma^\mu q_\tau^\nu T^{\sigma\tau},$$

when the coordinates are transformed according to

$$x'^\mu = \sum_{\sigma=1}^{4} q_\sigma^\mu x^\sigma.$$

20.3   Other Group Representations in Physics

Pauli's 1927 theory of electron spin and Dirac's 1928 relativistic wave equation led to transformation laws under rotations and Lorentz transformations of a different kind from the familiar transformation laws for vectors and tensors. That led in turn to the theory of new objects called spinors, which take their place in relativistic quantum mechanics alongside scalars, vectors, and tensors, and are discussed in Chapter 22. The transformation laws for spinors give so-called two-valued representations of SO(3) and $\mathscr{L}_p$, which, however, are true representations of the covering groups SU(2) and SL(2, $\mathbb{C}$) discussed in Sections 19.7 and 19.8. Why the latter groups should appear at all in physical problems seemed a paradox and is still left quite unclear in most books on quantum mechanics. The resolution of the paradox, due mainly to Hermann Weyl, is described in Chapter 22. It involves the so-called ray representations, which are not true representations (as defined in this chapter), but are nevertheless appropriate for quantum mechanical phenomena. In Weyl 1928 it is shown that the ray representations of a group are precisely determined by the true representations of its covering groups; hence, the representations of SU(2) and SL(2, $\mathbb{C}$) play a role. The theory shows further that since the manifolds of SU(2) and SL(2, $\mathbb{C}$) are simply connected, these groups are the so-called universal covering groups of SO(3) and $\mathscr{L}_p$, respectively, (see Chapters 24 and 27) and that, in consequence, there are no multivalued representations of SO(3) and $\mathscr{L}_p$ of multiplicity >2. It is thus concluded, on the basis of group theory, that vectors, tensors, and spinors provide the only possible transformation laws of quantum mechanical phenomena.

In classical physics a distinction is made between polar vectors (like momentum and electric field) and axial vectors (like angular momentum and magnetic field). The former change sign under an inversion $\mathbf{x} \to -\mathbf{x}$, while the latter do not. Hence there are two (in fact only two) ways of extending the transformation law for vectors to the full orthogonal group O(3).

The symmetry or antisymmetry of a many-particle wave function under interchange (or more generally permutation) of identical particles gives a simple representation of the relevant permutation group. More complicated representations of the permutation groups appear in the theory of parastatistics.

20.4 Infinite-Dimensional Representations

Infinite-dimensional representations, for example of the Lorentz and Poincaré groups, are encountered in applications like quantum field theory. Normally, such a representation is a homomorphism of the group G onto a group of bounded linear transformations $\rho(g)$ in a Banach or Hilbert space X. If G is a continuous matrix group, $\rho(g)$ is required to be a strongly continuous function of g; that is, it is required that for every u in X,

$$\|\rho(g)u - \rho(g_0)u\| \to 0,$$

as the matrix elements of g converge to those of g_0.

While all irreducible representations of compact groups like SO(3) are finite-dimensional, there are infinite-dimensional representations of the Lorentz, Poincaré, and rigid motion groups such that no finite-dimensional subspace of the Banach or Hilbert space X is invariant under all the transformations $\rho(g)$, $g \in G$, which therefore cannot be broken down into finite-dimensional subrepresentations. In the next chapter we describe infinite-dimensional representations of the 2-dimensional rigid motion group M_2, which is fairly typical, and which involves Bessel functions. We refer to Boerner 1955, Gel'fand, Minlos, and Shapiro 1963, Gel'fand, Graev, and Vilenkin 1966, Vilenkin 1968, Warner 1972, and Barut and Raczka (1977) for further material on the subject (which is a fairly vast one, as the sizes of those books indicates) and in particular for the representations of the Lorentz and Poincaré groups.

In Chapter 27 we give an example of a continuous group (Lie group) that has no finite-dimensional faithful representation, and hence cannot be realized as a matrix group at all.

20.5 A Simple Case: SO(2)

The general method of finding representations can be illustrated by the (rather trivial) case of the group SO(2) of rotations about the origin in the plane. Under this group any circle centered at the origin is invariant. Let X^∞ be the infinite-dimensional space of infinitely differentiable functions $f(\varphi)$ defined on the unit circle. (For the present qualitative discussion, we shall not introduce a norm or topology in this space.) Then, with the rotation g_α through an angle α [an element of SO(2)] we associate the linear transformation

$$\rho_\alpha: f(\varphi) \to f(\varphi - \alpha) \qquad (20.5\text{-}1)$$

in the space X^∞. The correspondence $g_\alpha \to \rho_\alpha$ is a representation of SO(2) on X_∞. (In this example, the transformation $f(\varphi) \to f(\varphi + \alpha)$ would do equally well—see **Note** in the next section.)

An important technique for finding subrepresentations is the use of the so-called infinitesimal operators of a representation, such as the operator T obtained in the present instance by differentiating the operator ρ_α with respect to α at $\alpha = 0$, namely

$$(Tf)(\varphi) = -f'(\varphi), \qquad (20.5\text{-}2)$$

so that T may be regarded as the limit of $(1/\alpha)(\rho_\alpha - \rho_0)$ as $\alpha \to 0$. Evidently, any subspace of X^∞ that is invariant under all the transformations ρ_α is invariant under T. We look for invariant subspaces of smallest possible dimension. Hence we look for a function $f(\varphi)$ such that the minimal subspace that contains $f(\varphi)$ and is invariant under T contains no other functions, or as few such as possible. For a one-dimensional subspace Tf should be a multiple of f, say λf; that leads to the eigenvalue problem

$$-f'(\varphi) = \lambda f(\varphi). \qquad (20.5\text{-}3)$$

In the present example, since the functions in X^∞ are required to be single-valued on the unit circle, the eigenvalues and eigenfunctions are

$$\lambda = im \qquad (m = 0, \pm 1, \pm 2, \ldots),$$
$$f(\varphi) = f_m(\varphi) = e^{-im\varphi}. \qquad (20.5\text{-}4)$$

For each α, the transformation ρ_α given by (20.5-1) also causes each eigenfunction f_m of T to be merely multiplied by a constant, namely

$$(\rho_\alpha f_m)(\varphi) = e^{im\alpha} f_m(\varphi).$$

That is, for each m the 1- (complex) dimensional subspace

$$X_m = \{Ae^{-im\varphi}: A \in \mathbb{C}\} \qquad (20.5\text{-}5)$$

is invariant not only under T but also under all the transformations ρ_α. By Fourier's theorem each function in X^∞ can be expanded in these functions, hence the subspaces X_m span X^∞.

It will now be shown that the *only* finite-dimensional representations of SO(2) on X^∞ are those that can be built up from the ones obtained above, for it will be shown that any finite-dimensional invariant subspace X' of X^∞ is the direct sum of finitely many of the subspaces (20.5-5). Namely, let $f(\varphi)$ be any function in X' and write it as a Fourier series

$$f(\varphi) = \sum_{(m)} c_m e^{im\varphi}. \tag{20.5-6}$$

It will be proved that the subspace X' not only contains this sum but contains each term $c_m e^{im\varphi}$ individually; hence, it contains the subspace X_{-m} for each m such that $c_m \neq 0$. Namely, since X' is invariant under all the transformations (20.5-1) it contains all translates $f(\varphi - \alpha)$ of the given function $f(\varphi)$; hence it contains any function of the form

$$\sum_{k=1}^{K} h_k f(\varphi - \alpha_k),$$

where the h_k and α_k are constants. By choosing this sum as a Riemann sum that approximates an integral and then going to the limit, it is seen that X' contains any function of the form

$$\int h(\alpha) f(\varphi - \alpha) d\alpha,$$

hence, in particular, the function

$$\frac{1}{2\pi} \int_{-\pi}^{\pi} e^{im\alpha} f(\varphi - \alpha) d\alpha = c_m e^{im\varphi}.$$

Therefore, X' contains X_{-m} if $c_m \neq 0$, as was to be proved. (Since X' was assumed finite-dimensional, it is now seen that c_m is $\neq 0$ only for finitely many m).

For non-Abelian groups, the minimal invariant subspaces are generally or more than one dimension.

20.6 Representations of Matrix Groups on X^∞

The ideas in the preceding section can be generalized. If G is a group of $n \times n$ matrices, then its elements g determine linear transformations in an n-dimensional space $V^n (= \mathbb{R}^n$ or $\mathbb{C}^n)$. Let X^∞ denote the space of all infinitely differentiable functions $f(\mathbf{x})$ on V^n or on any surface in V^n that is invariant under G. With each element g of G we associate a linear transformation

$$\rho(g): f(\mathbf{x}) \rightarrow f'(\mathbf{x}) = f(g^{-1}\mathbf{x}) \tag{20.6-1}$$

of X^∞ into itself (the prime does not denote differentiation). If the transformation $\rho(g) = \rho(g_1)$ is followed by another transformation $\rho(g_2)$, namely

$$\rho(g_2): f'(\mathbf{x}) \rightarrow f''(\mathbf{x}) = f'(g_2^{-1}\mathbf{x}),$$

the result is

$$\rho(g_2)\rho(g_1): f(\mathbf{x}) \to f''(\mathbf{x})$$
$$= f'(g_2^{-1}\mathbf{x})$$
$$= f(g_1^{-1}(g_2^{-1}\mathbf{x}))$$
$$= f((g_2 g_1)^{-1}\mathbf{x}),$$

so that

$$\rho(g_2)\rho(g_1) = \rho(g_2 g_1). \qquad (20.6\text{-}2)$$

Therefore, the association $g \to \rho(g)$ is a representation of G on X^∞.

Note. The appearance of g^{-1} (rather than g itself) in (20.6-1) is necessary to obtain the correct order of the factors in (20.6-2). It corresponds to the minus sign in (20.5-1); for an Abelian group the order is of course irrelevant, hence $f(\varphi + \alpha)$ would have been equally satisfactory in (20.5-1).

Replacing $f(\mathbf{x})$ by $f(g^{-1}\mathbf{x})$ is equivalent to performing the mapping $\mathbf{x} \to g\mathbf{x}$ in V^n and carrying the values of f along during the mapping, just as the substitution $f(t) \to f(t - a)$ carries the values of f to the right on $\mathbb{R}$, for $a > 0$, by a distance a.

If G is a continuous group and the group elements depend on parameters $\alpha, \beta, \ldots$, that is $g = g_{\alpha, \beta, \ldots}$, where $g_{0, 0, \ldots}$ is the identity elements of G, then the infinitesimal operators corresponding to the operator T in the preceding section are given by differentiating $\rho(g_{\alpha, \beta, \ldots})$ with respect to each parameter $\alpha, \beta, \ldots$ in turn and then setting $\alpha = \beta = \cdots = 0$. The infinitesimal operators of SO(3) are given in Section 20.9.

20.7 Homogeneous Spaces

If the space V^n contains a smooth surface S of some lower dimension that is invariant under all transformations of G, like the unit circle in the case of SO(2), then X^∞ can be replaced by a space of functions defined on S rather than on all V^n. In the example above, S could be any circle centered at the origin. If G is SO(3), S can be any sphere centered at the origin of V^3. In each of these cases, the action of the group on S has the following properties:

a. The group G is *effective* for S; that is, only the identity e of G gives the identity mapping in S; in other words, if $g \neq e$, then there is at least one $\mathbf{x}$ in S such that $g\mathbf{x} \neq \mathbf{x}$.

b. The group is *transitive* for S; that is, if $\mathbf{x}$ and $\mathbf{y}$ are any two points of S, there is a g in G that carries $\mathbf{x}$ into $\mathbf{y}$, i.e., $\mathbf{y} = g\mathbf{x}$.

If S is invariant under G, and a and b hold, then S is called a *homogeneous space* for G.

Suppose, for example, that G consists of all transformations in real 3-space of the form

$$\mathbf{x} \to \mathbf{x}' = \begin{pmatrix} a & b & e \\ c & d & f \\ 0 & 0 & 1 \end{pmatrix} \mathbf{x},$$

where $ad - bc \neq 0$. Any plane $x_3 = $ constant is invariant under G and satisfies condition b, but the plane $x_3 = 0$ fails to satisfy a, because every point of the plane $x_3 = 0$ is mapped onto itself under the mappings

$$\mathbf{x} \to \begin{pmatrix} 1 & 0 & e \\ 0 & 1 & f \\ 0 & 0 & 1 \end{pmatrix} \mathbf{x}.$$

Hence, only the planes $x_3 = $ constant $\neq 0$ are homogeneous spaces for this group.

A general procedure for finding irreducible representations of a continuous group G of linear transformations in V^n consists of finding a homogeneous space S for G in V^n, then letting ρ be the representation $f(\mathbf{x}) \to f(g^{-1}\mathbf{x})$ of G on a space X^∞ of functions on S, usually $L^2(S)$, then finding a complete set of infinitesimal operators, and then finding minimal invariant subspaces of X^∞ with the help of the infinitesimal operators. The special functions associated with the symmetries described by G are the elements of the invariant subspaces of X^∞.

The procedure is described in more detail in Sec. 20.9 below for the case of the rotation group SO(3). In that case, the invariant subspaces that will be found are all finite-dimensional. It will be seen in the next chapter that the same is true for any compact group. In that chapter, the question whether all irreducible representations are found in this way is also answered, for compact groups.

The theory for noncompact groups, where infinite-dimensional irreducible representations can occur, is considerably beyond the scope of this book, and we shall be content with a discussion of some of the main features of three examples, in the next chapter: the rigid motion group M_2, where Bessel functions appear, and the Lorentz group and SL(2, $\mathbb{C}$), where spinors appear.

20.8 Regular Representations

For Lie groups, the group manifold itself can always serve as a homogeneous space. Consider $G = $ SO(3), for example. It is recalled that the manifold of SO(3) is a certain 3-dimensional algebraic surface S in a space of 9 real dimensions; each point of S represents an element g of G. For fixed h in G, the *left translation* $g \to hg$ maps S onto itself. Let X^∞ be the space of all C^∞ functions $f(g)$ on S, or the Hilbert space $L^2(S)$. For any h in G, the mapping

$$\rho(h): f(g) \to f(h^{-1}g)$$

is a linear transformation in X^∞, and the association $h \to \rho(h)$ is a representation of G. Similarly, the association with h of the mapping $f(g) \to f(gh)$ (note that h itself appears here, rather than its inverse) is also a representation of G. Representations of this type, the so-called *left* and *right regular representations*, are discussed in the next chapter.

EXERCISE

Show that the set of left translations is effective and transitive for the group manifold.

20.9 Representations of the Rotation Group SO(3)

The methods outlined in Section 20.6 and 20.7 are here applied to the rotation group. Our approach contrasts with the one often taken, in that we assume nothing in advance about the spherical harmonic functions, $Y_l^m(\theta, \varphi)$, but derive those functions and their properties from group theory.

Let $g_\omega = g_{\omega_x, \omega_y, \omega_z}$ be the matrix of a rotation through an angle $\|\omega\|$ about an axis in the direction of ω. That is, $g_\omega = R(\omega)$, where $R(\cdot)$ is defined by (19.6-1). Let X^∞ be the space of all infinitely differentiable functions $f(\mathbf{x})$ in $\mathbb{R}^3$. For each $g = g_\omega$, an operator $\rho(g)$ on X^∞ is defined, according to (20.6-1), by the equation

$$(\rho(g)f)(\mathbf{x}) = f(g^{-1}\mathbf{x}). \tag{20.9-1}$$

These operators $\rho(g)$ constitute a representation of SO(3).

The infinitesimal operators of this representation are obtained as follows: Since the functions in X^∞ are differentiable, the operator

$$\frac{1}{\omega} [\rho(g_{\omega,0,0}) - \rho(g_{0,0,0})]$$

on X^∞ has a limit, called L_1, as $\omega \to 0$. [Recall that $\rho(g_{0,0,0})$ is the identity operator $f \to f$ on X^∞.] Any finite-dimensional subspace X_1 of X^∞ that is invariant under all $\rho(g)$ is also invariant under (i.e., is transformed into itself by) L_1. If $g = g_{\omega,0,0}$ and $f(\mathbf{x}) = f(x, y, z)$, then

$$f(g^{-1}\mathbf{x}) = f(x, y \cos \omega + z \sin \omega, -y \sin \omega + z \cos \omega),$$

and it follows that

$$L_1 \overset{\text{def}}{=} \frac{d}{d\omega} \rho(g_{\omega,0\,0}) \Big|_{\omega=0} = z \frac{\partial}{\partial y} - y \frac{\partial}{\partial z}. \tag{20.9-2}$$

Similarly,

$$L_2 \overset{\text{def}}{=} \frac{d}{d\omega} \rho(g_{0,\omega,0}) \Big|_{\omega=0} = x \frac{\partial}{\partial z} - z \frac{\partial}{\partial x},$$

$$L_3 \overset{\text{def}}{=} \frac{d}{d\omega} \rho(g_{0,0,\omega}) \Big|_{\omega=0} = y \frac{\partial}{\partial x} - x \frac{\partial}{\partial y}. \tag{20.9-3}$$

Except for a factor $i\hbar$, these are the quantum-mechanical operators for the components of angular momentum—see Schiff 1955, Chapter IV. They obey the commutation rules

$$[L_i, L_j] = L_k \qquad (ijk = 1\,2\,3, 2\,3\,1, 3\,1\,2). \qquad (20.9\text{-}4)$$

Note. The infinitesimal operators of *any* representation of SO(3) obey these rules, because the infinitesimal group elements themselves obey them: Namely, if

$$T_i = \left.\frac{\partial g_{\omega_1, \omega_2, \omega_3}}{\partial \omega i}\right|_{\omega = 0} \qquad (i = 1, 2, 3),$$

then, according to (19.9-1),

$$T_1 = \begin{pmatrix} 0 & 0 & 0 \\ 0 & 0 & -1 \\ 0 & 1 & 0 \end{pmatrix}, \quad T_2 = \begin{pmatrix} 0 & 0 & 1 \\ 0 & 0 & 0 \\ -1 & 0 & 0 \end{pmatrix}, \quad T_3 = \begin{pmatrix} 0 & -1 & 0 \\ 1 & 0 & 0 \\ 0 & 0 & 0 \end{pmatrix};$$

hence $[T_i, T_j] = T_k$, where $ijk = 1\,2\,3$, $2\,3\,1$, or $3\,1\,2$. Since products are mapped onto products in the representation $g \to \rho(g)$, the rules (20.9-4) follow for any representation.

Now let S be the unit sphere $x^2 + y^2 + z^2 = 1$ (a homogeneous space for the rotation group), and let $X^\infty(S)$ be the space of all C^∞ functions on S.

Starting with the operators L_i, the invariant subspaces can be found by the so-called raising and lowering operators introduced by Dirac (see his *Quantum Mechanics*, 1958, Chapter 6) for the quantization of the harmonic oscillator and of angular momentum. (See also W. Miller 1973.) In polar coordinates

$$L_1 = \sin \varphi \, \frac{\partial}{\partial \theta} + \cot \theta \cos \varphi \, \frac{\partial}{\partial \varphi},$$

$$L_2 = -\cos \varphi \, \frac{\partial}{\partial \theta} + \cot \theta \sin \varphi \, \frac{\partial}{\partial \varphi}, \qquad (20.9\text{-}5)$$

$$L_3 = -\frac{\partial}{\partial \varphi}.$$

The operators $L^\pm$, defined as $L_1 \pm iL_2$, are given by

$$L^\pm = e^{\pm i\varphi}\left(\mp i \frac{\partial}{\partial \theta} + \cot \theta \frac{\partial}{\partial \varphi}\right), \qquad (20.9\text{-}6)$$

and it is noted that

$$[L^+, L^-] = -2iL_3 = 2i \frac{\partial}{\partial \varphi}. \qquad (20.9\text{-}7)$$

Suppose that $f(\theta, \varphi)$ is a function in $X^\infty(S)$, not identically zero. We wish to find the minimal invariant subspace X_1 containing $f(\theta, \varphi)$ and to choose $f(\theta, \varphi)$ so as to make that subspace as small as possible, in a sense. If $f(\theta, \varphi) = \sum g_m(\theta)e^{im\varphi}$, and if, for some m^*, $g_{m^*}(\theta)$ is not $\equiv 0$, then, by the argument of Section 20.5, the subspace contains all multiples of the single term $g_{m^*}(\theta)e^{im^*\varphi}$. The operators L^+ and L^-, when applied to a function of the form $f(\theta)e^{im\varphi}$, give functions of the form $f_1(\theta)e^{i(m+1)\varphi}$ and $f_2(\theta)e^{i(m-1)\varphi}$ (for this reason, L^+ and L^- are called *raising* and *lowering* operators), and these functions must be in X_1, since X_1 is invariant; hence, X_1 contains functions of the form

$$\psi_m(\theta, \varphi) = g_m(\theta)e^{im\varphi} \tag{20.9-8}$$

for $m = m^* + 1, m^* + 2$, etc., and for $m = m^* - 1, m^* - 2, \ldots$, etc.

It will appear that the functions $g_m(\theta)$ can be so chosen as to make X_1 finite-dimensional; to achieve this, $L^+\psi_m$ must be zero for some m, say, $m = l$, and $L^-\psi_m$ must be zero for some $m \le l$, say $m = l'$; it will be seen below that $l' = -l$. From the first of these conditions, $g_l'(\theta) - l\cot\theta g_l(\theta) = 0$, according to the formula (20.9-6) for L^+. The solution of this differential equation is $g_l(\theta) = $ const. $(\sin\theta)^l$, and since the functions in X^∞ have no singularities on the unit sphere, it follows that $l \ge 0$; hence,

$$\psi_l(\theta, \varphi) = C(e^{i\varphi}\sin\theta)^l, \tag{20.9-9}$$

where C is a constant to be determined later. Starting with this function, a sequence of functions $\psi_{l-1}, \psi_{l-2}, \ldots$, is obtained by repeated use of the operator L^-, which transforms a function of the form $g(\theta)e^{im\varphi}$ into one of the form $h(\theta)e^{i(m-1)\varphi}$. All these functions are in X_1. We shall now show first that no new functions are obtained from these by the raising operator, i.e., that $L^+\psi_{m-1}$ is the same function as ψ_m, except for normalization, and second that $L^-\psi_{-l} = 0$, so that the sequence terminates at $m = -l$. We use induction on decreasing m, starting with $m = l$: Assume that, for some m, $L^+\psi_m \propto \psi_{m+1}$, i.e., that $L^-L^+\psi_m \propto \psi_m$, and note that this last is in any case true for $m = l$, because $L^+\psi_l = 0$. According to (20.9-7),

$$(L^+L^- - L^-L^+)\psi_m = -2iL_3\psi_m = -2m\psi_m, \tag{20.9-10}$$

and it follows that $L^+L^-\psi_m$ is also $\propto \psi_m$; hence $L^+\psi_{m-1} \propto \psi_m$, and the induction follows.

We now determine the functions ψ_m more explicitly. Let the proportionalities referred to be written as

$$L^+\psi_m = -i\alpha_m\psi_{m+1}, \qquad L^-\psi_{m+1} = -i\beta_m\psi_m. \tag{20.9-11}$$

Since each ψ_m contains an arbitrary factor, these equations determine only the product $\alpha_m\beta_m$, by the equation $L^-L^+\psi_m = -\alpha_m\beta_m\psi_m$. Hence, we can take $\beta_m = \alpha_m$, for all m. It then follows from (20.9-10) that

$$-\alpha_{m-1}^2 + \alpha_m^2 = -2m, \qquad \text{for all } m < l,$$

and this equation holds also for $m = l$, if α_l is set $= 0$. It follows by an induction on decreasing m that

$$\alpha_m^2 = (l + m + 1)(l - m) \qquad (m = l, l - 1, \ldots, -l);$$

hence we can choose

$$\alpha_m = \sqrt{(l + m + 1)(l - m)} \qquad (m = l, l - 1, \ldots), \qquad (20.9\text{-}12)$$

where the positive square root is understood. In particular, $\alpha_{-l-1} = 0$; hence $L^- \psi_{-l} = 0$, as stated.

The conclusion is that equations (20.9-9, 11, 12) determine all the functions ψ_m $(-l \le m \le l)$ up to the constant C. These functions span a $(2l + 1)$-dimensional subspace X^{2l+1} of $X^\infty(S)$ that is invariant under L_1, L_2, L_3 and also, as will be seen at the end of Section 20.14, under the transformations $\rho(g)$ for all g in SO(3); there is one such subspace for each $l = 0, 1, 2, \ldots$. The transformations $\rho(g)$, when restricted to the subspace X^{2l+1}, are called $\rho^l(g)$; they constitute a finite-dimensional irreducible representation of G, and it will be shown in Section 21.13 that these are the only irreducible representations, up to equivalence.

20.10 Tesseral Harmonics; Legendre Functions

It will be shown in this section that the properties of the spherical harmonics follow from the representation theory of the rotation group, and that the tesseral harmonics form a basis for the representation of SO(3).

If the functions $\psi_m(\theta, \varphi)$ of the preceding section are taken as a basis in X^{2l+1}, then the transformations $\rho(g)$, when restricted to X^{2l+1}, are given by $(2l + 1) \times (2l + 1)$ matrices. Before these matrices can be computed, the functions ψ_m must be discussed further; they will be denoted henceforth by $Y_l^m(\theta, \varphi)$, to acknowledge the dependence on l. They are called *tesseral (surface) harmonics*. A *surface harmonic* is a function $f(\theta, \varphi)$ such that $r^p f(\theta, \varphi)$ satisfies Laplace's equation in x, y, z, for some integer p, and it will be seen that $r^l Y_l^m(\theta, \varphi)$ satisfies Laplace's equation. A *tessera* (which comes through Latin from a Greek word meaning "four-cornered") is a curvilinear rectangle such as the ones into which the sphere is divided by the zeros or nodal lines of Re Y_l^m or Im Y_l^m, which occur on certain circles of lattitude $\theta = $ const. and certain meridians $\varphi = $ const.

An inner product is defined in the space $X^\infty(S)$ of functions on the unit sphere S, as follows:

$$(f_1, f_2) = \int_0^{2\pi} \int_0^\pi \overline{f_1(\theta, \varphi)} f_2(\theta, \varphi) \sin \theta \, d\theta \, d\varphi. \qquad (20.10\text{-}1)$$

The completion of $X^\infty(S)$ with respect to the norm $\|f\| = (f,f)^{1/2}$ is the Hilbert space $L^2(S)$. The operators $\rho(g)$, $g \in$ SO(3), are unitary in $L^2(S)$, because they are defined in all $L^2(S)$ and are invertible, and [since the integral (20.10-1) is invariant under rotations] because

$$(\rho(g)f_1, \rho(g)f_2) = (f_1, f_2) \tag{20.10-2}$$

for all f_1 and f_2. It will be shown that the functions Y_l^m are orthogonal with respect to the inner product (20.10-1). If the constant C in (20.9-9) is suitably chosen (it can depend on l), they are also normalized. It will be proved that they form a complete orthonormal set of functions on the sphere.

The φ integration alone shows immediately that $Y_{l_1}^{m_1}$ and $Y_{l_2}^{m_2}$ are orthogonal, if $m_1 \neq m_2$, because $\overline{Y_{l_1}^{m_1}} Y_{l_2}^{m_2}$ contains a factor $e^{i(m_2 - m_1)\varphi}$. It is evident from (20.9-2, 3) that the operators L_i are antisymmetric, i.e., $(L_i f, g) = -(f, L_i g)$, and from this it follows that $L^- L^+$ is symmetric. Furthermore, from (20.9-11),

$$L^- L^+ Y_l^m = -(\alpha_l^m)^2 Y_l^m, \tag{20.10-3}$$

where, with a slight improvement of notation,

$$(\alpha_l^m)^2 = (l + m + 1)(l - m), \tag{20.10-4}$$

according to (20.9-12). Therefore, the equation

$$(L^- L^+ Y_{l_1}^m, Y_{l_2}^m) = (Y_{l_1}^m, L^- L^+ Y_{l_2}^m)$$

is equivalent to

$$(\alpha_{l_1}^m)^2 (Y_{l_1}^m, Y_{l_2}^m) = (\alpha_{l_2}^m)^2 (Y_{l_1}^m, Y_{l_2}^m);$$

hence, since $\alpha_{l_1}^m \neq \alpha_{l_2}^m$ for $l_1 \neq l_2$, it is seen that the Y_l^m are orthogonal.

We now show how to choose the constant C in (20.9-9) so as to normalize the functions Y_l^m. The adjoint of the operator L^+ is $-L^-$; hence

$$(L^+ Y_l^m, Y_l^{m+1}) = (Y_l^m, -L^- Y_l^{m+1}).$$

It follows from (20.9-11), since $\beta_m = \alpha_m = \alpha_l^m$, that

$$(-i\alpha_l^m Y_l^{m+1}, Y_l^{m+1}) = (Y_l^m, i\alpha_l^m Y_l^m),$$

from which it is seen that $\|Y_l^m\|^2$ is independent of m, for given l. From the equation (20.9-9) for $\psi_l = Y_l^l$,

$$\|Y_l^l\|^2 = 2\pi |C|^2 \int_0^\pi \sin^{2l+1}\theta \, d\theta$$

$$= 4\pi |C|^2 \frac{2 \cdot 4 \cdots (2l)}{1 \cdot 3 \cdots (2l + 1)}$$

$$= 4\pi |C|^2 \frac{(2^l l!)^2}{(2l + 1)!}. \tag{20.10-5}$$

Therefore, if the constant C is chosen as

$$C = C_l = \frac{(-1)^l}{2^l l!} \sqrt{\frac{(2l + 1)!}{4\pi}}, \tag{20.10-6}$$

then all the functions Y_l^m are normalized.

With Y_l^1 given by (20.9-9) and (20.10-6), and the other Y_l^m given in terms of Y_l^1 by the recurrence relation (20.9-11), which says that $L^- Y_l^{m+1} = -i\alpha_l^m Y_l^m$, we define new functions $P_l^m(w)$, called the *associated Legendre functions*, for $-1 \leq w \leq 1$ by the equation

$$Y_l^m(\theta, \varphi) = (-1)^m \sqrt{\frac{2l+1}{4\pi} \frac{(l-m)!}{(l+m)!}} \, P_l^m(\cos\theta) e^{im\varphi} \qquad (-l \leq m \leq l).$$

$$(20.10\text{-}7)$$

Notes. (1) The factors $(-1)^l$ in (20.10-6) and $(-1)^m$ in (20.10-7) are arbitrary, but conventional. (2) Historically, $P_l^m(w)$ was first defined by equation (20.11-6) below, and $Y_l^m(\theta, \varphi)$ was then defined by (20.10-7). (3) The symbol $P_l^m(w)$ is used by different authors to denote slightly different functions. The usage here agrees with that of Talman 1968 and others, for all m ($-l \leq m \leq l$), and agrees with the original definition of Ferrers (see Whittaker and Watson 1927), for $m \geq 0$.

20.11 Associated Legendre Functions

The properties of the functions $P_l^m(w)$ are derived in this section from considerations of group representations.

Starting from the recurrence relations for the Y_l^m and the formula (20.9-6) for the operators $L^\pm$, a short calculation gives recurrence relations for the P_l^m, which are

$$\sqrt{1-w^2}\, P_l^{m\prime}(w) + \frac{mw}{\sqrt{1-w^2}}\, P_l^m(w) = P_l^{m+1}(w) \qquad (-l \leq m < l), \quad (20.11\text{-}1)$$

$$\sqrt{1-w^2}\, P_l^{m\prime}(w) - \frac{mw}{\sqrt{1-w^2}}\, P_l^m(w) = -(l+m)(l-m+1)P_l^{m-1}(w)$$

$$(-l < m \leq l), \quad (20.11\text{-}2)$$

where the prime indicates differentiation with respect to w. If the first of these equations is differentiated once more, and P_l^{m+1} and $P_l^{m+1\prime}$ are eliminated, using the second equation, the result is the associated Legendre differential equation

$$(1-w^2)P_l^{m\prime\prime} - 2wP_l^{m\prime} + \left[l^2 + l - \frac{m^2}{1-w^2}\right]P_l^m = 0, \quad (20.11\text{-}3)$$

for $P_l^m(w)$. The first recurrence relation (20.11-1) can be written as

$$(1-w^2)^{-(m+1)/2}P_l^{m+1}(w) = \frac{d}{dw}\left[(1-w^2)^{-m/2}P_l^m(w)\right],$$

from which it is deduced, by an evident induction, that

$$(1-w^2)^{-m/2}P_l^m(w) = \left(\frac{d}{dw}\right)^{l+m}\left[(1-w^2)^{l/2}P_l^{-l}(w)\right]. \quad (20.11\text{-}4)$$

It will be shown below that

$$P_l^{-l}(w) = \frac{(-1)^l}{2^l l!} (1 - w^2)^{l/2},$$ (20.11-5)

from which the so-called Rodrigues formula follows:

$$P_l^m(w) = \frac{1}{2^l l!} (1 - w^2)^{m/2} \left(\frac{d}{dw}\right)^{l+m} (w^2 - 1)^l \qquad (m = -l, -l+1, \ldots, l).$$

(20.11-6)

[For the present purpose, it would have been more reasonable to fix Y_l^{-l} initially, rather than Y_l^l, and then determine the other Y_l^m by means of the raising operator L^+, rather than the lowering operator L^-. However, the general relation between Y_l^m and Y_l^{-m}, which is now needed, has independent interest.] Complex conjugation interchanges L^+ and L^-; hence, the conjugates of equations (20.9-11) are

$$L^- \overline{\psi_m} = i\alpha_m \overline{\psi_{m+1}}, \, L^+ \overline{\psi_{m+1}} = i\alpha_m \overline{\psi_m},$$

from which it is seen that the quantities $(-1)^m \overline{\psi_m}$ satisfy the same equations as the quantities ψ_{-m}; hence

$$Y_l^{-m} = k(-1)^m \overline{Y_l^m},$$

where k is a constant, which will soon be seen to be $=1$. Since C in (20.10-6) is real, equations (20.9-9) and (20.10-7) show that $P_l^l(w)$ is real; then (20.11-1) shows that all the $P_l^m(w)$ are real; hence, by (20.10-7), Y_l^0 is real. The above equation, with $m = 0$ then shows that $k = 1$. Therefore,

$$Y_l^{-m}(\theta, \varphi) = (-1)^m \overline{Y_l^m(\theta, \varphi)}.$$ (20.11-7)

Since $Y_l^l = C(e^{i\varphi} \sin \theta)^l$, with C given by (20.10-6), the above equation gives Y_l^{-l} explicitly, from which equation (20.11-5) for P_l^{-l} follows from (20.10-7). (Some authors define Y_l^{-m} to be the complex conjugate of Y_l^m, after defining the latter for $m \geq 0$. The procedure followed here has some advantages; for example, the matrices $\rho_{m'm}^l$ of the irreducible representations of the rotation group, given below, are symmetric.)

Clearly $P_l^m(w)$ is a polynomial, for even m. $P_l^0(w)$, usually denoted by $P_l(w)$, is the *Legendre polynomial of degree l*.

EXERCISES

1. Show that $\int_0^1 (1 - w^2)^l \, dw = [2l/(2l + 1)] \int_0^1 (1 - w^2)^{l-1} \, dw$, and use this result to show, by an obvious induction, that the integral in (20.10-5) has been correctly evaluated.

2. Express the operators $L^\pm$ in terms of the variables w and φ, where $w = \cos \theta$, and derive the recurrence relationships (20.10-8, 9) from (20.9-11).

3. For the special case $m = 0$, verify that the Rodrigues formula (20.11-6) gives a solution of Legendre's differential equation, which is (20.11-3) with $m = 0$. [The solution is $P_l(w)$.] You are welcome to do the same for $m \neq 0$; it is just more work.

4.   Since P_l^m and P_l^{-m} satisfy the same equation (20.11-3) (this equation is unaltered by replacing m by $-m$), which can have at most one solution regular at $w = \pm 1$, they must be proportional. Find the proportionality constant. *Further warning about notation*: some authors define P_l^{-m} to be $= P_l^m$.

5.   Show that, as an alternative to (20.11-6), the equation

$$P_l^m(w) = (-1)^m \frac{(l+m)!}{(l-m)!} \frac{1}{2^l l!} (1 - w^2)^{-m/2} \left(\frac{d}{dw}\right)^{l-m} (w^2 - 1)^l \qquad (20.11\text{-}8)$$

holds, for $-1 \le m \le l$.

20.12   Matrices of the Irreducible Representations of SO(3); the Euler Angles

For given l, the functions Y_l^m $(m = l, l - 1, \ldots, -l)$ are taken as basis vectors in the space X^{2l+1} of the $(2l + 1)$-dimensional representation of SO(3) found in the preceding sections. For any function $f = f(\theta, \varphi)$, $\rho(g)f$ is the function obtained by carrying the values of $f(\theta, \varphi)$ around the sphere by the rotation g. Hence, the matrix $\rho^l(g)$ of the transformation $\rho(g)$, when restricted to X^{2l+1}, has components $\rho_{m'm}^l$ given by

$$(\rho(g)Y_l^m)(\theta, \varphi) = \sum_{m'=-l}^{l} \rho_{m'm}^l(g) Y_l^{m'}(\theta, \varphi) \qquad (m = l, l - 1, \ldots, -l).$$

$$(20.12\text{-}1)$$

It is convenient to express the rotation by its Euler angles α, β, γ, and to write $\rho(\alpha, \beta, \gamma)$ instead of $\rho(g)$. Then, g is the result of the following rotations in succession:

1.   A rotation through the angle γ about the z axis,
2.   A rotation through the angle β about the x axis,
3.   A rotation through the angle α about the z axis.

(See Exercises 3, 4, and 5 in Section 21.5.) The matrix ρ^l is decomposed accordingly as

$$\rho^l(\alpha, \beta, \gamma) = \rho^l(\alpha, 0, 0)\rho^l(0, \beta, 0)\rho^l(0, 0, \gamma).$$

The first and third factors are diagonal matrices; the transformation $\rho(\alpha, 0, 0)$ merely replaces φ in a function by $\varphi - \alpha$ and hence multiplies Y_l^m by $e^{-i\alpha m}$; that is,

$$\rho_{m'm}^l(\alpha, 0, 0) = e^{-i\alpha m'} \delta_{m'm}.$$

Therefore, $\rho_{m'm}^l(\alpha, \beta, \gamma)$ can be written in the form

$$\rho_{m'm}^l(\alpha, \beta, \gamma) = e^{-i\alpha m'} P_{m'm}^l(\cos \beta) e^{-i\gamma m}. \qquad (20.12\text{-}2)$$

The functions $P_{m'm}^l(w)$ are closely related to the Jacobi polynomials; their properties are discussed at length in Gel'fand, Minlos, and Shapiro 1963 and in Vilenkin 1968, to which the reader is referred for details. (The definition of $P_{m'm}^l$ given below agrees with that in Gel'fand et al. and gives the complex conjugate of the function defined by Vilenkin.) The $P_{m'm}^l$ are defined by the equation

$$P_{m'm}^l(w) = C(1 + w)^{-(m+m')/2}(1 - w)^{(m-m')/2}\left(\frac{d}{dw}\right)^{l-m'}$$

$$\times [(w - 1)^{l-m}(w + 1)^{l+m}], \qquad (20.12\text{-}3)$$

where

$$C = i^{m'-m}2^{-l}\left(\frac{(l + m')!}{(l - m)!(l + m)!(l - m')!}\right)^{1/2}. \qquad (20.12\text{-}4)$$

(It would perhaps be more logical to incorporate the factor $i^{m'-m}$ explicitly in $\rho_{m'm}^l$, rather than in $P_{m'm}^l$, which would then be a real function, for $-1 \le w \le 1$, but it is not customary to do so.)

For $m = 0$ (and for $m' = 0$), these functions are proportional to the associated Legendre functions. Comparison of (20.12-3) with (20.11-8) shows that

$$P_{m'0}^l(w) = (-i)^{m'}\sqrt{\frac{(l - m')!}{(l + m')!}} P_l^{m'}(w) \qquad (20.12\text{-}5)$$

[in particular, $P_{00}^l(w) = P_l(w)$]; hence,

$$\rho_{m'0}^l(\alpha, \beta, 0) = \sqrt{\frac{4\pi}{2l + 1}} Y_l^{m'}\left(\beta, \alpha - \frac{\pi}{2}\right). \qquad (20.12\text{-}6)$$

[This is the same as $\rho_{m'0}^l(\alpha, \beta, \gamma)$, because $\rho_{m'0}^l$ is independent of γ.]

EXERCISES

1. Verify (20.12-1) for g = rotation about the z axis.
2. Verify (20.12-1) for g = rotation by π about the x axis.
3. For $l = 1$, show that the $P_{m'm}^l$ are given in matrix format by

$$(P_{m'm}^l) = \begin{vmatrix} \dfrac{1 + w}{2} & -i\sqrt{\dfrac{1 - w^2}{2}} & \dfrac{w - 1}{2} \\ -i\sqrt{\dfrac{1 - w^2}{2}} & w & -i\sqrt{\dfrac{1 - w^2}{2}} \\ \dfrac{w - 1}{2} & -i\sqrt{\dfrac{1 - w^2}{2}} & \dfrac{1 + w}{2} \end{vmatrix},$$

where the rows are numbered by $m' = -1, 0, 1$ and the columns by $m = -1, 0, 1$. Note that this matrix is unitary.

4. Identify the left member of (20.12-1) with $Y_l^m(\theta', \varphi')$, multiply the equation through by r, call $z' = r \cos \theta'$, $x' \pm iy' = r \sin \theta' e^{\pm i\varphi'}$, and similarly for x, y, z. Again, take $l = 1$. Then, using the result of Exercise 3, show that, for the case g = rotation

through the angle β about the x axis (i.e., $\alpha = \gamma = 0$), the transformation (20.12-1) reduces to

$$x' = x,$$
$$y' = y \cos \beta + z \sin \beta,$$
$$z' = -y \sin \beta + z \cos \beta.$$

5. Show that $P^l_{m'm} = P^l_{mm'}$.

6. Show that the appearance of $\alpha - \pi/2$, rather than α itself, in (20.12-6) could be avoided if the second step in the definition of the Euler angles were taken to be a rotation through the angle β about the y axis rather than the x axis.

20.13 The Addition Theorem for the Tesseral Harmonics

In equation (20.12-1), which tells how the functions Y^m_l, for given l, transform among themselves under a rotation g, we set $m = 0$, and we take g to be the rotation with Euler angles α, β, 0:

$$(\rho(\alpha, \beta, 0) Y^0_l)(\theta, \varphi) = \sum_{m'=-l}^{l} \rho^l_{m'0}(\alpha, \beta, 0) Y^{m'}_l(\theta, \varphi). \qquad (20.13\text{-}1)$$

Now the operator $\rho(\alpha, \beta, 0)$ carries the values of a function around the sphere by the rotation g, which takes the polar axis into the direction with coordinates $\beta, \alpha - \pi/2$. Therefore, the left member above is equal to

$$Y^0_l(\theta_{12}, 0),$$

where θ_{12} is the angle between the direction $\beta, \alpha - \pi/2$ and the direction θ, φ [recall that $Y^0_l(\theta, \varphi)$ does not depend on φ]. We rename the first of these directions θ_1, φ_1 and the second θ_2, φ_2, and we use the trigonometrical identity

$$\cos \theta_{12} = \cos \theta_1 \cos \theta_2 + \sin \theta_1 \sin \theta_2 \cos(\varphi_1 - \varphi_2). \qquad (20.13\text{-}2)$$

We also substitute (20.12-6) into (20.13-1); the result is

$$Y^0_l(\theta_{12}, -) = \sqrt{\frac{4\pi}{2l+1}} \sum_{m=-l}^{l} \overline{Y^m_l(\theta_1, \varphi_1)}\, Y^m_l(\theta_2, \varphi_2), \qquad (20.13\text{-}3)$$

which is the required addition theorem. It can also be written as

$$P_l(\cos \theta_{12}) = P_l(\cos \theta_1) P_l(\cos \theta_2)$$

$$+ 2 \sum_{m=1}^{l} \frac{(l-m)!}{(l+m)!}\, P^m_l(\cos \theta_1) P^m_l(\cos \theta_2) \cos m(\varphi_1 - \varphi_2).$$

$$(20.13\text{-}4)$$

It is noted in passing that the special case $l = 1$ gives the trigonometric formula (20.13-2) back again, because $P_1(w) = w$, $P^1_1(w) = \sqrt{1 - w^2}$.

20.14 Completeness of the Tesseral Harmonics

It will be shown that the tesseral harmonics form a complete set for the expansion of functions defined on the sphere. Since (20.11-3) is a differential equation of Sturm–Liouville type in the interval $(-1, 1)$ (with singular endpoints of the limit-point type at each end), the completeness can be established by the methods of Section 10.6 in Volume I. Another approach is followed here, based on potential theory.

Rodrigues' formula (20.11-6) shows that the function $P_l^m(w)$ contains only even powers of w for $l + m$ even, and only odd powers for $l + m$ odd, together with a factor $\sqrt{1 - w^2}$, for m odd. It follows from (20.1-7), by repeated use of the equations

$$r \cos \theta = z,$$

$$r \sin \theta e^{i\varphi} = x + iy,$$

$$r^2 = x^2 + y^2 + z^2,$$

that $r^l Y_l^m(\theta, \varphi)$ is a homogeneous polynomial of degree l in x, y, z; i.e., each term is of the form: constant times $x^i y^j z^k$, where $i + j + k = l$.

It will now be shown that these polynomials satisfy Laplace's equation. The Laplacian, when written in polar coordinates, can be expressed in terms of the operators L_i, by use of (20.9-5, 6), as follows:

$$\nabla^2 = \frac{1}{r^2} \frac{\partial}{\partial r} r^2 \frac{\partial}{\partial r} + \frac{1}{r^2 \sin \theta} \frac{\partial}{\partial \theta} \sin \theta \frac{\partial}{\partial \theta} + \frac{1}{r^2 \sin^2 \theta} \frac{\partial^2}{\partial \varphi^2}$$

$$= \frac{1}{r^2} \left[\frac{\partial}{\partial r} r^2 \frac{\partial}{\partial r} + L^- L^+ + L_3^2 - iL_3 \right]. \qquad (20.14\text{-}1)$$

Now Y_l^m is an eigenfunction of $L^- L^+$ and of L_3, with eigenvalues $-(\alpha_l^m)^2$ and $-im$, respectively. Therefore,

$$\nabla^2 r^l Y_l^m(\theta, \varphi) = \frac{1}{r^2} [l(l + 1) - (l + m + 1)(l - m) - m^2 - m] r^l Y_l^m(\theta, \varphi) = 0,$$

where (20.9-12) has been used.

The general homogeneous polynomial $p(x, y, z)$ of degree l has $\frac{1}{2}(l + 1)(l + 2)$ terms $x^i y^j z^k$, because if $i = 0$ there are $l + 1$ possible values of j, if $i = 1$ there are l possible values, etc.; in each case k is then fixed as $l - i - j$; hence the number of terms is $1 + 2 + \cdots + (l + 1) = \frac{1}{2}(l + 1)(l + 2)$. Now suppose that $p(x, y, z)$ satisfies Laplace's equation. Since $\nabla^2 p$ is a homogeneous polynomial of degree $l - 2$, its vanishing imposes $\frac{1}{2}(l - 1)l$ conditions on the coefficients of p, and it is easy to see that these conditions are independent. Hence, the harmonic polynomials of degree l form a space of

$$\frac{(l + 1)(l + 2)}{2} - \frac{(l - 1)l}{2} = 2l + 1$$

dimensions. This space is precisely spanned by the polynomials

$$r^l Y_l^m \ (m = l, l - 1, \ldots, -l),$$

for there are $2l + 1$ of them, and they are obviously independent because of the orthogonality of the Y_l^m. It is concluded *that any harmonic polynomial can be expressed in terms of the functions* $r^l Y_l^m(\theta, \varphi)$.

According to the theory of the Dirichlet problem in potential theory, if $f(\theta, \varphi)$ is any continuous function on the unit sphere, there is a function $\psi(x, y, z)$ that satisfies $\nabla^2 \psi = 0$, for $x^2 + y^2 + z^2 < 1$, and is continuous, for $x^2 + y^2 + z^2 \leq 1$, and takes the values $f(\theta, \varphi)$ on the sphere $x^2 + y^2 + z^2 = 1$. (In fact, ψ can be expressed in terms of f by the Poisson integral formula.) ψ is analytic in the ball and can be expressed as a power series in x, y, and z. The terms of this expansion, of a given degree l constitute a harmonic polynomial of degree l, and hence can be expressed in terms of the functions $r^l Y_l^m(\theta, \varphi)$; that is,

$$\psi(x, y, z) = \sum_{l=0}^{\infty} \sum_{m=-l}^{l} A_l^m r^l Y_l^m(\theta, \varphi).$$

It can be shown that, for a continuous boundary function f, the solution ψ of the Dirichlet problem converges to f, uniformly in angle, as $r \to 1$. Hence, it converges in L^2; therefore,

$$f(\theta, \varphi) \quad = \sum_{(l, m)} A_l^m Y_l^m(\theta, \varphi), \qquad (20.14\text{-}2)$$

in the sense of mean convergence. Since the continuous functions are dense in $L^2(S)$, it follows that the tesseral harmonics form a complete orthonormal set of functions on the sphere. It follows also that any distribution $f(\theta, \varphi)$ in $L^2(S)$ can be expanded as (20.14-2), where the coefficients are given by

$$A_l^m = \int_0^{2\pi} \int_0^{\pi} \overline{Y_l^m(\theta, \varphi)} f(\theta, \varphi) \sin \theta \; d\theta \; d\varphi,$$

and where the series converges to $f(\theta, \varphi)$ in the mean.

A gap in the discussion of the representations of SO(3) in Section 20.9 can now be filled in. It is recalled that X^∞ was defined as the space of all C^∞ functions on the sphere, and X^{2l+1} was the subspace spanned by

$$Y_l^m \; (m = l, l - 1, \ldots, -l).$$

It was proved that X^{2l+1} is invariant under (is transformed into itself by) the infinitesimal operators L_1, L_2, L_3 of the representation ρ, and it can now be proved that X^{2l+1} is invariant also under the transformations $\rho(g), g \in SO(3)$: Namely, $\rho(g) Y_l^m$ is a function of θ and φ obtained by rotating the sphere by the rotation g and carrying the values of $Y_l^m(\theta, \varphi)$ along; hence, $r^l \rho(g) Y_l^m$ is also a homogeneous harmonic polynomial of degree l in x, y, z; hence, it can be expressed linearly in the polynomials $r^l Y_l^{m'}(\theta, \varphi)$, $m' = l, l - 1, \ldots, -l$. Therefore, X^{2l+1} is invariant.

The completeness of the set of tesseral harmonics shows that the space $L^2(S^2)$, where S^2 is the unit sphere in $\mathbb{R}^3$, is the direct sum (with respect to the L^2 norm) of the spaces X^{2l+1} ($l = 0, 1, 2, \ldots$).

CHAPTER 21

Group Representations II:
General; Rigid Motions; Bessel Functions

Unitary representation; reduction; decomposition; direct sum; complete
reduction; Schur's Lemma; compact and noncompact groups; invariant
integration; Haar measure; right and left translations; invariant integration in
SU(2); area of the n-sphere; regular representations; invariant integration in
SO(3); completeness theorems of Peter and Weyl and Vilenkin; wave
functions of the symmetric top; rigid motion groups; Bessel functions;
recurrence relations, differential equations, and generating equations;
expansion of plane wave; characters; completeness of representations of
SO(3).

Prerequisites: Chapters 18–20.

The irreducible representations of SO(2) found in Section 20.5 were one-
dimensional; those of SO(3) found in Section 20.9 were multidimensional,
with dimension $2l + 1$, $l = 0, 1, 2, \ldots$; those of the rigid motion groups, to
be discussed below, are infinite-dimensional. These differences reflect
various group properties, as will be shown.

21.1 Equivalence; Unitary Representations

The first step in classifying representations is to decide when two representations are equivalent, so that only one of them need be described. Two
representations $\rho(g)$ and $\rho'(g)$ on a (finite-dimensional) space X are *equivalent*
if the first can be transformed into the second by a suitable change of the
coordinates in X, i.e., if there is a matrix A such that $\rho(g) = A^{-1}\rho'(g)A$,
for all g. More generally, if ρ and ρ' are representations on different spaces
X and X' of the same dimension, and if there is a matrix A such that the
association $\mathbf{x} \rightarrow \mathbf{x}' = A\mathbf{x}$ is a one-to-one mapping of X onto X', and if $\rho(g)$
and $A^{-1}\rho'(g)A$ are the same matrix, for every g, then ρ and ρ' are equivalent
representations.

In the infinite-dimensional case, as noted in Section 20.4, $\rho(g)$ is a bounded linear transformation in a Banach or Hilbert space for each g. If ρ and ρ' are representations in spaces X and X' of the same kind, say both separable Hilbert spaces, and if there is a bounded linear transformation A from X onto X' with an inverse A^{-1}, and if $\rho(g)$ and $A^{-1}\rho'(g)A$ are the same operator for every g, then ρ and ρ' are equivalent representations.

This equivalence relation is reflexive, symmetric, and transitive, hence divides the set of all representations of a given group into equivalence classes. It is natural to try to choose from each equivalence class a representation with convenient properties. In most cases it is possible to choose a *unitary* representation, i.e., one in which each $\rho(g)$ is a unitary matrix or unitary operator, for all g in G; this is advantageous because of the theorem in the next section.

It should be noted that the equivalence of two representations ρ and ρ' is not merely a matter of the isomorphism of the two matrix groups $\{\rho(g)\}$ and $\{\rho'(g)\}$: The matrices have to be of the same order in the two representations, and the representations have to be so related that $\rho(g) = A^{-1}\rho'(g)A$, for all g, for some fixed A. The representations $\rho^l(l = 0, 1, 2, \ldots)$ of SO(3) found in Section 20.3 are all inequivalent (they have different dimensions), but the resulting matrix groups $\{\rho^l(g)\}$ are all isomorphic, for $l \neq 0$: In fact, they are all isomorphic to SO(3). The following example shows that two representations may have the same dimension but be inequivalent: Let G be the torus group T_2, consisting of 2×2 diagonal unitary matrices, i.e., matrices of the form

$$g = \begin{pmatrix} \alpha & 0 \\ 0 & \beta \end{pmatrix}, \qquad |\alpha| = 1, \qquad |\beta| = 1.$$

The two representations on $\mathbb{C}$ given by

$$\rho(g): z \to \alpha z, \qquad \rho'(g): z \to \beta z$$

are inequivalent; no transformation of the z plane can take $\rho(g)$ into $\rho'(g)$ for all g. For $g = \begin{pmatrix} 1 & 0 \\ 0 & -1 \end{pmatrix}$, $\rho(g)$ is the identity mapping in the plane, and $\rho'(g)$ is not.

21.2 The Reduction of Representations

The next step in the classification is to decompose a given representation, as far as possible, into irreducible components. If ρ is a representation on a finite-dimensional space X, and if a subspace X_1 is invariant under all $\rho(g)$— i.e., if $\mathbf{x} \in X_1$ implies $\rho(g)\mathbf{x} \in X_1$, for all g—then ρ is called *reducible*, as in the preceding chapter (otherwise, ρ is *irreducible*). If another invariant subspace X_2 can be found such that $X = X_1 \oplus X_2$ (by which is meant that X_1 and X_2 have no vector in common except $\mathbf{x} = 0$ and that any $\mathbf{x}$ in X can be written as $\mathbf{x}_1 + \mathbf{x}_2$, where $\mathbf{x}_1 \in X_1$ and $\mathbf{x}_2 \in X_2$), then ρ is called *fully reducible* or *decomposable*. In this case, if $\mathbf{e}^1, \ldots, \mathbf{e}^n$ is a basis in X such that $\mathbf{e}^1, \ldots, \mathbf{e}^m$ is a basis in X_1 while $\mathbf{e}^{m+1}, \ldots, \mathbf{e}^n$ is a basis in X_2, then, with respect to this basis, all the matrices $\rho(g)$ have the form

$$\rho(g) = \begin{pmatrix} \boxed{\begin{array}{c} m \times m \end{array}} & (0) \\[1em] (0) & \boxed{\begin{array}{c} (n-m) \\ \times \\ (n-m) \end{array}} \end{pmatrix} ; \qquad (21.2\text{-}1)$$

i.e., all matrix elements that connect the two subspaces are zero. (If ρ is reducible but not decomposable, then it is possible to choose the basis so as to make all matrix elements zero in the lower left rectangular block shown, but not also in the upper right block.) If, for each g, $\rho_1(g)$ and $\rho_2(g)$ denote the $m \times m$ and $(n-m) \times (n-m)$ matrices shown, respectively, then each of the mappings $g \rightarrow \rho_1(g)$ and $g \rightarrow \rho_2(g)$ is a representation of G, and the representation ρ is their *direct sum*; in symbols, $\rho = \rho_1 \dotplus \rho_2$.

If X is an infinite-dimensional Banach or Hilbert space, X_1 is understood as a closed linear manifold in X; there is no loss of generality here, because the operators $\rho(g)$ are all bounded, hence the closure of an invariant linear manifold is invariant. Again, if $X = X_1 \oplus X_2$, and if X_1 and X_2 are invariant under all $\rho(g)$, then we write

$$\rho = \rho_1 \dotplus \rho_2,$$

where ρ_1 and ρ_2 are the restrictions of the representation ρ to X_1 and X_2, respectively. (One of ρ_1, ρ_2 may be finite-dimensional.)

It may be that X_1 or X_2 contains further invariant subspaces such that ρ_1 and ρ_2 (or both) can be further decomposed, and so on. Then, with respect to a suitable basis in X, the matrices $\rho(g)$ contain a number of square blocks straddling the main diagonal, and all elements outside those blocks are zero. Each square block gives a representation of G, and ρ is the direct sum of those representations. When this process has been carried as far as possible, it may turn out that the resulting representations into which ρ has been decomposed are all irreducible. In this case, ρ is called *completely reducible*. Then one can find the structure of all representations of G by finding all minimal invariant subspaces of a sufficiently large initial space X, as was done in the preceding chapter for SO(2) and SO(3).

Theorem. *Any finite-dimensional unitary representation of a group G is completely reducible, i.e., either is irreducible or can be written as a direct sum of irreducible representations.*

PROOF. The result follows from the fact that if ρ is unitary and X_1 is invariant, then, as is easily seen, $X_1^\perp$ is also invariant.

21.3 Schur's Lemma and Its Corollaries

The next theorem depends on the famous lemma proved by I. Schur in 1905, an apparently trivial but quite powerful bit of elementary linear algebra.

Lemma (Schur).   *If ρ_1 and ρ_2 are irreducible representations of G on V^k and V^l, respectively, and if A is an $l \times k$ matrix such that*

$$A\rho_1(g) = \rho_2(g)A, \quad \text{for all } g \text{ in } G, \tag{21.3-1}$$

then either $k = l$, and A has an inverse A^{-1} (hence ρ_1 and ρ_2 are equivalent), or A is the zero matrix. In the first case, A is uniquely determined, up to a scalar multiplier, by (21.3-1).

PROOF.   Let S be the null space of A, consisting of all vectors $\mathbf{x}$ in V^k such that $A\mathbf{x} = 0$. S is invariant under all $\rho_1(g)$, for if $A\mathbf{x} = 0$, then $A\rho_1(g)\mathbf{x} = 0$, by (21.3-1). Since ρ_1 is irreducible, either S is all of V^k, in which case $A = 0$, or S consists of the zero vector alone. In the latter case, no nonzero vector is orthogonal to all the rows of A; hence the number l of rows is $\geq k$, and the mapping $\mathbf{x} \to \mathbf{y} = A\mathbf{x}$ is invertible because any $k \times k$ submatrix in A is nonsingular and can be used to solve the equation $y = A\mathbf{x}$ for $\mathbf{x}$. Let S' be the image of V^k in V^l under the mapping $\mathbf{x} \to A\mathbf{x}$; then S' is k-dimensional and is invariant under all $\rho_2(g)$, because if y is $A\mathbf{x}$ for some $\mathbf{x}$, then $\rho_2(g)y$ is $A\mathbf{x}'$, for $\mathbf{x}' = \rho_1(g)\mathbf{x}$. The representation ρ_2 is irreducible; hence $S' = V^l$, $k = l$, and A^{-1} is the matrix of the inverse mapping. We now prove the uniqueness of A: Suppose that B is another nonzero matrix such that $B\rho_1(g) = \rho_2(g)B$, for all g; then it must be proved that $A = \text{const. } B$. Clearly,

$$\rho_2(g)AB^{-1} = AB^{-1}\rho_2(g), \quad \text{for all } g.$$

It follows from this equation that if $\mathbf{v}$ is an eigenvector of AB^{-1} (every matrix has at least one eigenvector), then $\rho_2(g)\mathbf{v}$ is also an eigenvector corresponding to the same eigenvalue, say λ. That is, the one-dimensional subspace containing $\mathbf{v}$ is invariant under $\rho_2(g)$ for all g. Since ρ_2 is irreducible, the complete eigenspace corresponding to λ must be all of V^k; that is, $AB^{-1}\mathbf{v} = \lambda\mathbf{v}$, for all $\mathbf{v}$; hence $B^{-1} = \lambda A^{-1}$ or $A = \lambda B$, as required.

Comment. No use has been made of the fact that G is a group. The same conclusions hold if the sets $\{\rho_1(g)\}$ and $\{\rho_2(g)\}$ are any two irreducible sets of square matrices. A set $\{M_i\}$ of $k \times k$ matrices is *irreducible* if there is no nontrivial proper subspace of V^k that is invariant under all mappings $\mathbf{x} \to M_i\mathbf{x}$.

The following was established during the proof of Schur's lemma:

Corollary.   *Any matrix that commutes with all the matrices of an irreducible representation (or an irreducible set of matrices) is a multiple of the identity.*

A further corollary is the following:

Theorem.   *Every finite-dimensional irreducible representation ρ of an Abelian (commutative) group is one-dimensional.*

PROOF.   Each $\rho(h)$ commutes with all $\rho(g)$; hence each $\rho(h)$ is a multiple of the identity matrix; hence, ρ would be reducible if it were not one-dimensional.

This explains why only one-dimensional representations of SO(2) were found in Section 20.5.

21.4 Compact and Noncompact Groups

It was pointed out in Section 19.5 that an $n \times n$ matrix (generally complex) can be represented by a point in a space V of $2n^2$ (real) dimensions, and that the set of all matrices constituting any group like GL(n), SL(n), U(n),O(n), etc., is an algebraic surface $\mathscr{S}$ in V. For the groups mentioned (they are *continuous* groups or *Lie* groups), $\mathscr{S}$ is always a closed point set, but it may not be bounded. For O(n), U(n), SO(n), SU(n), the surface $\mathscr{S}$ is bounded [for example, equation (19.5-1) shows that, for the group O(3), no point of $\mathscr{S}$ can have a coordinate R_{jk} larger than 1 in absolute value], whereas, for GL(n), SL(n), M_n, and the Lorentz groups, $\mathscr{S}$ is unbounded (extends to infinity in V). In the first case, the group is called *compact* ($\mathscr{S}$ is a compact set in V), and in the second case *noncompact*. The theory of compact groups is much simpler than that of the noncompact ones; that shows an interesting interplay between different parts of mathematics: The compactness of a group manifold is a matter of geometry, while the simplifications referred to are mainly algebraic.

The following theorems, stated without proof, show that for compact groups it suffices to consider unitary representations.

Theorem 1.   *Let ρ be a finite-dimensional representation of a compact group G. Then ρ is equivalent to a unitary representation; that is, there is a fixed nonsingular matrix A such that the matrix $A\rho(g)A^{-1}$ is unitary for all g in G.*

This can be restated as follows: If a new inner product $(\cdot, \cdot)_1$ is defined in the representation space X by the equation $(\xi, \eta)_1 \overset{\text{def}}{=} (A\xi, A\eta)$, where $(\cdot, \cdot)$ is the ordinary inner product, given by $(\mathbf{x}, \mathbf{y}) = \mathbf{x}^*\mathbf{y}$, then

$$(\rho(g)\xi, \rho(g)\eta)_1 = (\xi, \eta)_1, \quad \text{for all } g \text{ in } G, \text{ all } \xi, \eta \text{ in } X; \quad (21.4\text{-}1)$$

in other words, the matrices $\rho(g)$ are unitary with respect to the new inner product. The same holds generally:

Theorem 2.   *Let ρ be a representation of a compact group G on a Hilbert space $\mathfrak{H}$; i.e., each $\rho(g)$ is an invertible bounded linear operator in $\mathfrak{H}$, and $\rho(g_1 g_2) = \rho(g_1)\rho(g_2)$. Then there is a new inner product $(\cdot, \cdot)_1$ in $\mathfrak{H}$ such that*

$$(\rho(g)u, \rho(g)v)_1 = (u, v)_1, \quad \text{for all } g \text{ in } G, \text{ all } u, v \text{ in } \mathfrak{H}; \quad (21.14\text{-}2)$$

Theorem 3.   *All irreducible unitary representations of a compact group G are finite-dimensional.*

By contrast, the Lorentz groups have infinite-dimensional irreducible unitary representations.

21.5 Invariant Integration; Haar Measure

Let G be a compact group of $n \times n$ matrices; the manifold $\mathscr{S}$ of G is a compact (closed, bounded) surface (of some dimension $m \leq 2n^2$) in a space V of $2n^2$ (real) dimensions (see Section 19.5). It can be proved that there is a positive continuous function $w(g)$ defined on $\mathscr{S}$—here, the symbol g is used both for a group element and for the corresponding point of $\mathscr{S}$—with the remarkable property that if f is any continuous function on $\mathscr{S}$ and h is any fixed group element, then

$$\int_{\mathscr{S}} f(hg)w(g)d\mathscr{A}(g) = \int_{\mathscr{S}} f(g)w(g)d\mathscr{A}(g)$$

$$(21.5\text{-}1)$$

for all h in G, all continuous f on $\mathscr{S}$,

where $d\mathscr{A}(g)$ is the m-dimensional volume element, or "element of area" on $\mathscr{S}$.

If G is noncompact, so that the surface $\mathscr{S}$ extends to infinity in V, then the same is true—i.e., there is a weight function $w(g)$ on $\mathscr{S}$ such that (21.5-1) holds—provided that f is a function such that the integrals converge.

The mapping $g \rightarrow hg$ of G onto itself, for fixed h, is called a *left translation* in G. The equation above shows that the integral of a continuous function on $\mathscr{S}$ with respect to the weight function w is invariant under all left translations in the group. The integral in (21.5-1) is called a *left-invariant integral*. A proof that such a function w exists can be found in Wigner's book *Gruppentheorie* (1931) under the heading "Hurwitzsches Integral." See also Nachbin 1965.

There is a similar invariant integral for right translations. It can be proved that, in particular, if the group G is compact, then the above integral (with the *same* weight function w) is also invariant under right translations and under inversions; that is

$$\int f(g)w(g)d\mathscr{A}(g) = \int f(gh)w(g)d\mathscr{A}(g) = \int f(g^{-1})\dot{w}(g)d\mathscr{A}(g). \quad (21.5\text{-}2)$$

The proof can be found in Weyl's book, *The Theory of Groups and Quantum Mechanics* (1932), Chapter III, Section 12.

Exercises

1. Show that any matrix g in SU(2) can be written as

$$g = \begin{pmatrix} a & b \\ -\bar{b} & \bar{a} \end{pmatrix},$$

where a and b are complex numbers subject only to the restriction $|a|^2 + |b|^2 = 1$. Hence, if $a = x_1 + ix_2$ and $b = x_3 + ix_4$, the manifold S of SU(2) is homeomorphic to the 3-sphere S^3, i.e., the unit sphere $x_1^2 + x_2^2 + x_3^2 + x_4^2 = 1$ in $\mathbb{R}^4$.

2.   Show that a left translation $g \to hg$ in SU(2), where h is a fixed group element, induces a rotation of S^3 about its center, and hence that if $d\mathscr{A}(g)$ is the element of 3-dimensional area on S^3, the weight function $w(g)$ in (20.7-3) can be taken to be constant, and the integral $\int f(g)d\mathscr{A}(g)$ is invariant under left translations (also right translations) in the group.

3.   Let a and b in Exercise 1 be written as

$$a = \left(\cos\frac{\beta}{2}\right)\exp\left(i\,\frac{\alpha + \gamma}{2}\right), \qquad b = i\left(\sin\frac{\beta}{2}\right)\exp\left(i\,\frac{\alpha - \gamma}{2}\right),$$

where

$$0 \le \alpha < 2\pi,$$

$$0 \le \beta \le \pi,$$

$$-2\pi \le \gamma < 2\pi;$$

then g is written as $g(\alpha, \beta, \gamma)$, and the variables α, β, γ are called the *Euler angles* of g. [Under the homomorphism of SU(2) onto SO(3) given in Section 19.7, they become the Euler angles of the rotation $R(g)$, with γ restricted to the range $0 \le \gamma < 2\pi$—note that replacing γ by $\gamma + 2\pi$ replaces g by $-g$ and leaves $R(g)$ unaltered.] Show that if the Euler angles are taken as intrinsic coordinates in the group SU(2), then the element of area on S^3 is given by

$$d\mathscr{A}(g) = \frac{\sin\beta}{8}\,d\alpha\,d\beta\,d\gamma.$$

4.   Show that

$$g(\alpha, \beta, \gamma) = g(\alpha, 0, 0)g(0, \beta, 0)\,g(0, 0, \gamma). \tag{21.5-3}$$

5.   Show that, under the homomorphism of SU(2) onto SO(3) given in Section 19.7, if $R(g(\alpha, \beta, \gamma))$ is called $R(\alpha, \beta, \gamma)$, then $R(\alpha, 0, 0)$ is $= R(0, 0, \alpha)$ and is a rotation through α about the z axis, while $R(0, \beta, 0)$ is a rotation through β about the x axis. Give the geometrical interpretation of the result of Exercise 4 as the law of composition of an arbitrary rotation in terms of successive rotations about the z, x, and z axes, respectively.

6.   Derive the formula for the area A_n of the n-sphere (the unit sphere in E^{n+1}) from the evident equation

$$\left(\int_{-\infty}^{\infty} e^{-x^2}\,dx\right)^{n+1} = \int_0^{\infty} e^{-r^2} A_n r^n\,dr,$$

using the gamma function, and verify directly that the formula of Exercise 3 is correctly normalized. Conclude, on the basis of the 2-to-1 homomorphism of SU(2) onto SO(3) that the (3-dimensional) area of the surface that has been identified with the manifold of SO(3)—see Section 19.5—is equal to π^2. Show that the volume of an n-dimensional ball of radius R is

$$V_n = \frac{2\pi^{n/2}}{n\Gamma(n/2)}\,R^n.$$

The system $\{Y_l^m\}$ of tesseral harmonics is not the only orthogonal function system that comes from the representations of SO(3). The tesseral harmonics are orthogonal on the unit 2-sphere, which was taken as the homogeneous space for the representation. However, it was pointed out in Section 20.8 that the group manifold can also be taken as the homogeneous space; then, a larger class of orthogonal functions appears—they are functions of the Euler angles α, β, γ, which can be taken as intrinsic coordinates in SO(3). The theorem below deals with such function systems in general.

It is customary to denote the expression $w(g)d\mathscr{A}(g)$ that appears in the left-invariant integration over the group manifold simply by dg, and to write equation (21.5-1) as

$$\int_G f(hg)dg = \int_G f(g)dg. \tag{21.5-4}$$

Let $C_0(G)$ denote the space of continuous functions of compact support on the group manifold, or, as one says, on G (if G itself is compact, then this includes all continuous functions on G). Then $\int_G f(g)dg$, $f \in C_0(G)$, is a continuous linear functional on $C_0(G)$ and is a measure (see Chapter 13). Hence, one sometimes speaks of a *left-invariant measure* on G—or *Haar measure*, because it was discussed in a paper by A. Haar in 1933.

Let $L^2(G)$ denote the Hilbert space of quadratically integrable distributions defined on the group manifold, with the inner product

$$(f_1, f_2) = \int_G \overline{f_1(g)} f_2(g)dg, \tag{21.5-5}$$

where, as above, dg is the left-invariant measure (G not necessarily compact). The *left-regular representation* of G is the association with each h in G of the mapping

$$\rho(h): f(g) \to f(h^{-1}g) \tag{21.5-6}$$

of $L^2(G)$ onto itself. (The right-invariant integral leads similarly to a *right-regular representation*.) Since the inner product is based on the left-invariant integral, it is seen that

$$(\rho(h)f_1, \rho(h)f_2) = \int_G \overline{f_1(h^{-1}g)} f_2(h^{-1}g)dg$$

$$= \int_G \overline{f_1(g)} f_2(g)dg$$

$$= (f_1, f_2), \qquad \text{for all } f_1, f_2 \text{ in } L^2 \text{ and all } h \text{ in } G.$$

That is, the representation ρ is unitary.

It can be proved that, if G is a compact group, then all irreducible representations can be obtained by the decomposition of the (left- or right-) regular representation ρ. That is, if ρ_1 is any irreducible representation, then there is a subspace X_1 of $L^2(G)$ such that the restriction of ρ to X_1 is equivalent to ρ_1.

Often, even if G is noncompact, one can find irreducible representations by operators as in (21.5-6) on *some* space of functions defined on the group, although it may be necessary to go outside the Hilbert space $L^2(G)$ to find the functions. We shall not attempt to discuss the general case but merely give an example: in Section 21.10, the functions that appear in connection with irreducible representations of the noncompact group M_2 of rigid motions in the plane are not square integrable over G.

We consider now the calculation of $w(g)$. Suppose that $\theta_1, \ldots, \theta_n$ are intrinsic coordinates in the manifold of a group G and that we wish to find the weight function $w(\theta)$ such that $w(\theta)d\theta_1 \ldots d\theta_n$ is the element of invariant measure dg.

We rewrite equation (21.5-4) as

$$\mathscr{I} = \int_G f(hg')dg' = \int_G f(g)dg.$$

We consider the case where the function $f(g)$ is zero except for elements g in a small neighborhood $\mathscr{N}$ of the identity element of the group, and $f(g) = 1$ for those elements. They occupy a small volume V in the coordinate space near $\theta = 0$; hence from the right member of the above equation we have $\mathscr{I} \approx Vw(0)$. The left member comes from group elements hg' in the neighborhood $\mathscr{N}$ of the identity, hence from g' in a neighborhood of h^{-1}, having volume V', so that $\mathscr{I} \approx V'w(h^{-1})$; hence we must determine V' to determine $w(h^{-1})$. If we call $g' = h^{-1}k$, then k varies in $\mathscr{N}$. We denote by $\hat{\theta}(g)$ the coordinates of any group element g. Hence, if θ and θ' denote the coordinates of k and $g' = h^{-1}k$, we have

$$\theta = \hat{\theta}(k), \qquad \theta' = \hat{\theta}(h^{-1}k) = \theta'(\theta).$$

As θ ranges through the volume V, θ' ranges through V'; hence V' is given in terms of the Jacobian as

$$V' \approx \left. \frac{\partial(\theta'_1, \ldots, \theta'_n)}{\partial(\theta_1, \ldots, \theta_n)} \right|_0 V,$$

where the subscript 0 indicates that the Jacobian is to be taken at $\theta = 0$. We conclude that

$$w(\hat{\theta}(h^{-1})) = C\left[\left. \frac{\partial(\theta'_1, \ldots, \theta'_n)}{\partial(\theta_1, \ldots, \theta_n)} \right|_0 \right]^{-1}, \tag{21.5-7}$$

where C is $= w(0)$. Since h was arbitrary, h^{-1} is also arbitrary; hence this equation determines $w(\theta)$ for all θ. If the group is compact, C can be chosen so as to normalize $\int_G dg$ to 1.

EXERCISE

7. Let G be the rotation group SO(3), let $\theta_x, \theta_y, \theta_z$ be the intrinsic coordinates introduced in Section 19.6, and let θ be the vector with components $\theta_x, \theta_y, \theta_z$. Specifically, let θ represent the group element k in the above discussion, where $\|\theta\| \ll 1$, and let θ'

represent $h^{-1}k$. To first order in small quantities [see (19.6-1)],

$$k = \begin{pmatrix} 1 & -\theta_z & \theta_y \\ \theta_z & 1 & -\theta_x \\ -\theta_y & \theta_x & 1 \end{pmatrix}.$$

We can take h^{-1} as a rotation through an angle α about the x axis, since $w(h^{-1})$ is independent of the direction of the rotation-axis, i.e., we can take

$$h^{-1} = \begin{pmatrix} 1 & 0 & 0 \\ 0 & \cos\alpha & -\sin\alpha \\ 0 & \sin\alpha & \cos\alpha \end{pmatrix}.$$

Show that the coordinates $\theta'_x, \theta'_y, \theta'_z$ of the element $h^{-1}k$ are, to first order,

$$\theta'_x = \alpha + \theta_x, \qquad \theta'_y = \alpha\left[\theta_y \frac{1+\cos\alpha}{2\sin\alpha} - \frac{\theta_z}{2}\right], \qquad \theta'_z = \alpha\left[\theta_z \frac{1+\cos\alpha}{2\sin\alpha} + \frac{\theta_y}{2}\right],$$

hence that the Jacobian is

$$\frac{\partial(\theta'_x, \theta'_y, \theta'_z)}{\partial(\theta_x, \theta_y, \theta_z)}\bigg|_{\theta=0} = \frac{\alpha^2}{2(1-\cos\alpha)}, \tag{21.5-8}$$

hence that the normalized weight function is given by

$$w(\theta') = \frac{1-\cos\alpha}{4\pi^2\alpha^2}, \qquad \alpha = \|\theta'\|. \tag{21.5-9}$$

Hint: The angle of rotation and the direction of the axis are given by (19.2-7 and 8) for a given rotation matrix.

21.6 Complete System of Representations of a Compact Group

We come now to one of the most important theorems on compact groups.

Theorem.   *Let ρ^k ($k = 1, 2, \ldots$) be a complete set of inequivalent irreducible unitary representations of a compact group G; let d_k be the dimension of ρ^k, and let $\rho^k_{mn}(g)$ be the matrix elements of the transformation $\rho^k(g)$. Then the functions*

$$\sqrt{d_k}\,\rho^k_{mn}(g), \qquad \begin{cases} k = 1, 2, \ldots, \\ 1 \leq m, n \leq d_k, \end{cases}$$

form a complete orthonormal system on G with respect to the inner product based on invariant integration on G.

Remarks. (1) Recall that right- and left-invariant integration are the same on a compact group. (2) It has been assumed that $\int_G dg = 1$. (3) The matrix elements are assumed to refer to an orthonormal set of vectors, so that the matrices $(\rho^k_{mn}(g))$ of the unitary transformations $\rho^k(g)$ are unitary matrices.

For the proof of the theorem, see Vilenkin 1968. The orthonormality of the functions is a straightforward matter; their completeness is somewhat deeper and was proved by F. Peter and H. Weyl in 1927.

For $G = SO(3)$, the theorem says that the functions

$$\sqrt{2l + 1}\, e^{im'\alpha} P^l_{m'm}(\cos \beta) e^{-imy}, \qquad \begin{cases} l = 0, 1, 2, \ldots \\ -l \leq m', m \leq l \end{cases}$$

form a complete orthonormal system on the manifold of $SO(3)$ with respect to the inner product

$$(f_1, f_2) = \frac{1}{8\pi^2} \int_0^{2\pi} \int_0^{\pi} \int_0^{2\pi} \overline{f_1(\alpha, \beta, \gamma)}\, f_2(\alpha_2, \beta, \gamma) \sin \beta \, d\alpha \, d\beta \, d\gamma;$$

see (20.12-2) and Exercise 3 in the preceding section.

The expansion of functions in $L^2(G)$ in (generalized) Fourier series in the functions described in the theorem is called harmonic analysis on the group. Harmonic analysis on noncompact groups involves generalized Fourier integrals; see, for example, Gel'fand, Graev, and Vilenkin 1966 for harmonic analysis on $SL(2, \mathbb{C})$ and Warner 1972 for harmonic analysis on semisimple Lie groups.

A similar theorem holds, under certain circumstances, for the functions obtained on other homogeneous spaces, with respect to a suitably chosen inner product. See Vilenkin 1968, Section 4.5 of Chapter 1. An example that has already been established above is that the functions $Y_l^m(\theta, \varphi)$ form a complete orthonormal set on the unit 2-sphere.

21.7 Homogeneous Spaces as Configuration Spaces in Physics

Considerations of physics often indicate which choice of homogeneous space will yield the functions of interest in a given problem, for a given symmetry group. For the quantum mechanical motion in a central force field, the coordinates are r, θ, φ; hence there are only two angle variables, and the 2-sphere is the appropriate homogeneous space. This choice leads to the functions $Y_l^m(\theta, \varphi)$, which can be taken as giving the angular dependence of the wave function. For the motion of a rigid body about its center of mass, on the other hand, there are three angular variables, the Euler angles α, β, γ, and the configuration space may be regarded as identical with the manifold of the group $SO(3)$. It was shown in 1929 by A. Sommerfield that the quantum mechanical problem of the symmetric top (a rigid body with two equal moments of inertia), which is of interest for the theory of molecular spectra, can be solved by use of the Jacobi polynomials. Indeed, the functions (20.12-2) can be taken as the wave functions of the symmetric top.

21.8 M_2 and Related Groups

A rigid motion in real n-space V^n consists of a rotation $\mathbf{x} \to R\mathbf{x}$ $[R \in SO(n)]$ followed by a translation $\mathbf{x} \to \mathbf{x} + \boldsymbol{\xi}$, where $\boldsymbol{\xi}$ is a constant vector. (If the same rotation is *preceded* by a translation $\mathbf{x} \to \mathbf{x} + \boldsymbol{\xi}_1$, the result is the same, for suitably chosen $\boldsymbol{\xi}_1$, namely $\boldsymbol{\xi}_1 = R^{-1}\boldsymbol{\xi}$.) *Note*: The words "rotation" and "motion" are possibly slightly misleading, because nothing depends on the time t. A rotation is merely a fixed change of orientation, and a translation is merely a fixed displacement.

The relations among some of the groups of interest in physics are

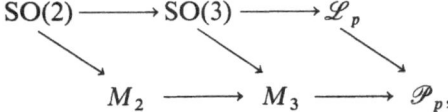

where an arrow indicates the subgroup relationship and where $\mathscr{P}_p$ is the proper Poincaré group (consisting of proper Lorentz transformations combined with displacements in space and time). Groups involving spatial inversion and time reversal are also of interest but would complicate the diagram. M_2 will be considered in some detail in this chapter.

An element of M_2 is a mapping $g = g_{\xi, \eta, \theta}$ of the x, y plane onto itself given by

$$g: \begin{pmatrix} x \\ y \end{pmatrix} \to \begin{pmatrix} x' \\ y' \end{pmatrix} = \begin{pmatrix} x \cos\theta - y \sin\theta + \xi \\ x \sin\theta + y \cos\theta + \eta \end{pmatrix}. \tag{21.8-1}$$

Note. In goup theory generally, and, in particular, in the definition of "representation," a "linear transformation" is understood to be homogeneous; hence, (21.8-1) will be called simply a "mapping." It is easily verified that a 3-dimensional faithful representation of M_2 is given by the association

$$g \leftrightarrow \begin{pmatrix} \cos\theta & -\sin\theta & \xi \\ \sin\theta & \cos\theta & \eta \\ 0 & 0 & 1 \end{pmatrix}. \tag{21.8-2}$$

EXERCISE

Verify that under this association the composition $g_2 g_1$ of two mappings of the form (21.8-1) is associated with the product of the corresponding matrices of the form (21.8-2).

21.9 Representations of M_2

To find other representations, let X^∞ denote the space of all infinitely differentiable functions $f(x, y)$ defined for all x and y. Evidently, the plane is a homogeneous space for M_2. A representation of M_2 on X^∞ is obtained by transforming each $f(x, y)$ into

$$(\rho(g)f)(x, y) = f([x - \xi]\cos\theta + [y - \eta]\sin\theta,$$

$$-[x - \xi]\sin\theta + [y - \eta]\cos\theta), \qquad (21.9\text{-}1)$$

according to the rule (20.6-1). Three operators L_1, L_2, and L_3 ("infinitesimal operators") are defined by

$$(L_1 f)(x, y) = \frac{\partial}{\partial\xi}(\rho(g)f)(x, y)\bigg|_{\xi=\eta=\theta=0},$$

$$(L_2 f)(x, y) = \frac{\partial}{\partial\eta}(\rho(g)f)(x, y)\bigg|_{\xi=\eta=\theta=0}, \qquad (21.9\text{-}2)$$

$$(L_3 f)(x, y) = \frac{\partial}{\partial\theta}(\rho(g)f)(x, y)\bigg|_{\xi=\eta=\theta=0}.$$

From (21.9-1) it is seen that

$$L_1 = -\frac{\partial}{\partial x}, \qquad L_2 = -\frac{\partial}{\partial y}, \qquad L_3 = y\frac{\partial}{\partial x} - x\frac{\partial}{\partial y}. \qquad (21.9\text{-}3)$$

If r and φ are polar coordinates in the x, y plane, and if L^+ and L^- are defined as $L_1 + iL_2$ and $L_1 - iL_2$, respectively, it is found that

$$L_3 = -\frac{\partial}{\partial\varphi}, \qquad L^{\pm} = e^{\pm i\varphi}\left(\frac{\partial}{\partial r} \pm \frac{i}{r}\frac{\partial}{\partial\varphi}\right), \qquad (21.9\text{-}4)$$

$$L^+L^- = L^-L^+ = \nabla^2.$$

21.10 Some Irreducible Representations

The representation (21.9-1) is highly reducible, and we look for an invariant subspace X_1 of X^∞ that is as small as possible in some sense. X_1 must be transformed into itself by each of L_1, L_2, L_3. By Fourier series expansion with respect to φ, any $f \in X^\infty$ can be expanded in the functions

$$\psi_m(x, y) = i^{-m}e^{im\varphi}g_m(r), \qquad (21.10\text{-}1)$$

where g_m is some C^∞ function on $(0, \infty)$. The reason for the factor i^{-m} will be apparent soon. Each ψ_m is transformed into a multiple of itself by L_3, and we wish to start with one of the ψ_m and find out what is the smallest collection of additional functions that must be included to get an invariant subspace. From (21.9-4) it is seen that $L^+\psi_m$ is of the form $e^{i(m+1)\varphi}h(r)$ and hence that $L^-L^+\psi_m$ is of the form $e^{im\varphi}\tilde{g}(r)$. In the interest of obtaining the smallest possible invariant subspace X_1, it is now *assumed* that $\tilde{g}(r)$ is $=g_m(r)$ times a constant, say $-\alpha_m^2$, and we investigate whether the g_m can be so chosen that this assumption is valid. Hence, we wish to choose the $g_m(r)$ so that

$$L^+\psi_m = i\alpha_m\psi_{m+1}, \qquad L^-\psi_{m+1} = i\alpha_m\psi_m. \qquad (21.10\text{-}2)$$

In this way, a sequence $\{\psi_m\}_{-\infty}^{\infty}$ of functions is generated, assuming that none of the α's vanish; however, none can vanish (unless all do), because L^+ and L^- commute, from which it follows that $\alpha_m^2 = \alpha_{m-1}^2$; hence all the α's can be taken as equal. Then, from (21.9-4) it is seen that

$$\nabla^2 \psi_m = -\alpha^2 \psi_m, \tag{21.10-3}$$

for each m. This requires, of course, that $g_m(r)$ be proportional to the Bessel function $J_m(\alpha r)$; Bessel functions are discussed in the next section.

Conclusion. X_1 is the subspace of X^{∞} consisting of the (nonsingular) solutions of $\nabla^2 \psi + \alpha^2 \psi = 0$: it will be called X_{α}. Each nonzero value of α leads in this way to an irreducible representation of M_2. Under a transformation $\mathbf{x} \to \mu \mathbf{x}$, where μ is real, $\nabla^2 \to \mu^{-2} \nabla^2$; hence we can assume, without loss of generality, that $|\alpha| = 1$. Furthermore, α and $-\alpha$ determine the same subspace; hence the relevant values of α may be taken as $e^{i\beta}$, $0 \le \beta < \pi$. For each such α, the representation of M_2 on X_{α} is given by the transformations (21.9-1).

21.11 Bessel Functions

We normalize the system of functions $\{\psi_m\}_{-\infty}^{\infty}$ by setting $\psi_0(0, 0) = 1$. Then, for each m, the *Bessel function of order* m can be defined as $J_m(z) = g_m(z/\alpha)$, so that (21.10-1) becomes

$$\psi_m(x, y) = i^{-m} e^{im\varphi} J_m(\alpha r).$$

The equations (21.10-2) then take the form

$$\left(\frac{d}{dz} - \frac{m}{z}\right) J_m(z) = -J_{m+1}(z),$$

$$\left(\frac{d}{dz} + \frac{m+1}{z}\right) J_{m+1}(z) = J_m(z), \tag{21.11-1}$$

which are the recurrence relations for the Bessel functions. Elimination of J_{m+1} gives

$$\left(\frac{d^2}{dz^2} + \frac{1}{z}\frac{d}{dz} + 1 - \frac{m^2}{z^2}\right) J_m(z) = 0, \tag{21.11-2}$$

which is Bessel's differential equation.

The Bessel functions $J_m(z)$ are completely determined by these equations and the starting conditions $J_0(0) = 1$ and $J_m(0) = 0$ for $m \ne 0$. From now on, we assume familiarity with these functions, and, in particular, with the integral representation

$$J_m(z) = \frac{1}{2\pi} \int_{-\pi}^{\pi} e^{iz\sin t + imt} \, dt. \tag{21.11-3}$$

21.12 Matrices of the Representations

For some purposes, it is more convenient to have the transformations represented by (infinite) matrices; the matrix elements can be obtained as follows: Let α be a vector whose (generally complex) components are $\alpha \cos \chi$ and $\alpha \sin \chi$, where α is as above and χ is a real angle. Then, a solution of the reduced wave equation $\nabla^2 u + \alpha^2 u$ is

$$e^{i\boldsymbol{\alpha} \cdot \mathbf{x}} = e^{i\alpha r \cos(\varphi - \chi)} \tag{21.12-1}$$

(as before, $x = r \cos \varphi$, $y = r \sin \varphi$), and the general solution is

$$f(r, \varphi) = \int_{-\pi}^{\pi} e^{i\alpha r \cos(\varphi - \chi)} \hat{f}(\chi) d\chi, \tag{21.12-2}$$

where $\hat{f}$ is an arbitrary function in some class of admissible functions, whose exact properties are not important here. [In the case where α is real, if $\hat{f}(r, \varphi)$ is interpreted as a wave function, then $\alpha \cos \chi$ and $\alpha \sin \chi$ are the momentum variables, and $\hat{f}(\chi)$ is closely related to the momentum representation of $f(r, \varphi)$.] If the plane wave $\exp\{i\boldsymbol{\alpha} \cdot \mathbf{x}\}$ is subjected to the transformation (21.9-1), it becomes $\exp\{i\boldsymbol{\alpha} \cdot (\mathbf{x} - \boldsymbol{\xi})_\theta\}$, where the subscript θ indicates that the vectors $\mathbf{x}$ and $\boldsymbol{\xi}$ have been rotated (counterclockwise) through the angle θ. If the displacement components are written as $\xi = \zeta \cos \omega$, $\eta = \zeta \sin \omega$, then it is seen from (21.9-1) that

$$(\rho(g)f)(r, \varphi) = \int_{-\pi}^{\pi} e^{i\alpha r \cos(\varphi - \chi)} \hat{f}'(\chi) d\chi,$$

where

$$\hat{f}'(\chi) = e^{-i\alpha\zeta \cos(\omega - \chi)} \hat{f}(\chi + \theta) \tag{21.12-3}$$

(the prime does not denote differentiation); a change of variable $\chi - \theta \rightarrow \chi$ has been made in the integral, without altering the limits, because the integrand has period 2π. Hence, the group element g of M_2, which consists of a clockwise rotation through θ followed by a translation by $\boldsymbol{\xi}$, induces the transformation (21.12-3) in the space of functions $\hat{f}$ of period 2π. $\hat{f}(\chi)$ and $\hat{f}'(\chi)$ are now expanded in the Fourier series $\sum c_m e^{im\chi}$ and $\sum c'_m e^{im\chi}$, respectively; it is found that

$$c'_m = \sum_{-\infty}^{\infty} {}_{(n)} \hat{\rho}_{mn} c_n,$$

where

$$\hat{\rho}_{mn} = \hat{\rho}_{mn}^{(\alpha)}(g) = e^{im\theta} e^{i(m-n)(\omega - \pi/2)} \frac{1}{2\pi} \int_{-\pi}^{\pi} e^{i\alpha\zeta \sin t + i(m-n)t} \, dt$$

$$= e^{im\theta} e^{i(m-n)(\omega - \pi/2)} J_{n-m}(\alpha\zeta), \tag{21.12-4}$$

where (21.11-3) has been used. It is seen that the Bessel functions appear not only in the characterization of the invariant subspaces of X^∞, but also in the dependence of the matrix elements $\hat{\rho}_{mn}$ on the parameters ξ, η, θ or ζ, ω, θ in the group M_2.

Irreducible representations of the group M_3 of rigid motions in Euclidean space E^3 can be obtained in a similar way, taking E^3 as the homogeneous space. Let X_α denote the space of all functions $u(x, y, z)$ that satisfy the 3-dimensional reduced wave equation $\nabla^2 u + \alpha^2 u = 0$ in all of E^3, for fixed α. Then the representation of M_3 on X_α given by the association with each g in M_3 of the transformation

$$\rho(g): f(\mathbf{x}) \rightarrow f(g^{-1}\mathbf{x}) \qquad (21.12\text{-}5)$$

of X_α onto itself is irreducible. From the infinitesimal operators of this representation, a basis in X_α is found, consisting of the functions

$$Y_l^m(\theta, \varphi) r^{-1/2} J_{l+1/2}(\alpha r), \qquad \begin{cases} l = 0, 1, 2, \ldots \\ m = -l, -l+1, \ldots, l. \end{cases} \qquad (21.12\text{-}6)$$

Therefore, the so-called spherical Bessel functions

$$j_l(z) = \sqrt{\frac{\pi}{2z}} J_{l+1/2}(z) \qquad (21.12\text{-}7)$$

have their origin in the irreducible representations of M_3.

Representations of M_n are discussed in Vilenkin 1968.

In contrast with the irreducible representations of compact groups, which are finite-dimensional and depend on a discrete parameter [e.g., $l = 0, 1, 2, \ldots$ for SO(3)], the irreducible representations of the rigid motion groups are infinite-dimensional and depend on a continuous parameter α.

The irreducible representations of the Lorentz groups are of two kinds: finite-dimensional ones that depend on a discrete parameter, and infinite-dimensional ones that depend on a continuous parameter. The former are obtained from finite-dimensional representations of SL(2, $\mathbb{C}$) (see next chapter) and appear in the relativistic quantum mechanics of various particles. Both are needed, in a complete set of irreducible representations; hence, one must expect that the infinite-dimensional ones may also play a role in physics.

21.13 Characters

The concept of the *character* $\chi(g)$ of a representation plays an important role in representation theory. For compact groups, it is the key to establishing the completeness of a set of irreducible representations, hence to deciding whether all representations have been found. If ρ is a representation on a finite-dimensional space X^n, so that the $\rho(g)$ are matrices with elements $\rho_{jk}(g)$, then $\chi(g) = \text{tr } \rho(g) = \sum_{j=1}^n \rho_{jj}(g)$. Hence, χ is a scalar-valued function on the group G. If G is compact, the characters χ^1 and χ^2 of two inequivalent irreducible representations are orthogonal with respect to the inner product (21.5-5):

$$\int_G \overline{\chi^1(g)} \chi^2(g) dg = 0, \qquad (21.13\text{-}1)$$

whereas $\chi^1(g) \equiv \chi^2(g)$ if the representations are equivalent. Furthermore,

$$\int_G |\chi(g)|^2 \, dg = 1 \tag{21.13-2}$$

for an irreducible representation, if the Haar measure is so normalized that the measure (volume) of the whole group is $= 1$.

The character $\chi(g)$ depends only on the conjugacy class of the group element g; that is, $\chi(g) = \chi(hgh^{-1})$ for all g, all h. The set of all the characters χ^i of irreducible representations of a compact group form a complete set of functions for the expansion of functions $f(g)$ that depend only on the conjugacy class of g.

In the rotation group SO(3), all rotations through a given angle ω belong to the same conjugacy class, regardless of the direction of the axis of rotation; hence, the characters are functions of ω only. The classical proof of the completeness of the set of irreducible representations ρ^l ($l = 0, 1, 2, \ldots$) referred to at the end of Section 20.9 consists in showing (see Wigner 1931) that the corresponding characters $\chi^l(g)$ form a complete set of functions for the expansion of functions of ω. It then follows, since there is no nonzero function of ω orthogonal to all the $\chi^l(g)$, that there is no irreducible representation inequivalent to all the ρ^l.

EXERCISES

1. Show that if two finite-dimensional representations ρ^1 and ρ^2 are equivalent [i.e., if they have the same dimension and if there is a matrix $A \neq 0$ such that $A\rho^1(g) = \rho^2(g)A$ for all g], then $\chi^1(g) \equiv \chi^2(g)$.

2. Show that two rotations g_1 and g_2 through a given angle ω about two different axes are conjugate, i.e., that there is a rotation h such that $g_1 = hg_2 h^{-1}$.

3. Show that the characters of the irreducible representations ρ^l of SO(3) are $\chi^l = \sin(l + \frac{1}{2})\alpha/\sin\frac{1}{2}\alpha$ ($l = 0, 1, \ldots$), and show that they satisfy the orthonormality relation (21.13-1), where dg is $= ((1 - \cos\alpha)/(4\pi^2\alpha^2))d^3\theta$, according to Exercise 7 in Section 21.5. *Hint*: It suffices to consider rotations about the z axis, for which the matrices $\rho^l(g)$ are diagonal; see (20.12-2) and the preceding equation.

To show the completeness of the characters χ^l for SO(3), we must show that if $\psi(\alpha)$ is any continuous function such that $(\chi^l, \psi) = 0$ for all l, then $\psi(\alpha) \equiv 0$. Since $1 - \cos\alpha = 2(\sin\frac{1}{2}\alpha)^2$ and $d^3\theta = 4\pi\alpha^2 \, d\alpha$, this is equivalent to showing that if

$$2 \int_0^\pi \sin(l + \tfrac{1}{2})\alpha \, \sin\tfrac{1}{2}\alpha \, \psi(\alpha)d\alpha = 0$$

for all l, then $\psi(\alpha) \equiv 0$. If we call $\frac{1}{2}\alpha = t$ and $\sin\frac{1}{2}\alpha \, \psi(\alpha) = \chi(t)$, this is equivalent to showing that if

$$\int_0^{\pi/2} \chi(t)\sin(2l + 1)t \, dt = 0$$

for all l, then $\chi(t) \equiv 0$. This is, however, the case, for if $\chi(t)$ is extended to the entire interval $-\pi \leq t \leq \pi$ by requiring it to be an odd function about $t = 0$ and an even one about $t = \pm \pi/2$, then the functions $\sin(2l + 1)t$ just suffice for the Fourier series for $\chi(t)$. Since the characters χ^l form a complete set of functions, the representations ρ^l ($l = 0, 1, 2, \ldots$) are *all* the irreducible representations of SO(3). This yields the answer to the question raised in Section 20.2 for the case of rotations of the Cartesian coordinate axes in 3-space: All possible nonrelativistic transformation laws of physical quantities are provided by the representations ρ^l of SO(3).

CHAPTER 22

Group Representations and Quantum Mechanics

Ray space and ray representations in quantum mechanics; extensions of local
representations; effect of double connectedness of SO(3) and simple
connectedness of SU(2); double-valued representations; spinors.

Prerequisites: Chapters 18–21 and elementary quantum mechanics.

The purpose of this chapter is to elucidate one particular point in the application of group theory to quantum mechanics, namely the occurrence of double-
valued or spin representations of the rotation and Lorentz groups.

22.1 Representations in Quantum Mechanics

It has been seen that, in classical physics, various sets of quantities transform,
under rotations of the coordinate axes, so as to give representations of the
rotation group. (The same applies in classical physics to other symmetry
groups, such as the groups of rigid motion, the crystal symmetry groups, the
Lorentz group, and so on.)

In quantum mechanics, on the other hand, some quantities transform, under
rotations of the coordinate axes, like the components of spinors and thus give
representations of SU(2) rather than of the rotation group SO(3). This was
shown by Dirac (in somewhat different language) in his paper (1928) on the
relativistic wave equation, and it was also implicit in Pauli's theory of the
electron spin, published a year earlier. More generally, spinor components
transform under a Lorentz transformation $\mathscr{L}_p$ so as to give representations
of SL(2, $\mathbb{C}$) rather than $\mathscr{L}_p$. This seemed rather surprising at the time, even
though Dirac showed that all observable quantities transform like scalars,
vectors, and tensors, i.e., according to the representations of SO(3) and $\mathscr{L}_p$. It
was seen in Sections 19.7 and 19.8 that the homomorphisms of SU(2) and
SL(2, $\mathbb{C}$) onto SO(3) and $\mathscr{L}_p$, respectively, are 2-to-1; hence a representation
of the first group can associate two different matrices, M and $-M$, with each

element g of the second group, i.e., with each of the transformations of space-time. This association is sometimes called a *two-valued representation* of the second group. How they arise is discussed in this chapter. It will be seen that the role of SU(2) and SL(2, $\mathbb{C}$) is to determine the so-called ray representations of the physically relevant groups SO(3) and $\mathscr{L}_p$.

Each possible state of a quantum mechanical system corresponds not to a single vector ψ in a Hilbert space $\mathfrak{H}$, but to a ray $\{\alpha\psi\}$ consisting of all numerical multiples of ψ. If all vectors are normalized ($\|\psi\| = 1$, $\|\alpha\psi\| = 1$), then α has unit modulus ($|\alpha| = 1$), but its phase (arg α) is arbitrary. This arbitrariness affects the interpretation of representation theory, as will be seen.

22.2 Rotations of the Axes

A state of a system may be regarded as determined in principle by the simultaneous measured values $\{a, b, \ldots\}$ of a complete set of commuting observables (self-adjoint operators) $\{A, B, \ldots\}$. Hence, the set $\{a, b, \ldots\}$ of numbers determines a ray $\{\alpha\psi\}$ in a Hilbert space $\mathfrak{H}$. The observables correspond in principle to an experimental arrangement, or apparatus, for measuring them. Suppose that the entire apparatus is rotated about some fixed point p into a new orientation by a rotation g [an element of SO(3)]. It then determines a new set of similar observables $\{A', B', \ldots\}$. A given state of the system now corresponds to a new set $\{a', b', \ldots\}$ of numbers, which determine a new ray $\{\alpha\psi'\}$ in $\mathfrak{H}$. Under the action of the rotation g, then, each ray $\{\alpha\psi\}$ is mapped into another ray $\{\alpha\psi'\}$. These mappings provide a *ray representation* of the group, as discussed in the next section.

Suppose that a normalized vector ψ is somehow chosen in each ray in $\mathfrak{H}$. Then the rotation g determines a one-to-one mapping among those vectors in $\mathfrak{H}$ that have been thus chosen. We assume, as an axiom of quantum mechanics, that the ψ's can be so chosen that the mapping is linear, and hence can then be defined in all $\mathfrak{H}$ by linearity. Since the representing ψ's were all normalized, the mapping is a unitary transformation U or $U(g)$. It is not unique, however, for given g, because of the arbitrariness of the phases of the representing ψ's. The degree of arbitrariness of U is described by the following lemma, whose proof is left as an exercise.

Lemma.   *Let U_1 and U_2 be two unitary transformations in $\mathfrak{H}$ such that, for every ψ, the two transformed vectors $U_1\psi$ and $U_2\psi$ determine the same ray. That is, there is a complex-valued function $\beta(\psi)$ such that $U_1\psi = \beta(\psi)U_2\psi$, for all ψ. Then $\beta(\psi) = const. = \beta$, where $|\beta| = 1$, i.e., $U_1 = \beta U_2$.*

Unitary transformations U and βU, where β is a constant and $|\beta| = 1$, are called *equivalent*: $U \cong \beta U$. We have seen that each rotation g corresponds to an equivalence class $\{\beta U : |\beta| = 1\}$ of unitary transformations having different phases arg β.

Now suppose that for each g in SO(3) a single unitary transformation $U(g)$ is somehow chosen from the corresponding equivalence class. If $\psi' = U(g)\psi$ and $\psi'' = U(h)\psi'$, then the resulting transformation matrix for the mapping $\psi \to \psi''$, i.e., $U(h)U(h)$, is not necessarily $= U(hg)$, but is $\cong U(hg)$. Hence, for each pair h, g of rotations there is a phase factor $\gamma(h, g)$ such that

$$U(h)U(g) = \gamma(h, g)U(hg), \qquad (22.2\text{-}1)$$

where $|\gamma(h, g)| = 1$. Possibilities for the choice of the function $\gamma(h, g)$ are discussed below.

22.3 Ray Representations

The set S of all rays is called a *ray space*. It is not a linear space, because if r is a ray, and c a number, cr is not defined, and if r_1 and r_2 are two rays, $r_1 + r_2$ is not defined. If we drop the requirement of normalization, each ray is a one-dimensional subspace of $\mathfrak{H}$. From that point of view, the only reasonable definitions would make cr the same ray as r, even for $c \neq 1$, and would make $r_1 + r_2$ a *two*-dimensional subspace for $\mathfrak{H}$, and hence not an element of S. However, each element of S corresponds to a state of the physical system, and the correspondence is one-to-one. S is a topological, in fact a metric, space in a very natural way. If r_1 and r_2 are two rays, their distance can be defined as

$$d(r_1, r_2) = \inf\{\|\psi_1 - \psi_2\| : \psi_1 \in r_1, \psi_2 \in r_2, \|\psi_1\| = \|\psi_2\| = 1\}.$$

The physical properties of the two corresponding states (expectations of observables) are nearly the same, if $d(r_1, r_2)$ is small.

Each rotation in space (each change of orientation of the apparatus) induces a transformation is S, as described above. It is not a linear transformation, since S is not a linear space, but it is continuous with respect to the metric in S. The mapping of the elements of SO(3) onto the corresponding transformations in S is an isomorphism: It is one-to-one, and the product of any two elements in SO(3) maps onto the compositions of the corresponding transformations in S, etc. Each such transformation in S corresponds to an equivalence class of unitary transformations in $\mathfrak{H}$. Generally, a homomorphism of a group G onto a group of equivalence classes of unity transformations in a vector space V is called a *ray representation* of G on V. As above, two unitary transformations U_1 and U_2 in V are *equivalent* if $U_1 = \beta U_2$ for some constant β.

The ray representations of SO(3) on $\mathfrak{H}$ are the physically relevant expressions of spherical symmetry. For calculational purposes, however, it is desirable to describe the transformations in S by something more tangible, like matrices. Hence, the question arises of selecting *one* unitary transformation $U(g)$ from each equivalence class in some convenient way.

If the phases of the $U(g)$ could be so chosen that the factor $\gamma(h, g)$ in (22.2-1) were $= 1$ for all h, all g, then the mapping $g \to U(g)$ would be an ordinary representation of SO(3) on $\mathfrak{H}$. That can't be done in general; what *can* be done will now be explained, after first restricting the problem to a finite-dimensional one.

22.4 A Finite-Dimensional Case

Suppose that the physical system has spherical symmetry and that there is a discrete energy state of energy E with finite multiplicity n. Then the corresponding eigenspace of the energy operator is an invariant subspace $\mathfrak{H}_E$ of $\mathfrak{H}$. Then rays in $\mathfrak{H}_E$ are transformed under rotations into other rays in $\mathfrak{H}_E$; hence $\mathfrak{H}_E$ is invariant under each of the operators $U(g)$, and the restriction of $U(g)$ to $\mathfrak{H}_E$ can be represented for each g by an $n \times n$ unitary matrix, which will also be denoted simply by $U(g)$. From this point on, the discussion will be restricted to the finite-dimensional case.

22.5 Local Representations

The physically reasonable assumption is now made that the phases of the unitary transformations can at least be so chosen that the matrix elements $U_{ij}(g)$ are continuous functions of g. Then let θ_x, θ_y, and θ_z be the intrinsic coordinates in SO(3) defined in Section 19.6, and let $\mathcal{N}$ and $\mathcal{N}_0$ be the sets of group elements for which $\|\boldsymbol{\theta}\| < \pi$ and $\|\boldsymbol{\theta}\| < \pi/2$, respectively. Although the group manifold as a whole is doubly connected, the regions $\mathcal{N}$ and $\mathcal{N}_0$ are simply connected neighborhoods of the identity. If g and h are in $\mathcal{N}_0$, gh is in $\mathcal{N}$. Now, g, h, and gh are rotation matrices, and the matrix elements of gh are continuous functions of the matrix elements of g and h; hence, as g and h vary continuously in $\mathcal{N}_0$, gh varies continuously in $\mathcal{N}$.

It will now be shown that the phases of the unitary matrices $U(g)$ described above can be so chosen in the neighborhood $\mathcal{N}$ that the function $\gamma(g, h)$ in (22.2-1) is $= 1$ for all g, h in $\mathcal{N}_0$. When that is done, the mapping $g \to U(g)$ is called a *local representation* of SO(3). (See Chapter 25.) Namely, since $U(g)$ is continuous in $\mathcal{N}$, the multivalued function $(\det U(g))^{1/n}$ splits into n independent continuous branches in $\mathcal{N}$, because $\mathcal{N}$ is simply connected. Clearly $U(e)$ is a multiple of the identity matrix I, and we write $U(e) = \beta^n I$, where $|\beta| = 1$. Then, $(\det U(e))^{1/n}$ is β times an nth root of unity, and a function $\alpha(g)$ can be defined as that branch of $(\det U(g))^{1/n}$ that is $= \beta$ for $g = e$. It is now asserted that if new unitary matrices $V(g)$ are defined in $\mathcal{N}$ by the equations

$$V(g) = \frac{1}{\alpha(g)} U(g), \qquad (22.5\text{-}1)$$

then

$$V(g)V(h) = V(gh) \quad \text{for all } g, h \text{ in } \mathcal{N}_0. \qquad (22.5\text{-}2)$$

To prove this, note that in any case

$$V(g)V(h) = \delta(g, h)V(gh),$$

where $\delta(g, h)$ is a continuous function [compare with (22.2-1)]. It is seen from (22.5-1) that $\det V(g) = 1$ for all g, hence $\delta(g, h)^n = 1$, hence $\delta(g, h)$ is some nth root of unity for all g, h; but $\delta(e, e) = 1$, hence $\delta(g, h) \equiv 1$, by continuity, and (22.5-2) follows.

22.6 Origin of the Two-Valued Representations

The remaining question is: When can the local representation $g \to V(g)$ be extended to a representation of all SO(3)? If H denotes the matrix group generated by the matrices $V(g)$, g in $\mathcal{N}$, then the mapping $g \to V(g)$ is a local homomorphism SO(3) into H. According to Theorem 3 in Section 25.13, a local homomorphism of a Lie group G into a Lie group H can be extended to a homomorphism of all of G into H if G is simply connected, but not necessarily otherwise. $G = $ SO(3) is, of course, not simply connected; however, from the mapping $g \to V(g)$ one can construct also a local homomorphism of SU(2) into H, and that homomorphism *can* be extended, because SU(2) is simply connected. Denote by $u \to g(u)$, where $u \in$ SU(2) and $g(u) \in$ SO(3), the homomorphism of SU(2) onto SO(3) that was constructed in Section 19.7. Then the mapping $u \to W(u) \overset{\text{def}}{=} V(g(u))$ is a local homomorphism of SU(2) into H, defined for those values of u for which $g(u) \in \mathcal{N}$. Its extension [which will also be denoted by $u \to W(u)$] is a representation of SU(2). Now, the homomorphism $u \to g(u)$ is 2-to-1, in fact $g(-u) = g(u)$, hence, the two equations $g = g(u)$, $W = W(u)$ give either (1) a representation $g \to W$ of SO(3)—this in case $W(u) = W(-u)$—or (2) an association of two unitary $n \times n$ matrices, say $V_1(g)$ and $V_2(g)$, $= -V_1(g)$, with each rotation matrix g, in such a way that each of the four products

$$V_i(g)V_j(h) \qquad (ij = 11, 12, 21, 22)$$

is equal to $V_1(gh)$ or to $V_2(gh)$. This association is called a *two-valued representation* of SO(3). Clearly every representation of SU(2) thus determines a two-valued representation, hence a ray representation, of SO(3).

Summary. Since the quantum-mechanical states correspond to rays in the Hilbert space rather than to vectors, a rotation g of the physical system corresponds not to a unique transformation among the vectors of a given invariant subspace, with matrix $U = U(g)$, but instead to a set $\{\alpha U : \text{all } \alpha \text{ in } \mathbb{C}$ such that $|\alpha| = 1\}$ of unitary transformations. It has been shown, however, that these sets are necessarily so interrelated that, by suitably choosing matrices from them, one can obtain a representation of SU(2). This can happen in either of two ways:

1. It may be possible to choose one matrix $U = U(g)$ from each set so as to give a representation of SO(3), and hence also a representation of SU(2) via the homomorphisms

$$\begin{array}{ccc} \text{SU(2)} & \to \text{SO(3)} & \to \{U(g)\}; \\ (2 \times 2) & (3 \times 3) & (n \times n) \end{array}$$

2. It may be necessary to choose two matrices U and $-U$ from each set and to associate them with the two elements u and $-u$ of SU(2), but with only one element g of SO(3), in such a way that they form an ordinary representation of SU(2) and a two-valued or *spin* representation of SO(3). It will be seen in Chapter 25 that there is no other group related to SO(3) in the way SU(2) is; hence there are no n-valued representations except two-valued ones.

Similarly, for a system that is invariant not merely under the rotation group SO(3) but also under the entire proper Lorentz group $\mathscr{L}_p$, the transformation of the wave functions corresponding to a given transformation g of $\mathscr{L}_p$ is not unique. Instead, there is a set $\{\alpha U : |\alpha| = 1\}$ of transformations corresponding to each g, and these sets are so correlated that one can choose transformations from them so as to give a representation of SL(2, ℂ) [which is related to $\mathscr{L}_p$ as SU(2) is to SO(3)]; this may be a representation of $\mathscr{L}_p$ itself, involving scalars, vectors, or general tensors, or it may be a two-valued representation (spin representation) of $\mathscr{L}_p$. In Dirac's theory of the electron, the transformation laws of the four components of the electron's wave function give a two-valued representation of $\mathscr{L}_p$ (see Dirac 1958, p. 258).

It is easy to see that a two-valued irreducible representation cannot be made into a single-valued one by somehow appropriately choosing one of the two matrices U and $-U$ that represent each given g of SO(3) (or $\mathscr{L}_p$); namely, if U_0 is a matrix that represents a rotation through π, in a two-valued irreducible representation, it can be shown that $U_0^2 = -I$, but U_0^2 represents the identity in SO(3), hence must be $= +I$ in any single-valued representation.

22.7 Representations of SU(2) and SL(2, ℂ)

The discussion so far is rather hypothetical until it can be shown that there actually exist representations of SU(2) that give two-valued representations of SO(3). To be sure, the identity representation of SU(2) by itself is one such, but there are many others. In the next few sections the irreducible representations of SU(2) will be discussed. They are all finite-dimensional, because SU(2) is compact, while SL(2, ℂ) and $\mathscr{L}_p$, which are not compact, have also infinite-dimensional irreducible representations, concerning which the reader is referred to Valenkin 1968, Naimark 1976, and Sugiura 1975. It turns out that certain *finite*-dimensional representations of SL(2, ℂ) remain irreducible when restricted to SU(2); they give rise to ordinary and spin representations of the Lorentz and rotation groups.

An element of SL(2, ℂ) is a unimodular transformation of $\mathbb{C}^2$ onto itself given by

$$u = \begin{pmatrix} \alpha & \beta \\ \gamma & \delta \end{pmatrix} : \begin{pmatrix} x_1 \\ x_2 \end{pmatrix} \rightarrow \begin{pmatrix} \alpha x_1 + \beta x_2 \\ \gamma x_1 + \delta x_2 \end{pmatrix}, \tag{22.7-1}$$

where $\alpha\delta - \gamma\beta = 1$. The matrix of the inverse transformation is

$$u^{-1} = \begin{pmatrix} \delta & -\beta \\ -\gamma & \alpha \end{pmatrix}.$$

Now the action of the group on $\mathbb{C}^2$ is effective (any transformation $u \neq e$ moves at least one point of $\mathbb{C}^2$) and transitive (given any two points **x** and **y**, there is always an element u in the group such that $\mathbf{y} = u\mathbf{x}$), that is, $\mathbb{C}^2$ is a homogeneous space for SL(2, ℂ). Therefore, let X^∞ be the space of all entire analytic functions $f(x_1, x_2)$ of two complex variables. Then, according to (20.6-1) an infinite-dimensional representation of SL(2, $\mathbb{C}$) is obtained by associating with u the transformation $\rho(u)$ in X^∞ given by

$$(\rho(u)f)(x_1, x_2) = f(\delta x_1 - \beta x_2, -\gamma x_1 + \alpha x_2). \qquad (22.7\text{-}2)$$

Certain elements of the subgroup SU(2) are now considered. Let $\omega_1, \omega_2, \omega_3$ be the intrinsic coordinates in SO(3) defined in Section 19.6, let $g_{\omega_1, \omega_2, \omega_3}$ be the corresponding rotation matrix [element of SO(3)], and let $\pm u_{\omega_1, \omega_2, \omega_3}$ be the elements SU(2) that are mapped onto $g_{\omega_1, \omega_2, \omega_3}$ by the homomorphism of Section 19.7. In particular, one can take

$$u_{\omega, 0, 0} = \begin{pmatrix} \cos \omega/2 & -i \sin \omega/2 \\ -i \sin \omega/2 & \cos \omega/2 \end{pmatrix},$$

$$u_{0, \omega, 0} = \begin{pmatrix} \cos \omega/2 & -\sin \omega/2 \\ \sin \omega/2 & \cos \omega/2 \end{pmatrix}, \qquad (22.7\text{-}3)$$

$$u_{0, 0, \omega} = \begin{pmatrix} e^{-i\omega/2} & 0 \\ 0 & e^{i\omega/2} \end{pmatrix},$$

because a direct calculation, using the equations of Section 19.7, shows that the corresponding transformations from x, y, z to x', y', z' are those given by the matrices

$$g_{\omega, 0, 0} = \begin{pmatrix} 0 & 0 & 0 \\ 0 & 0 & -\omega \\ 0 & \omega & 0 \end{pmatrix},$$

$$g_{0, \omega, 0} = \begin{pmatrix} 0 & 0 & \omega \\ 0 & 0 & 0 \\ -\omega & 0 & 0 \end{pmatrix}, \qquad (22.7\text{-}4)$$

$$g_{0, 0, \omega} = \begin{pmatrix} 0 & -\omega & 0 \\ \omega & 0 & 0 \\ 0 & 0 & 0 \end{pmatrix},$$

in agreement with (19.6-1). Infinitesimal group elements of SU(2) are obtained accordingly:

$$T_1 = \frac{\partial}{\partial \omega} u_{\omega, 0, 0}|_{\omega = 0} = \frac{1}{2} \begin{pmatrix} 0 & -i \\ -i & 0 \end{pmatrix},$$

$$T_2 = \frac{\partial}{\partial \omega} u_{0, \omega, 0}|_{\omega = 0} = \frac{1}{2} \begin{pmatrix} 0 & -1 \\ 1 & 0 \end{pmatrix}, \qquad (22.7\text{-}5)$$

$$T_3 = \frac{\partial}{\partial \omega} u_{0, 0, \omega}|_{\omega = 0} = \frac{1}{2} \begin{pmatrix} -i & 0 \\ 0 & i \end{pmatrix}.$$

The corresponding differential operators of the representation ρ are

$$L_1 = \frac{\partial}{\partial \omega} \rho(u_{\omega,0,0})|_{\omega=0} = \frac{i}{2}\left(x_2 \frac{\partial}{\partial x_1} + x_1 \frac{\partial}{\partial x_2}\right),$$

$$L_2 = \frac{\partial}{\partial \omega} \rho(u_{0,\omega,0})|_{\omega=0} = \frac{1}{2}\left(x_2 \frac{\partial}{\partial x_1} - x_1 \frac{\partial}{\partial x_2}\right), \qquad (22.7\text{-}6)$$

$$L_3 = \frac{\partial}{\partial \omega} \rho(u_{0,0,\omega})|_{\omega=0} = \frac{i}{2}\left(x_1 \frac{\partial}{\partial x_1} - x_2 \frac{\partial}{\partial x_2}\right).$$

The infinitesimal elements and operators satisfy the commutation relations

$$[T_i, T_j] = T_k, \qquad [L_i, L_j] = L_k \qquad (ijk = 123, 231, 312). \quad (22.7\text{-}7)$$

We note in passing that the matrices (22.7-5) can be regarded also as the infinitesimal group elements of the larger group SL(2, $\mathbb{C}$), for the following reason: First, it is easily verified that the matrices (22.7-3) are given in terms of the matrices T_i by the equations

$$u_{\omega,0,0} = \exp(\omega T_1),$$

$$u_{0,\omega,0} = \exp(\omega T_2),$$

$$u_{0,0,\omega} = \exp(\omega T_3).$$

By the methods of Chapter 25 (exponential mapping), it is seen that, more generally,

$$u_{\omega_1,\omega_2,\omega_3} = \exp(\omega_1 T_1 + \omega_2 T_2 + \omega_3 T_3). \qquad (22.7\text{-}8)$$

From (22.7-5) it is seen that the right member of this last equation is of the form $\exp(iA)$, where A is a general 2×2 Hermitian matrix of trace zero. If, now, ω_1, ω_2 and ω_3 are allowed to take complex values, then it is of the form $\exp B$, where B is a completely general 2×2 matrix of trace zero, and then $\exp B$ is a general 2×2 matrix of determinant $= 1$, i.e., a general element of the group SL(2, $\mathbb{C}$).

22.8 Irreducible Representations of SU(2)

For each value $0, \frac{1}{2}, 1, \frac{3}{2}, 2, \ldots$ of an index l, a subspace X^{2l+1} of X^∞ is defined as the space of all homogeneous polynomials in x_1 and x_2 of degree $2l$. From (22.7-2) it is seen that each operator $\rho(u)$ transforms any homogeneous polynomial into another homogeneous polynomial of the same degree; hence each subspace X^{2l+1} is invariant under $\rho(u)$, not only for all u in SU(2), but also for all u in SL(2, $\mathbb{C}$).

It will be shown that the representation of SU(2) given by (22.7-2) on each subspace X^{2l+1} (it will be called D^l) is irreducible; hence, the representation of

SL(2, $\mathbb{C}$) on X^{2l+1} is *a fortiori* also irreducible. It will be shown that any subspace of X^{2l+1} (other than the subspace consisting of the zero vector alone) which is invariant under SU(2) is *all* of X^{2l+1}. This is done by the now familiar method of raising and lowering operators: Any such subspace is invariant under the operators L_1, L_2, L_3 of equations (22.7-6), hence also under $L_1 \pm iL_2$.

The monomials

$$f_m(x_1, x_2) = x_1^{l-m}x_2^{l+m} \qquad (m = -l, -l+1, \ldots, l) \qquad (22.8\text{-}1)$$

are a basis in X^{2l+1}; f_m is an eigenfunction of the operator L_3 with eigenvalue im. Note that m assumes integer or half-odd-integer values according as l is an integer or half an odd integer. Any function g in X^{2l+1} can be written as $\sum c_m f_m$. By an argument like that of Section 20.5 it follows that if an invariant subspace of X^{2l+1} contains such a function g, then it contains all those monomials f_m, individually, for which $c_m \neq 0$. Namely, if the subspace contains the function g, then it contains the function $L_3 g$ (because the subspace is invariant under L_3), also the function $P(L_3)g$, where P is any polynomial, but P can be chosen as to annihilate all the terms of $\sum c_m f_m$ except one (P can be taken as the Lagrange interpolation polynomial which vanishes for all eigenvalues im of L_3 except one). Hence, the invariant subspace contains at least one of the functions f_m. But $L_1 + iL_2$ is a lowering operator, i.e., it converts f_m into a multiple of f_{m-1}, except that it converts f_{-l} into zero; and $L_1 - iL_2$ is a raising operator, i.e., it converts f_m into a multiple of f_{m+1}, except that it converts f_l into zero. Hence, the invariant subspace is all of X^{2l+1}, as was to be proved.

It will be shown in the next section that the D^l ($l = 0, \frac{1}{2}, 1, \frac{3}{2}, \ldots$) are the only irreducible representations of SU(2).

The two-valued representations of SO(3) come about as follows: If $u \in$ SU(2), then, under the homomorphism of SU(2) onto SO(3) of Section 19.7, u and $-u$ are both mapped onto an element g of SO(3). If D^l is one of the representations of SU(2) found above, then the mapping $g \to D^l(\pm u)$ is a single-valued representation of SO(3) on X^{2l+1} if $D^l(-u) = D^l(u)$ and is a double-valued representation if $D^l(-u) \neq D^l(u)$. It is only necessary to examine the case of $u = I_2$; then $D^l(u) = I_{2l+1}$. (Here, I_k denotes the $k \times k$ unit matrix). Under the mapping determined by the matrix $-I_2$, x_1 and x_2 go into $-x_1$ and $-x_2$; if $2l$ is even, the monomial f_m goes into itself; hence $D^l(-I_2) = I_{2l+1}$, and the mapping $g \to D^l(\pm u)$ is the ordinary representation of odd dimension $2l + 1$ found in Section 20.9. If $2l$ is odd, f_m goes into $-f_m$; hence $D^l(-I_2) = -I_{2l+1}$, and the mapping $g \to D^l(\pm u) = \pm D^l(u)$ is a double-valued or spin representation of even dimension.

In a similar fashion, the irreducible representations of SL(2, $\mathbb{C}$) given above [from which those of SU(2) were obtained by restriction] lead to finite-dimensional ordinary and spin representations of the Lorentz group $\mathscr{L}_p$. However, there are still other irreducible finite-dimensional representations of SL(2, $\mathbb{C}$), given in Section 22.11 below; they also determine ordinary and spin representations of $\mathscr{L}_p$, none of them unitary.

22.9 The Characters of SU(2)

The conjugacy classes of SU(2) are easily determined. First, matrices u_1 and u_2 in SU(2) have the same eigenvalues if and only if there is a unitary matrix u [element of U(2)] such that $u^*u_1u = u_2$. Since u can be multiplied by any complex number of modulus 1, we can assume det $u = 1$ without loss of generality, and then u is in SU(2). It follows that u_1 and u_2 are in the same conjugacy class in SU(2) if and only if they have the same eigenvalues. Therefore, each conjugacy class can be represented by a matrix of the form

$$u = \begin{pmatrix} u^{-i\alpha/2} & 0 \\ 0 & e^{i\alpha/2} \end{pmatrix}$$

for some α in $[0, 2\pi]$. For such u, the operator $D^l(u)$ simply multiplies the basis vector f_m (22.8-1) by $e^{im\alpha}$; hence $D^l(u)$ is a diagonal matrix, whose trace is

$$\chi^l(\alpha) = \frac{\sin(l + \frac{1}{2})\alpha}{\sin \frac{1}{2}\alpha}, \qquad (22.9\text{-}1)$$

just as for the case in which l is an integer, according to Exercise 3 at the end of Section 21.13. It was shown in that section that the functions 22.9-1, for $l = 0, \frac{1}{2}, 1, \frac{3}{2}, \ldots$ form a complete system for the expansion of functions depending only on the conjugacy class on the manifold of SU(2); hence, the representations D^l exhaust the irreducible representations of SU(2).

22.10 Functions of z and $\bar{z}$

The notation introduced in this section is convenient for discussing the representations of SL(2, $\mathbb{C}$) and is used in many branches of mathematics. Let $u(x, y)$ and $v(x, y)$ be two real C^∞ functions of two real variables x and y. We write $z = x + iy$, $f(z) = u(x, y) + iv(x, y)$, so that $f(z)$ is a complex-valued function of the complex variable z, not in general analytic. If $f(z)$ is analytic, then its derivative can be written in various forms, using the Cauchy–Riemann equations, namely

$$f'(z) = \frac{\partial}{\partial x}(u + iv) = -i\frac{\partial}{\partial y}(u + iv)$$

$$= \partial_z(u + iv),$$

where ∂_z is the operator

$$\partial_z = \frac{1}{2}\left(\frac{\partial}{\partial x} - i\frac{\partial}{\partial y}\right). \qquad (22.10\text{-}1)$$

An operator $\partial_{\bar{z}}$ is similarly defined as

$$\partial_{\bar{z}} = \frac{1}{2}\left(\frac{\partial}{\partial x} + i\frac{\partial}{\partial y}\right), \qquad (22.10\text{-}2)$$

and it is noted that $\partial_{\bar{z}} f(z) \equiv 0$, if $f(z)$ is analytic, by virtue of the Cauchy–Riemann equations. On the other hand, if $f(z)$ is a polynomial (or convergent power series) in $\bar{z}$, then $\partial_z f(z) = 0$. Furthermore, the operators (22.10-1, 2) are linear differential operators; hence the usual rule for differentiating a product holds. Therefore, if f is a polynomial (or a convergent power series) in both z and $\bar{z}$ [in which case one usually writes $f(z, \bar{z})$ to indicate that it is not necessarily analytic in either z or $\bar{z}$], then $\bar{z}$ can be regarded as a constant for the purpose of computing ∂_z, and z a constant for computing $\partial_{\bar{z}}$. That is, z and $\bar{z}$ may be regarded as independent variables for purposes of differentiation.

22.11 The Finite-Dimensional Representations of SL(2, $\mathbb{C}$)

The method of homogeneous polynomials used in Section 22.8 for SU(2) can also be used for SL(2, $\mathbb{C}$), but here a new aspect appears. Given a representation ρ of a group G, there are many ways in which another representation ρ' can be obtained. Among them is

$$\rho'(u) = \overline{\rho(u)}, \qquad \forall u \text{ in } G, \tag{22.11-1}$$

[which means that each matrix element $\rho_{mn}(u)$ is replaced by its complex conjugate], for then $\rho'(u_1 u_2) = \rho'(u_1)\rho'(u_2)$, etc. Another possibility is

$$\rho'(u) = (\rho(u)^T)^{-1}; \tag{22.11-2}$$

if the representation is unitary, this is the same as (22.11-1). If G is itself a group of matrices, then two other possibilities are

$$\rho'(u) = \rho(\bar{u}), \tag{22.11-3}$$

$$\rho'(u) = \rho((u^T)^{-1}). \tag{22.11-4}$$

If G is a unitary group, e.g., $U(n)$ or SU(n), then (22.11-3) and (22.11-4) are the same, but otherwise, they are generally different.

We now show that if G is SU(2), then the representation ρ' given by (22.11-3) is equivalent to ρ; hence, in this case, no new representations are obtained by these methods, and that is why these methods were not used in Section 22.8. Namely, call

$$\gamma = \begin{pmatrix} 0 & 1 \\ -1 & 0 \end{pmatrix},$$

so that generally

$$\gamma^{-1} \begin{pmatrix} a & b \\ c & d \end{pmatrix} \gamma = \begin{pmatrix} d & -c \\ -b & a \end{pmatrix}. \tag{22.11-5}$$

In particular, for any u in SU(2), $\bar{u} = \gamma^{-1}u\gamma$, which can be seen by writing u as $\begin{pmatrix} a & b \\ -\bar{b} & \bar{a} \end{pmatrix}$. Then, since γ is also in SU(2),

$$\rho'(u) = \rho(\gamma^{-1}u\gamma) = \rho(\gamma)^{-1}\rho(u)\rho(\gamma),$$

for all u; hence ρ and ρ' are equivalent representations.

When the representations (22.11-3) and (22.11-4) are extended from SU(2) to SL(2, ℂ) by writing

$$\rho'(m) = \rho(\bar{m}) \tag{22.11-6}$$

and

$$\rho'(m) = \rho((m^T)^{-1}), \tag{22.11-7}$$

respectively, for m in SL(2, ℂ), they are no longer identical, or even equivalent. The second one is equivalent to ρ, because $(m^T)^{-1} = \gamma^{-1}m\gamma$, by (22.11-5), since $\det m = 1$, while (22.11-6) is *inequivalent* to ρ, for if the equations

$$\rho(\bar{m}) = V^{-1}\rho(m)V \tag{22.11-8}$$

held for all m, then V would have to be $= \rho(\gamma)$ in order that this equation be satisfied for $m \in$ SU(2), in which case it would not be satisfied for $m \notin$ SU(2), since $\gamma^{-1}m\gamma$ is not in general $= \bar{m}$ for such m.

Clearly, then, SL(2, ℂ) has more representations, in some sense, than SU(2). To find them, we let X^∞ denote the set of all complex-valued functions of two complex variables x_1 and x_2 that are C^∞ in the real sense rather than entire analytic in contrast with Section 22.10, and we denote these functions by $f(x_1, x_2, \bar{x}_1, \bar{x}_2)$, following the procedure of Section 22.10. In place of (22.7-2), we write

$$(\rho(u)f)(x_1, x_2, \bar{x}_1, \bar{x}_2)$$
$$= f(\delta x_1 - \beta x_2, -\gamma x_1 + \alpha x_2, \bar{\delta}\bar{x}_1 - \bar{\beta}\bar{x}_2, -\bar{\gamma}\bar{x}_1 + \bar{\alpha}\bar{x}_2), \tag{22.11-9}$$

where u is the matrix

$$u = \begin{pmatrix} \alpha & \beta \\ \gamma & \delta \end{pmatrix}, \qquad \alpha\delta - \beta\gamma = 1, \tag{22.11-10}$$

i.e., any matrix in SL(2, ℂ). Now, in addition to the three matrices (22.7-3), which are in SU(2) and correspond to rotations in space, and which determine the infinitesimal operators L_1, L_2, and L_3 by (22.7-6), we have three additional matrices,

$$\begin{pmatrix} \cosh \omega/2 & \sinh \omega/h \\ \sinh \omega/2 & \cosh \omega/2 \end{pmatrix} \qquad (\omega = \varphi_x),$$

$$\begin{pmatrix} \cosh \omega/2 & -i \sinh \omega/2 \\ i \sinh \omega/2 & \cosh \omega/2 \end{pmatrix} \qquad (\omega = \varphi_y), \tag{22.11-11}$$

$$\begin{pmatrix} e^{\omega/2} & 0 \\ 0 & e^{-\omega/2} \end{pmatrix} \qquad (\omega = \varphi_z),$$

which correspond to Lorentz transformations and determine further infinitesimal operators K_1, K_2, and K_3. Hence, the infinitesimal operators of SL(2, $\mathbb{C}$) are obtained by substituting the matrices (22.7-6) and (22.11-11) into (22.11-3) and taking the derivative of each with respect to ω at $\omega = 0$. If we recall that $x_1, x_2, \bar{x}_1$, and $\bar{x}_2$ are to be regarded as independent for purposes of differentiation (see preceding section), we find that

$$L_1 = -\frac{i}{2}(x_2\,\partial_{x_1} + x_1\,\partial_{x_2}) + \frac{i}{2}(\bar{x}_2\,\partial_{\bar{x}_1} + \bar{x}_1\,\partial_{\bar{x}_2}),$$

$$L_2 = \frac{1}{2}(x_2\,\partial_{x_1} - x_1\,\partial_{x_2}) + \frac{1}{2}(\bar{x}_2\,\partial_{\bar{x}_1} - \bar{x}_1\,\partial_{\bar{x}_2}),$$

$$L_3 = -\frac{i}{2}(x_1\,\partial_{x_1} - x_2\,\partial_{x_2}) + \frac{i}{2}(\bar{x}_1\,\partial_{x_1} - \bar{x}_2\,\partial_{\bar{x}_2}),$$

$$\hspace{8cm}(22.11\text{-}12)$$

$$K_1 = -\frac{1}{2}(x_2\,\partial_{x_1} + x_1\,\partial_{x_2}) - \frac{1}{2}(\bar{x}_2\,\partial_{\bar{x}_1} + \bar{x}_1\,\partial_{\bar{x}_2}),$$

$$K_2 = \frac{i}{2}(x_2\,\partial_{x_1} - x_1\,\partial_{x_2}) - \frac{i}{2}(\bar{x}_2\,\partial_{\bar{x}_1} - \bar{x}_1\,\partial_{\bar{x}_2}),$$

$$K_3 = -\frac{1}{2}(x_1\,\partial_{x_1} - x_2\,\partial_{x_2}) - \frac{1}{2}(\bar{x}_1\,\partial_{\bar{x}_1} - \bar{x}_2\,\partial_{\bar{x}_2}).$$

The complete commutation relations are

$$\begin{aligned}
[L_i, L_j] &= L_k \\
[K_i, K_j] &= L_k \\
[K_i, L_i] &= 0 \\
[K_i, L_j] &= -K_k
\end{aligned} \qquad (ijk = 123,\ 231,\ \text{or } 312). \qquad (22.11\text{-}13)$$

We introduce also the operators

$$L^{\pm} = L_1 \pm iL_2, \qquad K^{\pm} = K_1 \pm iK_2. \qquad (22.11\text{-}14)$$

22.12 The Irreducible Invariant Subspaces of X^∞ for SL(2, $\mathbb{C}$)

As basis vectors for X^∞, or at least for the set of all polynomials in X^∞, we introduce the monomials

$$\psi = \psi_{lml'm'} = \frac{1}{C}x_1^{l-m}x_2^{l+m}\bar{x}_1^{l'-m'}\bar{x}_2^{l'+m'}, \qquad (22.12\text{-}1)$$

where

$$C^2 = (l-m)!\,(l+m)!\,(l'-m')!\,(l'+m')!,$$

where l and l' are any two of the numbers $0, \frac{1}{2}, 1, \frac{3}{2}, \ldots$, and where

$$m = l, l - 1, \ldots, -l,$$
$$m' = l', l' - 1, \ldots, -l'.$$

For given l and l', the space $X(l, l')$ spanned by the $\psi_{l m l' m'}$ is the space of all homogeneous polynomials of degree $2l$ in the variables x_1 and x_2 and of degree $2l'$ in $\bar{x}_1$ and $\bar{x}_2$. This space has (complex) dimension $(2l + 1)(2l' + 1)$. It is clear from (22.11-9) that each subspace $X(l, l')$ is mapped into itself under every $\rho(u)$, and hence is an invariant subspace. From (22.11-12) we see that

$$L_3 \psi = i(m - m')\psi,$$
$$K_3 \psi = (m + m')\psi;$$

i.e., m is an eigenvalue of $-\frac{1}{2}iL_3 + \frac{1}{2}K_3$, and m' is an eigenvalue of $\frac{1}{2}iL_3 + \frac{1}{2}K_3$. It follows, by the same kind of argument used in all previous cases, that if an invariant subspace contains any function (polynomial) in the subspace $X(l, l')$, then it contains every monomial $\psi_{l m l' m'}$ which appears in that polynomial with a nonzero coefficient. We find furthermore that

$$L^+ + iK^- = -2ix_1\, \partial_{x_2},$$
$$L^- + iK^+ = -2ix_2\, \partial_{x_1},$$
$$L^- - iK^+ = 2i\bar{x}_1\, \partial_{\bar{x}_2},$$
$$L^+ - iK^- = 2i\bar{x}_2\, \partial_{\bar{x}_1}.$$

Hence,

$$L^+ + iK^- \quad \text{lowers} \quad m,$$
$$L^- + iK^+ \quad \text{raises} \quad m,$$
$$L^- - iK^+ \quad \text{lowers} \quad m',$$
$$L^+ - iK^- \quad \text{raises} \quad m',$$

in the sense that $(L^+ + iK^-)\psi_{l m l' m'}$ is proportional to $\psi_{l m - 1 l' m'}$, etc. Hence, for given l and l', all the $\psi_{l m l' m'}$ are linked by the infinitesimal operators. We conclude that if an invariant subspace contains any function (polynomial) in $X(l, l')$, then it contains the entire subspace $X(l, l')$. That is, ρ, when restricted to $X(l, l')$, is irreducible; it is denoted by $\rho^{(l, l')}$; these are the only *finite-dimensional* irreducible representations of SL(2, $\mathbb{C}$).

22.13 Spinors

Spinors are sets of quantities related to SU(2) and SL(2, $\mathbb{C}$) in the same way that the tensors (including vectors and scalars) of prequantum physics are related to the physical groups SO(3) and $\mathscr{L}_p$. Their transformation laws constitute certain (generally reducible) representations of SU(2) and SL(2, $\mathbb{C}$), and hence certain one- or two-valued representations of the physical groups. Those spinors whose transformation laws constitute ordinary or one-valued representations of the physical groups are in fact tensors, slightly disguised (see Exercise 1, below); in this sense, tensors are special cases of spinors.

In order to describe a tensor, one associates a set of quantities (say, $T_{1\,1}$, $T_{1\,2}$, ..., $T_{4\,4}$) with each frame of reference, and then the relations between these sets, for different frames of reference, constitute the transformation laws. Once an initial frame of reference has been specified, this is equivalent to associating one such set of quantities with each element of the rotation or Lorentz group; the corresponding sets, for other frames of reference, are then given by the transformation laws. In the case of spinors, one associates a set of quantities for the initial frame of reference, with each element of the group SU(2) or SL(2, $\mathbb{C}$), and this amounts to associating two sets, differing, however, only in phase (in fact, only in sign), with each physical frame of reference. Since phases, hence signs, are physically irrelevant, we shall speak loosely of the association of a set of quantities with each frame of reference.

A *spinor of rank* 1, then, is the association of a pair of numbers ξ_1 and ξ_2 with each frame of reference, in such a way that under a Lorentz transformation corresponding to an element

$$m = \begin{pmatrix} m_{11} & m_{12} \\ m_{21} & m_{22} \end{pmatrix}$$

of SL(2, $\mathbb{C}$), the ξ's transformation according to the Law

$$\begin{aligned} \xi_1 \to \xi_1' &= m_{11}\xi_1 + m_{12}\xi_2, \\ \xi_2 \to \xi_2' &= m_{21}\xi_1 + m_{22}\xi_2. \end{aligned} \tag{22.13-1}$$

Higher rank spinors are defined in complete analogy with tensors: *A spinor of rank* r is the association with each frame of reference of a set of 2^r complex numbers $\xi_{\alpha_1 \ldots \alpha_r}$ (each index takes on the values 1 and 2), with the transformation law

$$\xi'_{\alpha_1 \ldots \alpha_r} = \sum_{1}^{2} {}_{(\beta_1, \ldots, \beta_r)} m_{\alpha_1\beta_1} \cdots m_{\alpha_r\beta_r} \xi_{\beta_1 \ldots \beta_r}. \tag{22.13-2}$$

If we were concerned only with rotations of the x, y, z axes, and not with Lorentz transformations, hence only with the group SU(2), this is all that would need to be said. It was seen in the preceding section, however, that in the study of the representations of SL(2, $\mathbb{C}$), the matrix $\bar{m}$ plays a parallel role with m. A *dotted spinor of rank* 1 is the association of a pair of complex numbers $\xi_{\dot{1}}$, $\xi_{\dot{2}}$ with each frame of reference, according to the transformation law

$$\begin{aligned} \xi_{\dot{1}}' &= \overline{m_{11}}\xi_{\dot{1}} + \overline{m_{12}}\xi_{\dot{2}}, \\ \xi_{\dot{2}}' &= \overline{m_{21}}\xi_{\dot{1}} + \overline{m_{22}}\xi_{\dot{2}}. \end{aligned} \tag{22.13-3}$$

Finally, a mixed spinor having r undotted and s dotted indices is the association of 2^{r+s} complex numbers with each frame of reference, with the transformation law

$$\xi'_{\alpha_1 \ldots \alpha_r \dot{\beta}_1 \ldots \dot{\beta}_s}$$

$$= \sum_{1}^{2} {}_{(\gamma_1, \ldots, \gamma_r, \delta_1, \ldots, \delta_s)} m_{\alpha_1 \gamma_1} \cdots m_{\alpha_r \gamma_r} \overline{m_{\beta_1 \delta_1}} \cdots \overline{m_{\beta_s \delta_s}}$$

$$\times \xi_{\gamma_1 \ldots \gamma_r \dot{\delta}_1 \ldots \dot{\delta}_s}. \tag{22.13-4}$$

EXERCISES

1. Let $\xi_{\alpha\dot{\beta}}$ be a mixed spinor of rank 2, and define quantities v_j ($j = 1, \ldots, 4$) by

$$v_1 = \xi_{1\dot{1}} + \xi_{2\dot{2}},$$

$$v_2 = \xi_{1\dot{1}} - \xi_{2\dot{2}},$$

$$v_3 = \xi_{1\dot{2}} + \xi_{2\dot{1}},$$

$$v_4 = i(\xi_{1\dot{2}} - \xi_{2\dot{1}}).$$

Show that quantities $v_1, \ldots, v_4$ transform like the components of a vector under rotations and Lorentz transformations. Show similarly that a mixed spinor of rank $2r$ having r dotted and r undotted indices determines a tensor of rank r.

2. A spinor is called *symmetric* if it is symmetric in the dotted indices (unchanged by any permutation of the dotted indices) and also symmetric in the undotted indices; show that the transformation law of such a spinor gives a representation of SL(2, $\mathbb{C}$), which is equivalent to the representation $\rho^{(l, l')}$ defined in the preceding section, where $2l$ and $2l'$ are the numbers of dotted and undotted indices, respectively.

CHAPTER 23

Elementary Theory of Manifolds

Locally n-dimensional space; sphere; torus; disk; Möbius strip; Klein bottle; identification of edges; coordinate charts; compatibility of charts; induced topology; Hausdorff separation axiom; manifold; curves; functions on a manifold; connectedness; simple connectedness; component; homotopic curves; homotopy classes of curves; fundamental group; double connectedness of SO(3); configuration space of a mechanical system; Cartesian product manifolds.

Prerequisites: Chapters 18 and 19.

The theory of manifolds is basic for the theory of Lie groups and Riemannian and Einsteinian geometries. The introduction of the manifold concept into general relativity around 1960, mainly by Martin Kruskal, put a new light on that subject and clarified the topological properties, both local and global, of space-time models. Statistical mechanics deals with flows on manifolds. Other applications of manifolds to physics appear from time to time, because of their basic geometric nature. Only finite-dimensional manifolds will be discussed. For more general manifolds, see Lang 1962.

23.1 Examples of Manifolds; Method of Identification

An n-dimensional manifold is, roughly speaking, a space that is locally topologically indistinguishable from Euclidean n-space E^n; that is, each point of the manifold lies in a region (connected open set) that is homeomorphic to a region in E^n. The formal definition is given in Section 23.4, below.

Any Euclidean space itself is trivially a manifold. A simple nontrivial example is the 2-sphere, i.e., the set S^2 of points in E^3 for which $x^2 + y^2 + z^2 = 1$, where x, y, and z are Cartesian coordinates. Any sufficiently small region $\mathfrak{U}$ on S^2 can be mapped onto a plane region in a one-to-one manner by any of the projections used by cartographers. (If stereographic projection is used, $\mathfrak{U}$ is all of S^2 minus one point.) Hence S^2 is a 2-dimensional manifold. The torus, that is, the surface of a ring, is another 2-dimensional manifold.

Any open subset of a manifold (e.g., the open disk $x^2 + y^2 < 1$ in the plane) is a manifold. The Möbius band (minus its edge) is a manifold. Boundary points must be omitted, because a boundary point does not lie in a part of the surface that can be mapped onto an *open* set in the plane. The ball and the solid torus (minus their surfaces) are 3-dimensional manifolds. For manifolds with boundary, which will not be discussed here, see Lang 1962.

The surface known as the *Klein bottle* is a 2-dimensional manifold. Imagine a surface in the form of a wine bottle with a long neck and a reentrant bottom, as in (a) in Figure 23.1. Then imagine that the neck is bent downward so as to penetrate through the side of the bottle, as at Q in (b), and then joined, inside the bottle, onto an opening in the reentrant bottom, thus forming a closed one-sided surface. This surface cannot be represented as immersed in V^3 without self-intersection, as at Q, but it can be so immersed in V^4. To see that, start with the surface immersed in V^3, as at (b); each point of it then has three coordinates x_1, x_2, x_3. It is only necessary to assign a fourth coordinate x_4 to each point in such a way that x_4 varies smoothly over the surface, but in a neighborhood of Q takes on one value on the bottle (say 0) and a different value on the neck (say 1). Then, no two different points of the surface have the same set of coordinates x_1, x_2, x_3, x_4.

Many 2-dimensional manifolds can be described by the method of identification or gluing together of edges. The Möbius band is obtained by starting with a rectangle $ABCD$, as in Figure 23.2, and then identifying each point of the edge AB (taking the points in order from A to B) with the corresponding point of the edge CD, so that A is identified with C, B with D, etc. The identified points are regarded as single points of the manifold, just

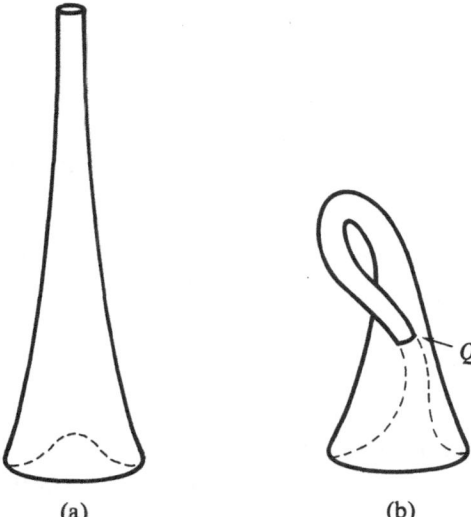

(a) (b) **Figure 23.1** The Klein bottle.

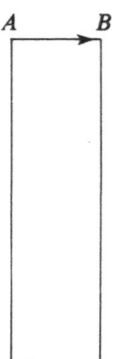

Figure 23.2

as if the rectangle were a narrow strip of paper, which has been bent into a circle, has had one end twisted through a half turn, and then has had the edges glued together.

If, in the above example, the edges AD and CB are also identified, in the same manner, the Klein bottle is the result. (This would require considerable stretching of the paper, to say nothing of the problem of self-intersection.)

Group manifolds were discussed in Section 19.5. The manifold of SO(3) was realized as a certain 3-dimensional algebraic surface in a 9-dimensional space. This surface is homeomorphic, in some neighborhood of each of its points, to a region in E^3, but as a whole is not homeomorphic to any region in E^3; it will be seen in Section 23.7 that it has a kind of connectivity that a region in E^3 cannot have.

The manifold of SO(3) can also be regarded as obtainable by a 3-dimensional version of the method of identification, used above for the Möbius band. If θ_x, θ_y, θ_z are the intrinsic coordinates introduced in Section 19.6, then each point of the ball $\|\boldsymbol{\theta}\| \leq \pi$ represents a single element of SO(3), and conversely, except that any two antipodal points on the surface, $\boldsymbol{\theta}$ and $-\boldsymbol{\theta}$, where $\|\boldsymbol{\theta}\| = \pi$, represent the same element of SO(3) and must be identified with each other. The identification cannot be achieved, in analogy with the Möbius band, by distorting the sphere in 3-dimensional space and gluing surfaces together, but evidently it can be achieved by suitably distorting the sphere in a 9-dimensional space.

According to Exercise 1 in Section 20.6, the manifold of SU(2) can be realized as the 3-sphere, i.e., the unit sphere in E^4. This manifold is simply connected but is also not homeomorphic to any region in E^3.

23.2 Coordinate Systems or Charts; Compatibility; Smoothness

Let $\mathfrak{M}_0$ be a space—i.e., a set of object called points. ($\mathfrak{M}_0$ may be a group, or the like, but is not yet assumed to have any topological structure—see Note in Section 23.4.) An n-dimensional coordinate chart in $\mathfrak{M}_0$ is the assignment of an n-tuple $\{x^1, \ldots, x^n\} \overset{\text{def}}{=} \mathbf{x}$ of real coordinates to each point P of a specified subset $\mathfrak{U}$ of $\mathfrak{M}_0$ in such a way that the assignment $P \to \mathbf{x}$ is a one-to-one mapping $\boldsymbol{\varphi}$ of $\mathfrak{U}$ onto a connected open set N in the coordinate space $\mathbb{R}^n$; one writes $\mathbf{x} = \boldsymbol{\varphi}(P)$, and one refers to the triple $\{\mathfrak{U}, \boldsymbol{\varphi}, N\}$ as a *coordinate chart* in $\mathfrak{M}_0$. The notation is, of course, redundant, since $\mathfrak{U}$ and $\boldsymbol{\varphi}$ determine N, but it is convenient (see Note in Section 23.4). The vector $\mathbf{x} = \boldsymbol{\varphi}(P)$ is sometimes called *the coordinate of P*.

For example, if θ and φ are polar coordinates on the sphere ($\theta = x^1$ and $\varphi = x^2$), then the mapping is from certain points of the sphere onto points of the open rectangle ($0 < \theta < \pi$, $-\pi < \varphi < \pi$) in the θ, φ plane $\mathbb{R}^2$. To make the mapping one-to-one, it is necessary to omit the north and south poles, $\theta = 0$ and $\theta = \pi$, respectively, and the international date line, $\varphi = \pm\pi$. To describe the entire sphere, one might use the method of identification, that is, extend the mapping to the boundary of the rectangle and then decree that the points $\theta = 0$, $-\pi \le \varphi \le \pi$ are all one point at the north pole, also the points $\theta = \pi$, $-\pi \le \varphi \le \pi$ at the south pole, and that, for each θ in $(0, \pi)$, the points with $\varphi = +\pi$ and $\varphi = -\pi$ are the same point. However, in order to be able to impose smoothness requirements and ensure that the surface, when glued together, really looks like a sphere, not like a kreplach or a sopaipilla, a different procedure is needed.

If $\{\mathfrak{U}_1, \boldsymbol{\varphi}_1, N_1\}$ and $\{\mathfrak{U}_2, \boldsymbol{\varphi}_2, N_2\}$ are two overlapping charts in $\mathfrak{M}_0$, they establish a relation between the two sets of coordinates for points P in the intersection $\mathfrak{U}_1 \cap \mathfrak{U}_2$, and this relation is one-to-one, because each of the mappings $P \to \boldsymbol{\varphi}_1(P)$ and $P \to \boldsymbol{\varphi}_2(P)$ is one-to-one. If we write $\mathbf{x} = \boldsymbol{\varphi}_1(P)$ and $\mathbf{y} = \boldsymbol{\varphi}_2(P)$, then the resulting relation between $\mathbf{x}$ and $\mathbf{y}$ and its inverse involve functions that will be denoted as follows:

$$x^i = \hat{x}^i(y^1, \ldots, y^n), \qquad i = 1, \ldots, n, \qquad (23.2\text{-}1)$$

$$y^i = \hat{y}^i(x^1, \ldots, x^n), \qquad i = 1, \ldots, n. \qquad (23.2\text{-}2)$$

These functions are assumed continuous and differentiable a certain number of times. Also, if $\mathfrak{U}_1 \cap \mathfrak{U}_2$ is not empty, each of the transformations (23.2-1) and (23.2-2) is assumed to apply throughout an *open* region in $\mathbb{R}^n$. That is, the two charts are called C^k-*compatible* if

1. the sets $\{\mathbf{x} = \boldsymbol{\varphi}_1(P) : P \in \mathfrak{U}_1 \cap \mathfrak{U}_2\}$ and $\{\mathbf{y} = \boldsymbol{\varphi}_2(P) : P \in \mathfrak{U}_1 \cap \mathfrak{U}_2\}$ are open sets in $\mathbb{R}^n$, and
2. the functions (23.2-1) and (23.2-2) are of class C^k.

If $\mathfrak{U}_1 \cap \mathfrak{U}_2$ is empty, the charts are automatically compatible.

Note. It follows from 2 that if either of the sets referred to in 1 is open in $\mathbb{R}^n$, the other is, too.

It is intuitively clear than any two reasonable charts in a reasonable space will always be compatible. The relationships are shown schematically in

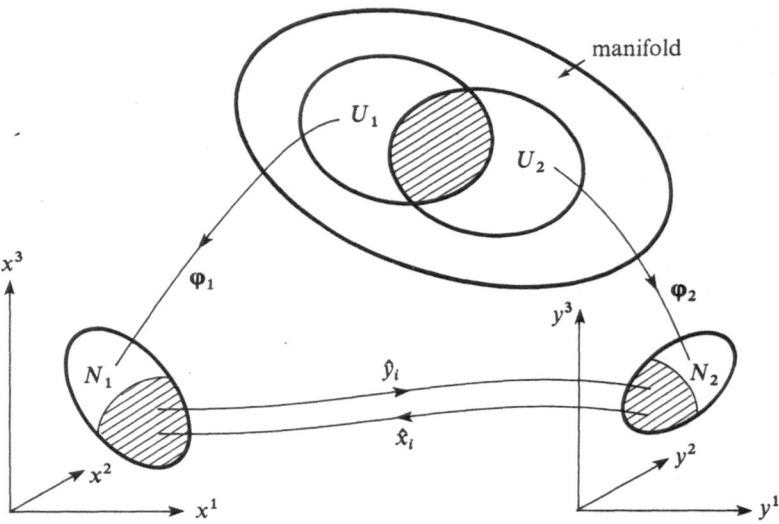

Figure 23.3   Schematic sketch of two charts in a manifold.

Figure 23.3, where the various mappings indicated are the following:

$$\varphi_1 : P \to \varphi_1(P),$$
$$\varphi_2 : P \to \varphi_2(P),$$
$$\hat{x}^i : y^i \to \hat{x}^i(y^1, \ldots, y^n),$$
$$\hat{y}^i : x^i \to \hat{y}^i(x^1, \ldots, x^n).$$

If the polar angles θ, φ on the sphere are the coordinates in the first system, a second system can be chosen so that the coordinates θ', φ' are polar angles with respect to different axes. For example, the north pole N' in the primed system ($\theta' = 0$) might be taken as the point ($\theta = \pi/2$, $\varphi = \pi/2$), in the old system, and the angle φ' about this new north pole so chosen that the new international date line is the portion ($\theta = \pi/2$, $-\pi/2 < \varphi < \pi/2$) of the old equator. See Figure 23.4. It is clear that these two coordinate systems together completely cover the sphere.

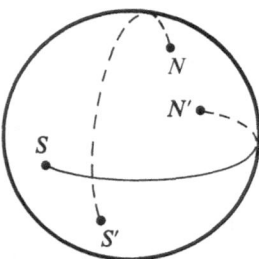

Figure 23.4

EXERCISES

1.   Find the transformations (23.2-1, 2) for this example, i.e., the relation between θ, φ and θ', φ'.

2.   Describe coordinate systems on the surface of the torus. Show that the torus can be covered by two charts, but if simply connected charts are required, three are needed.

23.3 Induced Topology

If $\{\mathfrak{U}, \varphi, N\}$ is a coordinate chart in a space $\mathfrak{M}_0$, if $\mathfrak{U}_0$ is any subset of $\mathfrak{U}$ whose image

$$\varphi(\mathfrak{U}_0) = \{\mathbf{x} = \varphi(P): P \in \mathfrak{U}_0\}$$

is an open set in the coordinate space $\mathbb{R}^n$, then $\mathfrak{U}_0$ is called an *open* set in $\mathfrak{M}_0$. In particular, $\mathfrak{U}$ itself is open. In $\mathbb{R}^n$, the intersection of finitely many open sets is open, and so is the union of an arbitrary collection of open sets. Hence, the open sets in $\mathfrak{M}_0$ determined by a chart have these same properties.

Conditions 1 and 2 for the compatibility of two charts ensure that these charts determine the same topology in their overlap. In particular, condition 1 ensures that the intersection of the open sets $\mathfrak{U}_1$ and $\mathfrak{U}_2$ is open. In the following example, two overlapping charts are given that satisfy condition 2 (trivially) but are not compatible because they do not satisfy condition 1: Let $\mathfrak{M}_0$ consist of those points (x, y) in $\mathbb{R}^2$ on the x and y axes; define two one-dimensional charts in $\mathfrak{M}_0$, one on each axis, as follows:

$$\mathfrak{U}_1 = \{(x, y): x = 0\}, \qquad \mathfrak{U}_2 = \{(x, y): y = 0\},$$

$$\varphi_1((x, y)) = y, \qquad \varphi_2((x, y)) = x,$$

$$N_1 = \mathbb{R}, \qquad N_2 = \mathbb{R}.$$

Then, $\mathfrak{U}_1 \cap \mathfrak{U}_2$ consists of a single point $(x = 0, y = 0)$, which is not an open set in either chart. More generally, two n-dimensional charts might intersect on a surface of smaller dimensions than n (and even in such a way that condition 2 is satisfied in the surface), and in this case their intersection would not be an open set in $\mathbb{R}^n$. Condition 1 prevents this sort of thing by requiring the intersection to be n-dimensional.

When a space $\mathfrak{M}_0$ is covered by a collection of compatible charts, its topology is completely determined by defining the open sets to be the open sets determined by the individual coordinate charts, as above, together with arbitrary unions of such sets. In particular, then, the entire space $\mathfrak{M}_0$ is an open set.

23.4 Definition of Manifold; Hausdorff Separation Axiom

Note. The present discussion differs from the usual discussion of manifolds in one respect. In the usual discussion, one starts with a space in which a topological structure has previously been defined; in fact, it is assumed to be a

Hausdorff space. (However, see Lang 1962.) One then requires the coordinate systems to be continuous with respect to that topology. On the other hand, the existence of coordinate systems restricts the topology considerably, in fact, in such a way that the space is locally Euclidean (this refers to topological, not to metric properties). For the purpose of this book it seems better to let the topological properties be entirely determined by the coordinate systems. Then, only the familiar topological concepts of Euclidean spaces are needed, except for one consideration: When a manifold is being constructed by piecing together two or more coordinate charts, care must be used to ensure that the Hausdorff separation axiom is satisfied—this question will be discussed below.

Note. The discussion starts with a space $\mathfrak{M}_0$, which is simply a collection (uncountably infinite) of otherwise undefined elements, called *points*. In some applications, the space $\mathfrak{M}_0$ is given in advance; for example, it may be a group. In Riemannian geometry or general relativity, on the other hand, one starts with a set of functions $g_{\mu\nu}(x^1, \ldots, x^n)$ defined for coordinates $x^1, \ldots, x^n$ lying in a certain domain N of the coordinate space $\mathbb{R}^n$; each point of N is then assumed to determine a point P of the Riemannian manifold or the physical space being constructed or described. Then, the part of the manifold or physical space thus described may be extended, by means of coordinate transformations like (23.2-1, 2), and so on, until one believes, on the basis of some criterion or other, that the complete manifold has been defined (see, for example, Kruskal's criterion of geodesic completeness described in Chapter 28). In this method, nothing is said in advance about the abstract space $\mathfrak{M}_0$ or its subsets $\mathfrak{U}$, $\mathfrak{U}'$, etc., until the description is complete. Each chart is specified by describing N, and nothing is said explicitly about $\mathfrak{U}$ and φ; hence we prefer to retain "N" in the designation $\{\mathfrak{U}, \varphi, N\}$ of a chart.

A manifold is required to satisfy the following axiom, which expresses an obvious property of Euclidean spaces:

Hausdorff Separation Axiom. If P and Q are any two distinct points, then there are neighborhoods $\mathfrak{U}$ and $\mathfrak{B}$ of P and Q, respectively, such that $\mathfrak{U} \cap \mathfrak{B} = 0$.

When two coordinate systems are pieced together to make a space, it is possible to violate this axiom, as the following one-dimensional example shows: The space $\mathfrak{M}_0$ consists of three copies of the line $\mathbb{R}$, and the points in $\mathfrak{M}_0$ are denoted by $\{x, \alpha\}$, where x is a real number, and α denotes one of the letters a, b, or c. Two charts are defined in $\mathfrak{M}_0$ as follows:

$$\mathfrak{U}_1 = \{\{x, a\}: x \geq 0\} \cup \{\{x, b\}: x < 0\},$$

$$\varphi_1 \quad (\{x, \alpha\}) = x, \qquad N_1 = \mathbb{R},$$

$$\mathfrak{U}_2 = \{\{x, c\}: x \geq 0\} \cup \{\{x, b\}: x < 0\},$$

$$\varphi_2 \quad (\{x, \alpha\}) = x, \qquad N_2 = \mathbb{R}.$$

Each chart, by itself, is homeomorphic to $\mathbb{R}$, but the points $P = \{0, a\}$ and $Q = \{0, c\}$ are not separated, because any two open intervals that contain P and Q, respectively, have points of the form $\{x, b\}$, $x < 0$, in common. It is clear that this phenomenon can easily be avoided in the practical construction of manifolds.

Definition. An *n-dimensional manifold* $\mathfrak{M}$ is a space $\mathfrak{M}_0$ together with a (finite or) countable set of compatible *n*-dimensional coordinate charts, which together cover $\mathfrak{M}_0$ in such a way that the resulting topology satisfies the Hausdorff separation axiom. It is understood that compatible coordinate systems can be added or deleted at will, so long as the space $\mathfrak{M}_0$ is kept covered at all times; the *intrinsic properties* of $\mathfrak{M}$ are those properties that are unaltered by such additions or deletions.

$\mathfrak{M}$ is called a C^k-*manifold* if the transformations (23.2-1, 2) between any two coordinate systems are of class C^k, that is, if the functions $\hat{x}^i(\cdots)$ and $\hat{y}^i(\cdots)$ have continuous partial derivatives (pure and mixed) of all orders up to and including order k. In this case, the addition of new coordinate systems is restricted to ones that satisfy this requirement. Similarly, $\mathfrak{M}$ is called a C^∞-*manifold*, if the transformations are of class C^∞; it is called a real analytic manifold if they are analytic. The manifolds of continuous groups (Lie groups) are real analytic. In general relativity, on the other hand, it is advisable to admit C^k manifolds with finite k, because the Einstein field equations are of hyperbolic type, and gravitational waves can in principle propagate discontinuities of various derivatives of the components $g_{\mu\nu}$ of the metric tensor.

EXERCISE

Intrinsic coordinates $\theta_x, \theta_y, \theta_z$ were defined in the rotation group SO(3) in Section 19.6. To serve in the above definition, they must be restricted to the *open* ball $\theta_x^2 + \theta_y^2 + \theta_z^2 < \pi$. Introduce additional coordinate systems so as to cover the manifold of SO(3). (A general method of doing this for groups is given in Chapter 27.)

23.5 Curves and Functions in a Manifold

Suppose $f(P)$ is a real- or complex-valued function defined for all points P in a manifold $\mathfrak{M}$. If $\{\mathfrak{U}, \varphi, N\}$ is any coordinate chart, then one can define a function $\hat{f}(\mathbf{x}) = \hat{f}(x^1, \ldots, x^n)$ by the equations $\hat{f}(\mathbf{x}) = f(P)$, $\mathbf{x} = \varphi(P)$ for all points $\mathbf{x}$ in the open set N into which $\mathfrak{U}$ is mapped by the mapping $P \to \mathbf{x} = \varphi(P)$; since this relation is one-to-one, $\hat{f}(\cdots)$ is well defined. If the resulting function $\hat{f}(\cdots)$ is continuous for every choice of the chart $\{\mathfrak{U}, \varphi, N\}$ in $\mathfrak{M}$, then $f(P)$ is called a *continuous function* on $\mathfrak{M}$. If $\mathfrak{M}$ is a C^k-manifold and the $\hat{f}$ are of class C^r ($r \le k$), then $f(P)$ is called a *function of class C^r* on $\mathfrak{M}$. Clearly, there can be no functions of class C^r on $\mathfrak{M}$, if $r > k$, other than constants, because allowable coordinate transformations could destroy the existence of derivatives of order $> k$. Continuous or C^r functions can also be defined in a portion of $\mathfrak{M}$.

If $P(t)$ is a one-parameter set of points in $\mathfrak{M}$, where t is a real variable, and if, for every chart $\{\mathfrak{U}, \varphi, N\}$, the functions $\hat{x}^i(t)$ given by

$$\mathbf{x}(t) = \varphi(P(t))$$

are continuous functions of t, for all t for which they are defined, then $P(t)$ is called a *curve* or *path* in $\mathfrak{M}$. If the $\hat{x}^i(t)$ are of class C^r, then $P(t)$ is said to be of class C^r. As t varies in an interval $[t_1, t_2]$, the function $P(t)$ describes a curve $\mathscr{C}$ going from the *initial point* $P(t_1)$ to the *terminal point* $P(t_2)$. It is assumed that all curves are either piecewise differentiable or at least rectifiable (i.e., that the image in any coordinate chart is such), unless otherwise specified. To make that possible, it is assumed that all manifolds considered are at least of class C^1.

Continuous or class C^r functions of two or more variables $P(t, s, \ldots)$ are similarly defined.

23.6 Connectedness; Components of a Manifold

If $\mathfrak{M}$ is such that, given any two points P_1 and P_2 in it, there is a curve, $\mathscr{C}$ in $\mathfrak{M}$ that goes from P_1 to P_2, then $\mathfrak{M}$ is called *pathwise* or *arcwise connected*.

If $\mathfrak{M}$ is arcwise connected and is furthermore such that, given any two curves $\mathscr{C}_1$ and $\mathscr{C}_2$ going from any point P_1 to any other point P_2, $\mathscr{C}_1$ can be continuously deformed into $\mathscr{C}_2$ in $\mathfrak{M}$; i.e., if there is a continuous function $P(t, s)$ such that for each s in an interval, say $[0, 1]$, $P(t, s)$ describes a curve from P_1 to P_2, as t varies, and this curve coincides with $\mathscr{C}_1$ for $s = 0$ and with $\mathscr{C}_2$ for $s = 1$, then $\mathfrak{M}$ is called *simply connected*.

If S is a set of points in a manifold $\mathfrak{M}$ and P is a point of $\mathfrak{M}$, and if every neighborhood of P (open set containing P), no matter how small, contains points of S, then P is called a *limit point* of S. If S contains all its limit points, it is called a *closed set*. The complement of a closed set is obviously an open set, and conversely.

An alternative definition of connectedness usually given by topologists is that a topological space is *connected* if it cannot be decomposed as $S_1 \cup S_2$, where S_1 and S_2 are nonempty disjoint ($S_1 \cap S_2 = \varnothing$) open sets, or, equivalently, nonempty disjoint closed sets. Any arcwise connected space is connected. (*Exercise*: prove this.) For a manifold, the converse is also true, so that then the two concepts are equivalent.

A *component* of a manifold $\mathfrak{M}$ is a maximal connected set in $\mathfrak{M}$; that is, if P_0 is a given point of $\mathfrak{M}$, the set

$$S = \{P: P \text{ can be joined to } P_0 \text{ by a curve}\}$$

is a component of $\mathfrak{M}$. It was seen in Section 19.5 that the manifold of the group $O(3)$ has two components.

As a set in $\mathfrak{M}$, a component S is both open and closed. First, it is open, because (1) any point P of S is an interior point of $\mathfrak{M}$ (all points of $\mathfrak{M}$ are interior points) and (2) a neighborhood of P that is the image of a ball in $\mathbb{R}^n$, in some chart containing P, obviously consists of points in $\mathfrak{M}$ that can be joined to P; hence this entire neighborhood belongs to S. Second, S is closed, because if P is a limit point of a sequence in S, then P can be joined to any point lying in a neighborhood of P of the kind just described, and hence can be joined to points of the sequence. Conversely, if a set S is connected and is both open and closed, then it is maximal, i.e., is a component of $\mathfrak{M}$; this fact will be used in the theory of Lie groups.

Warning. The concept of an open set in $\mathfrak{M}$ has nothing to do with the possible embedding of $\mathfrak{M}$ in a space of higher dimension. For example, if the unit sphere $x^2 + y^2 = z^2 = 1$ is regarded as a 2-dimensional manifold $\mathfrak{M}$, then a polar cap, i.e., the set of all points north of a given circle of latitude, is an open set in $\mathfrak{M}$ but not an open set in $\mathbb{R}^3$.

Note. Any component of $\mathfrak{M}$ is obviously itself a manifold.

23.7 Global Topology; Homotopic Curves; Fundamental Group

In the remainder of this chapter, only connected manifolds are considered. Two curves $\mathscr{C}_0$ and $\mathscr{C}_1$ in a manifold $\mathfrak{M}$, both having initial point P_0 and a terminal point P_1, are called *homotopic* if one of them can be deformed into the other by a continuous deformation in $\mathfrak{M}$, i.e., if there is a continuous function $P(t, s)$ $(0 \le t, s \le 1)$ such that, for every fixed s in $[0, 1]$, $P(t, s)$ traces a curve from P_0 to P_1 as t goes from 0 to 1, and this curve coincides with $\mathscr{C}_0$ for $s = 0$ and with $\mathscr{C}_1$ for $s = 1$. As was said in Section 19.5, a manifold is called *simply connected* if any two curves in it have the same initial and terminal points are homotopic.

It is clear that homotopy is an equivalence relation (it is reflexive, symmetric, and transitive); hence, for given initial and terminal points P_0 and P_1, the class of all curves that are homotopic to a given curve from P_0 to P_1 form an *equivalence class* or *homotopy class* in $\mathfrak{M}$. In the manifold

$$\{x, y : x^2 + y^2 > 1\}$$

consisting of the region of the x, y plane exterior to the unit disc, the first drawing in Figure 23.5 shows three paths from P to Q belonging to the same homotopy class, while the second drawing shows three paths, no two of which are homotopic. In general, if $\mathscr{C}$ denotes any curve in a manifold, $[\mathscr{C}]$ denotes the equivalence class of all curves homotopic to $\mathscr{C}$.

A partial law of composition of homotopy classes of curves is now defined. First, if $\mathscr{C}_1$ and $\mathscr{C}_2$ are curves going from P to Q and from Q to R, respectively, then $\mathscr{C}_1 \circ \mathscr{C}_2$ denotes the curve that follows $\mathscr{C}_1$ from P to Q and then follows

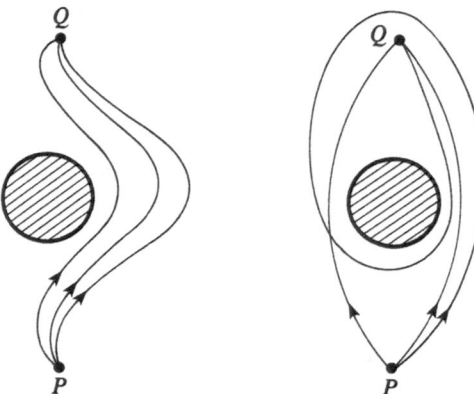

Figure 23.5   Homotopic and nonhomotopic curves.

$\mathscr{C}_2$ from Q to R. To put this in formulas, if the functions $P_1(t)$ ($0 \le t \le 1$) and $P_2(t)$ ($0 \le t \le 1$) describe these curves, then the function

$$P_3(t) \overset{\text{def}}{=} \begin{cases} P_1(2t), & \text{if } 0 \le t \le \frac{1}{2}, \\ P_2(2t - 1), & \text{if } \frac{1}{2} \le t \le 1. \end{cases} \tag{23.7-1}$$

describes the curve $\mathscr{C}_1 \circ \mathscr{C}_2$. The law of composition of homotopy classes is now specified by the formula

$$[\mathscr{C}_1] \circ [\mathscr{C}_2] = [\mathscr{C}_1 \circ \mathscr{C}_2].$$

This applies when the terminal point of the curves of the first class is the same as the initial point of the curves of the second class; otherwise, $[\mathscr{C}_1] \circ [\mathscr{C}_2]$ is undefined. It is easy to give a formal proof that the result is independent of the particular curves $\mathscr{C}_1$ and $\mathscr{C}_2$ chosen from the respective classes. The "product" $[\mathscr{C}_1] \circ [\mathscr{C}_2]$ consists of all curves homotopic to the curve $\mathscr{C}_1 \circ \mathscr{C}_2$, such as the curve $\mathscr{C}_3'$ in Figure 23.6.

This law of composition is associative but does not make the set of all homotopy classes into a group, because the composition is not defined for all pairs of classes, and nothing has been said about inverses. However, a group can be obtained as follows: A fixed base point B_0 is chosen, and consideration is restricted to curves that start from B_0 and return to B_0. (In the definition of homotopy, it was not excluded that the initial and terminal points might coincide.) The set of all homotopy classes of such curves is a group, called the *fundamental group* of the manifold, and is denoted by $\pi_1(\mathfrak{M})$. If $\mathscr{C}_0$ is a curve

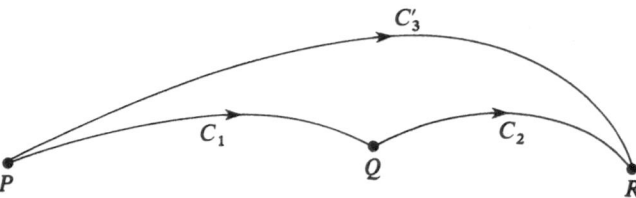

Figure 23.6

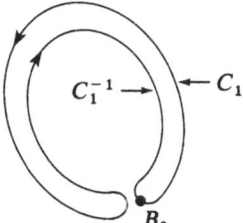

Figure 23.7

that can be contracted onto the base point B_0 by a continuous deformation in $\mathfrak{M}$ (such a curve is called *nullhomotopic*), then $\mathscr{C}_0 \circ \mathscr{C}_1$ can be deformed into $\mathscr{C}_1$; hence $[\mathscr{C}_0] \circ [\mathscr{C}_1] = [\mathscr{C}_1]$; that is, $[\mathscr{C}_0]$ is the identity element of the group. If $\mathscr{C}_1$ is any curve, we denote by $\mathscr{C}_1^{-1}$ the same curve traced backwards; i.e., if the function $P(t)$ $(0 \le t \le 1)$ describes $\mathscr{C}_1$, then the function $P(1 - t)$ describes $\mathscr{C}_1^{-1}$. Clearly, the curve $\mathscr{C}_1 \circ \mathscr{C}_1^{-1}$ can be contracted onto the basic point, i.e., is nullhomotopic—see Figure 23.7; therefore

$$[\mathscr{C}_1] \circ [\mathscr{C}_1^{-1}] = [\mathscr{C}_0] \quad \text{(identity)}; \tag{23.7-2}$$

that is,

$$[\mathscr{C}_1]^{-1} = [\mathscr{C}_1^{-1}]. \tag{23.7-3}$$

On the assumption that the manifold $\mathfrak{M}$ is connected, the fundamental group $\pi_1(\mathfrak{M})$ is independent of the choice of base point, for let B_1 be any other point of $\mathfrak{M}$, and let $\mathscr{C}_{01}$ be any fixed curve from B_0 to B_1. If $\mathscr{C}$ is any curve beginning and ending at B_1, then

$$\mathscr{C}_{01} \circ \mathscr{C} \circ \mathscr{C}_{01}^{-1}$$

is a curve beginning and ending at B_0—see Figure 23.8. The mapping

$$[\mathscr{C}] \rightarrow [\mathscr{C}_{01} \circ \mathscr{C} \circ \mathscr{C}_{01}^{-1}] \tag{23.7-4}$$

is an isomorphism of the fundamental group with B_1 as base point onto the fundamental group with B_0 as base point, for this mapping is obviously one-to-one and onto, and a product $[\mathscr{C}] \circ [\mathscr{C}'] = [\mathscr{C} \circ \mathscr{C}']$ is mapped under (23.7-4) onto

$$
\begin{aligned}
[\mathscr{C}_{01} \circ \mathscr{C} \circ \mathscr{C}' \circ \mathscr{C}_{01}^{-1}] &= [\mathscr{C}_{01} \circ \mathscr{C}] \circ [\mathscr{C}' \circ \mathscr{C}_{01}^{-1}] \\
&= [\mathscr{C}_{01} \circ \mathscr{C}] \circ ([\mathscr{C}_{01}]^{-1} \circ [\mathscr{C}_{01}]) \circ [\mathscr{C}' \circ \mathscr{C}_{01}^{-1}] \\
&= [\mathscr{C}_{01} \circ \mathscr{C} \circ \mathscr{C}_{01}^{-1}] \circ [\mathscr{C}_{01} \circ \mathscr{C}' \circ \mathscr{C}_{01}^{-1}].
\end{aligned}
$$

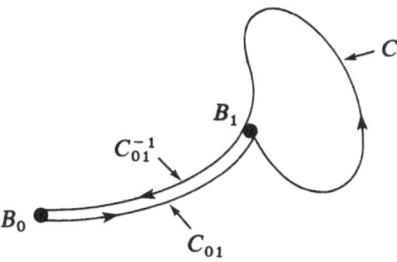

Figure 23.8

EXAMPLES

(1)   If $\mathfrak{M}$ is simply connected, then its fundamental group $\pi_1(\mathfrak{M})$ is the trivial group consisting of an identity element only.

(2)   Let $\mathfrak{M}$ be the surface of a cylinder (finite or infinite, but finite in the drawing of Figure 23.9). Let θ, z be cylindrical coordinates, and let values of θ, z be represented on a strip in the plane, as shown. Each point of $\mathfrak{M}$ is multiply represented on the strip, and in particular the base point B_0 of $\mathfrak{M}$ is represented by the points B'_0, $B'_{\pm 1}$, $B'_{\pm 2}$, etc. A curve in the strip, such as $\mathscr{C}$, going from B'_0 to any other image of B_0, say B'_k, is the image of a closed curve in $\mathfrak{M}$, and, conversely, every closed curve beginning and ending at B_0 in $\mathfrak{M}$ has such an image; furthermore, $\mathscr{C}$ can be continuously deformed, within the strip, keeping its ends fixed, into any other curve going from B'_0 to B'_k, such as $\mathscr{C}'$. Consequently, for each of the possible terminal points B'_k, there is precisely one homotopy class of curves beginning and ending at B_0 in $\mathfrak{M}$; k is the net number of windings about the cylinder made by a curve of the class. The composition of two such curves, say with terminal points B'_k and B'_l, is a curve with terminal point B'_{k+l}; hence $\pi_1(\mathfrak{M})$ is isomorphic to the additive group of integers, that is, to the infinite cyclic group C_∞. The annulus $a < x^2 + y^2 < b$, the punctured plane $x^2 + y^2 > 0$, and the Möbius strip all have fundamental groups isomorphic to C_∞.

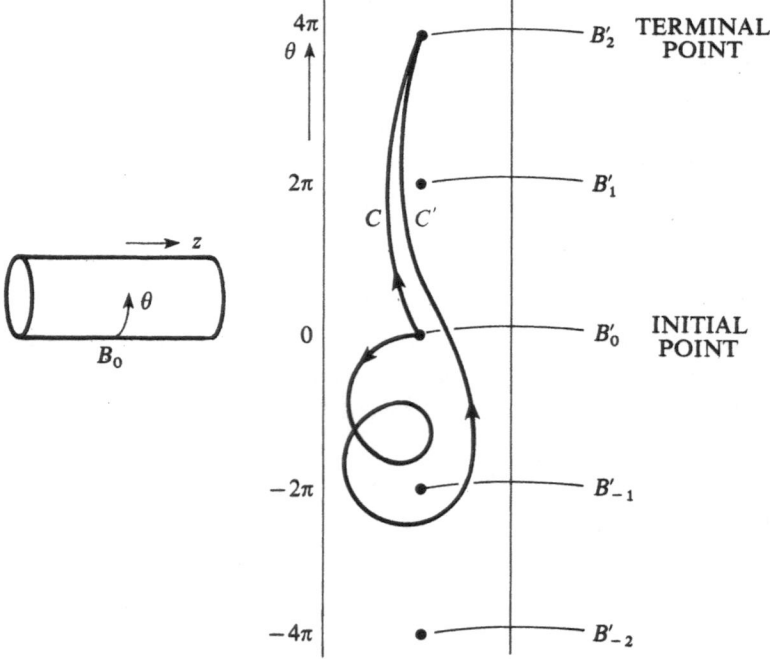

Figure 23.9

(3)   Let $\mathfrak{M}$ be the surface of the torus, which is given by the equations

$$z = a \sin \alpha,$$
$$x = (A + a \cos \alpha)\cos \beta,$$
$$y = (A + a \cos \alpha)\sin \beta,$$

where x, y, and z are Cartesian coordinates, a and A are constants ($A > a > 0$), and α and β are two angles or intrinsic coordinates on $\mathfrak{M}$. See Figure 23.10. If α and β are allowed to vary unrestrictedly, then the number pairs (α, β) and $(\alpha + 2\pi k, \beta + 2\pi l)$ represent the same point of $\mathfrak{M}$. Let the base point B_0 be given by $x = A + a, y = z = 0$; it is then represented by any of the points $(\alpha, \beta) = (2\pi k, 2\pi l)$ in a lattice in the α, β plane. Any curve from $(0, 0)$ to $(2\pi k, 2\pi l)$ in the plane represents a closed curve beginning and ending at B_0 in $\mathfrak{M}$ and can be continuously deformed into another curve from $(0, 0)$ to $(2\pi k, 2\pi l)$. Therefore, each integer pair (k, l) determines an element of the fundamental group $\pi_1(\mathfrak{M})$. Clearly, the composition of the elements determined by (k, l) and (k', l') is the element determined by $(k + k', l + l')$; that is, the fundamental group of the torus is isomorphic to the direct product $C_\infty \times C_\infty$, i.e., to the free abelian group on two generators.

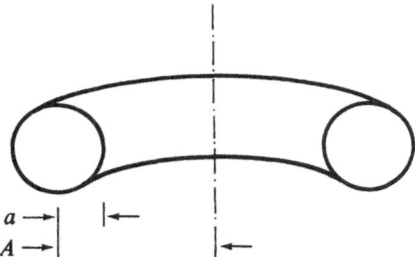

$a \rightarrow$

$A \rightarrow$

Figure 23.10   The 2-torus.

(4)   Consider the manifold $\mathfrak{M}$ consisting of the plane with two points a and b removed. In the theory of functions of a complex variable, a contour of integration is specified by writing an equation such as

$$J = \int^{(a+, a+, b-)} f(z)dz.$$

Here, the expression $(a+, a+, b-)$ indicates that the contour starts from some base point B (not coinciding with a or with b), encircles the point a twice positively (counterclockwise), then encircles the point b once negatively, and then returns to B, as in Figure 23.11. In function theory it is taken as geometrically evident that this procedure specifies the contour adequately, if $f(z)$ is analytic except for branch points at a and b; that is, any two contours which follow the above prescription

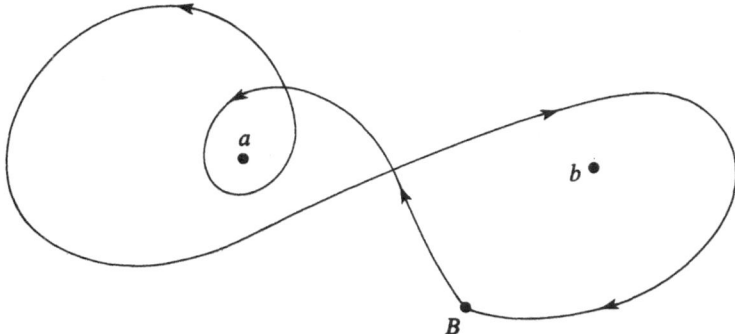

Figure 23.11   A contour in the complex plane.

can be deformed continuously into each other without crossing either branch point. In other words, the expression $(a+, a+, b-)$ determines a homotopy class of curves in $\mathfrak{M}$, i.e., an element of $\pi_1(\mathfrak{M})$. This point of view will be accepted here. The simplest nontrivial elements of the group $\pi_1(\mathfrak{M})$ are $(a+)$ and $(b+)$ and their inverses $(a-)$ and $(b-)$; these group elements will be denoted by α, β, α^{-1}, and β^{-1}. The general group element is of the form

$$\gamma_1^{\varepsilon_1}\gamma_2^{\varepsilon_2}\cdots\gamma_k^{\varepsilon_k},$$

where each γ_i is either α or β and each exponent ε_i is either $+1$ or -1. In other words, $\pi_1(\mathfrak{M})$ is isomorphic to the free group on two generators. Unlike the groups in the first three examples, this group is noncommutative. The same fundamental group is obtained for the figure-of-eight region shown shaded in Figure 23.12. If n distinct points are removed from the plane, then the resulting fundamental group is isomorphic to the free group on n generators.

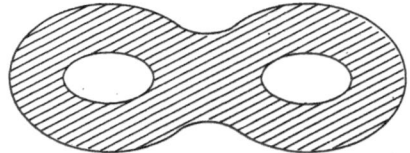

Figure 23.12

(5)   Let $\mathfrak{M}$ be the manifold of the rotation group SO(3). In Section 19.6 intrinsic coordinates in $\mathfrak{M}$ were introduced as the three components of a vector $\boldsymbol{\theta}$, which lies in the spherical ball $K = \{\boldsymbol{\theta}: \|\boldsymbol{\theta}\| \le \pi\}$ in the coordinate space. If opposite ends of each diameter of K are identified (regarded as the same point), then there is a one-to-one correspondence between the points of $\mathfrak{M}$ and those of K. A nontrivial element of $\pi_1(\mathfrak{M})$ is obtained by taking the base point B as the center K, and by considering a curve $\mathscr{C}$ that goes from B along a radius to a point A on the surface, then jumps to the antipodal point A', then returns to B, as shown in Figure 23.13. This curve $\mathscr{C}$ cannot be collapsed onto B by a continuous deformation, because (a) any curve that starts and ends at B and has such a jump has total length (in K) at least 2π, and (b) it is intuitively clear on grounds of continuity that continuous deformation cannot make the jump disappear. (This will be established more firmly in the next chapter.) Now consider a curve $\mathscr{C}$ that starts at B and returns to B after making a finite number of such jumps, let us say from A_1 to A'_1, from A_2 to A'_2, etc., where in each case the prime denotes the antipodal point. By continuous deformation of the curve, consecutive jumps can be made to disappear two at a time. Consider a portion of $\mathscr{C}$ that contains

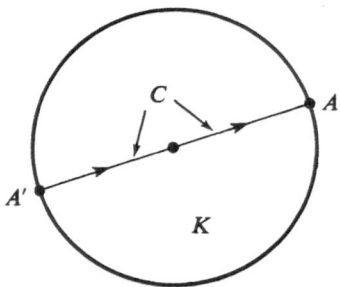

Figure 23.13   A nonnullhomotopic curve in the manifold of SO(3).

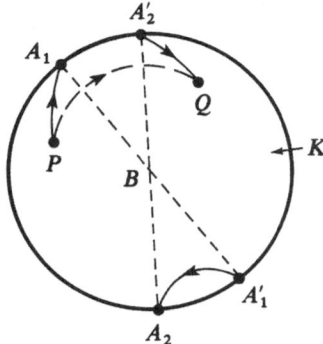

Figure 23.14   Homotopic curves from P to Q in the manifold of SO(3).

two consecutive jumps, as in Figure 23.14, where it consists of the parts PA_1, A'_1A_2, and A'_2Q. By moving the part A'_1A_2 to the surface of K and simultaneously drawing the points A_2 and A'_1 (also A_1 and A'_2) together, the part A'_1A_2 can be made to disappear, and what remains is a curve, like the dashed one, going from P to Q without a jump. If this procedure is continued, the curve $\mathscr{C}$ can be collapsed onto the base point B, if it had initially an even number of jumps, or onto a curve having a single jump, if it had initially an odd number. Therefore, the group $\pi_1(\text{SO}(3))$ is isomorphic to the cyclic group of order 2, consisting of just two elements. If initially $\mathscr{C}$ had infinitely many jumps, then many of these jumps would have to be very close together, so that the value of $\|\boldsymbol{\theta}\|$ would remain close to π between them, and a continuous deformation could be made to a curve that has a finite number of jumps. These results will all be obtained more simply and rigorously in the next chapter by means of the covering of SO(3) by SU(2).

It should be noted that a knowledge of the fundamental group does not completely fix the global topology of a manifold. For example, the sphere $x^2 + y^2 + z^2 = 1$ and the disk $x^2 + y^2 < 1$ are both simply connected 2-dimensional manifolds, but they are topologically different: If a single point is removed, the sphere remains simply connected, while the disk does not. For this reason, one might be tempted to introduce also the higher homotopy groups (see Hocking and Young 1961, Chapter 4) or other topological characterizations. However, it appears that the first homotopy group, the fundamental group, is precisely what is needed for many purposes, such as the theory of the covering of one manifold by another, which is the subject of the next chapter.

It will be shown in Chapter 27 that the fundamental group of the manifold of a Lie group is always Abelian (commutative).

23.8 Mechanical Linkages: Cartesian Products

Each configuration of the compound pendulum sketched in Figure 23.15 can be specified by the two angles α and β shown. For any integers k and l, the number pair $(\alpha + 2\pi k, \beta + 2\pi l)$ represents the same configuration as (α, β). Hence, according to Example 3 in the preceding section, there is a

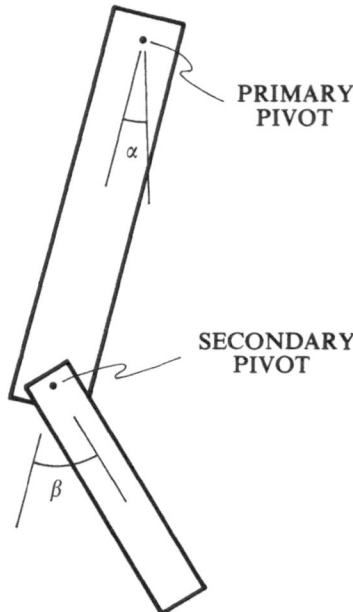

Figure 23.15   A compound pendulum.

one-to-one correspondence between the configurations of the pendulum and the points of a torus, of such a kind that as the pendulum moves, the corresponding point moves continuously on the torus.

If the secondary pivot is perpendicular to the primary one, as in Figure 23.16, then the tip of the secondary pendulum moves on a torus in space.

In either case, each part of the pendulum moves in a circle about its own pivot, and the resulting motion is the combined effect of these two circular motions. Correspondingly, the torus is regarded as the so-called *Cartesian product* of two circles. Generally, let $\mathfrak{M}$ and $\mathfrak{M}'$ be any two manifolds, of dimensions n and n'. Their Cartesian product, $\mathfrak{M} \times \mathfrak{M}'$, is an $(n + n')$-dimensional manifold defined as follows: (1) Each point of $\mathfrak{M} \times \mathfrak{M}'$ is an

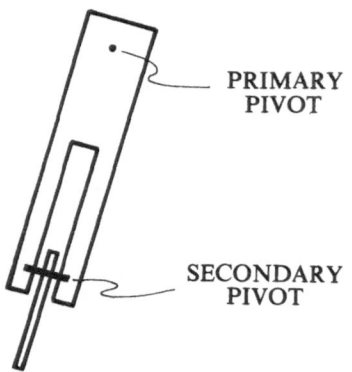

Figure 23.16   A compound pendulum.

ordered pair (P, P'), where P and P' are arbitrary points of $\mathfrak{M}$ and $\mathfrak{M}'$, respectively. That is, $\mathfrak{M} \times \mathfrak{M}'$, as a set, is the Cartesian product of $\mathfrak{M}$ and $\mathfrak{M}'$ in the set-theoretical sense. (2) If $\{\mathfrak{U}, \varphi, N\}$ and $\{\mathfrak{U}', \varphi', N'\}$ are any charts in $\mathfrak{M}$ and $\mathfrak{M}'$, respectively, then a chart $\{\mathfrak{U}'', \varphi'', N''\}$ is defined in $\mathfrak{M} \times \mathfrak{M}'$, as follows: $\mathfrak{U}''$ is the set of all points (P, P') such that P is in $\mathfrak{U}$ and P' is in $\mathfrak{U}'$, and $\varphi''((P, P'))$ is the $(n + n')$-component vector consisting of the components of $\varphi(P)$ together with those of $\varphi'(P')$, i.e.,

$$\varphi_i''((P, P')) = \begin{cases} \varphi_i(P), & \text{if } i = 1, 2, \ldots, n, \\ \varphi_{i-n}'(P'), & \text{if } i = n + 1, \ldots, n + n'. \end{cases}$$

It is evident that these definitions make $\mathfrak{M} \times \mathfrak{M}'$ into a manifold.

If the pivots of the compound pendulum are replaced by idealized ball-and-socket joints, then the configuration space is the Cartesian product of two 2-spheres, and hence is a 4-dimensional manifold. (Rotation of the two members about their own logitudinal axes has been neglected.)

Lastly, if a small spherical ball moves inside a hollow sphere in such a way as to maintain contact, then the configuration space is the Cartesian product of a sphere and the manifold of SO(3), and hence is a 5-dimensional manifold.

Clearly other examples of this kind can be constructed indefinitely.

CHAPTER 24

Covering Manifolds

Local homeomorphism; projection; p-sheeted covering; good
neighborhood; principles of lifting; universal covering manifold;
construction of mathematical models; manifolds covered by a given
manifold.

Prerequisites: Chapter 23 and parts of Chapters 18 and 19.

24.1 Definition and Examples

The 2-to-1 mapping ψ of SU(2) onto SO(3) found in Section 19.7 is more than
merely a group homomorphism. It is also a mapping of the manifold of
SU(2) onto the manifold of SO(3) of the kind known as a covering. It is
locally a homeomorphism, in the sense that if P is any point on the manifold
of SU(2), and Q is its image in the manifold of SO(3), then P has a neighborhood that is mapped homeomorphically by ψ onto a neighorhood of Q.
Furthermore, if Q is *any* point of the second manifold, then there are always
two points P in the first manifold, each of which has such a neighborhood.
ψ is a *two-sheeted covering* of SO(3) by SU(2).

A mapping $\psi: \mathfrak{M} \to \mathfrak{N}$ of a manifold $\mathfrak{M}$ to a manifold $\mathfrak{N}$ (they will be called
the "upper" and "lower" manifolds, respectively) is called a *covering* of $\mathfrak{N}$
by $\mathfrak{M}$ if it satisfies the following two requirements, of which the first says that
all of $\mathfrak{N}$ is covered, and the second tells just how it is covered: (a) ψ is an *onto*
mapping; i.e., for each point Q in the lower manifold ($\mathfrak{N}$) there is at least one
point P in the upper one ($\mathfrak{M}$) such that $\psi(P) = Q$; (b) each point Q of the
lower manifold is in some neighborhood $\mathfrak{B}$ whose preimage $\psi^{-1}(\mathfrak{B})$—this is
the set of all points of the upper manifold that are mapped onto points of
$\mathfrak{B}$—consists of one or more disjoint neighborhoods $\mathfrak{U}_1, \mathfrak{U}_2, \ldots$, or components
(one in each "sheet" of $\mathfrak{M}$), each of which is homeomorphic with $\mathfrak{B}$; that is,
for each j, the mapping $P \to \psi(P)$, restricted to $\mathfrak{U}_j$, is a one-to-one bicon-
tinuous mapping of $\mathfrak{U}_j$ onto $\mathfrak{B}$. A neighborhood $\mathfrak{B}$ in the lower manifold with
these properties is called a *good* neighborhood. (A mapping is called *bi-
continuous* if it and its inverse are both continuous.) If $x^1, \ldots, x^n$ are coordinates of P in the neighborhood $\mathfrak{U}_j$ in $\mathfrak{M}$, and if $y^1, \ldots, y^n$ are coordinates

of the corresponding point $\psi(P)$ in the neighborhood $\mathfrak{B}$ in $\mathfrak{N}$, then the x^i are continuous functions of the y^i, and conversely, throughout the respective neighborhoods. Obviously, $\mathfrak{M}$ and $\mathfrak{N}$ must have the same number of dimensions.

If such a mapping ψ exists, $\mathfrak{M}$ is called a *covering manifold* of $\mathfrak{N}$, and ψ is called a *covering* of $\mathfrak{N}$ by $\mathfrak{M}$ or a *projection* of $\mathfrak{M}$ onto $\mathfrak{N}$.

If the manifolds are connected, then the *multiplicity* of the covering (i.e., the number of points of $\mathfrak{M}$ that are mapped onto a single point on $\mathfrak{N}$) is constant throughout the manifolds, because this number is a positive integer or $+\infty$, and it is obviously constant in any neighborhood, and hence is constant throughout $\mathfrak{M}$ and $\mathfrak{N}$. If the multiplicity is p, then ψ is called a *p*-sheeted covering. If $\mathfrak{M}$ and $\mathfrak{N}$ are of class C^k, then the x_i as functions of the y_i are required to be of class C^k; that is, the mapping ψ is required to be of class C^k.

Note. The following 1-dimensional example shows that it would not have been equivalent to require merely that each point P of the upper manifold $\mathfrak{M}$ have a neighborhood that is mapped homeomorphically onto a neighborhood of the lower manifold $\mathfrak{N}$: Let $\mathfrak{N}$ be the unit circle in a plane, and let $\mathfrak{M}$ be an open interval of length greater than 2π wrapped around the unit circle. Then the points a and b of $\mathfrak{N}$ lying under the ends of $\mathfrak{M}$ (see Figure 24.1) do not satisfy the conditions of the definition, although, since $\mathfrak{M}$ is open, each of *its* points has a neighborhood that is mapped homeomorphically into $\mathfrak{N}$.

If $\mathfrak{M}$ denotes the Riemann surface of any algebraic function $F(z)$, with all branch points deleted, if $\mathfrak{N}$ denotes the complex plane, with the corresponding points deleted, and if ψ is the mapping that maps any point of $\mathfrak{M}$ onto the point of $\mathfrak{N}$ directly underneath it (i.e., the point with the same value of z attached), then ψ is a covering of $\mathfrak{N}$ by $\mathfrak{M}$. If P is any point of $\mathfrak{N}$, then P has a neighborhood $\mathfrak{B}$ which is simply connected and does not contain any of the branch points of $F(z)$. If a right cylinder is constructed with $\mathfrak{B}$ as base, then this cylinder intersects each sheet of the Riemann surface in a neighborhood $\mathfrak{U}$ which looks exactly like $\mathfrak{B}$. Hence, $\mathfrak{B}$ is a good neighborhood.

A given manifold $\mathfrak{N}$ may have many different covering manifolds $\mathfrak{M}$, and may have many different coverings by a given $\mathfrak{M}$. If $\mathfrak{N}$ is the unit circle $|z| = 1$ in the z plane, then $\mathfrak{N}$ can be covered by the real line by the mapping

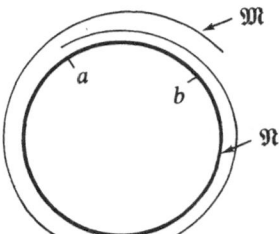

Figure 24.1

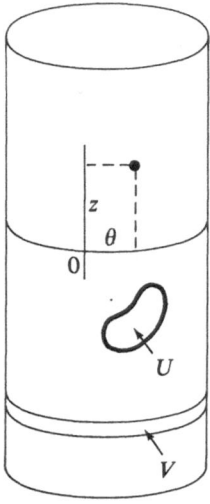

Figure 24.2

$x \to z = e^{ix}$ (the line is to be imagined as wrapped around the circle infinitely many times), or $\mathfrak{N}$ can be covered by the circle $\mathfrak{M}: |w| = 1$ in the w plane by any of the mappings $w \to z = w^n$, where n is any integer not $=0$ (the circle $\mathfrak{M}$ is to be imagined as stretched and wrapped n times around the circle $\mathfrak{N}$).

Consider the mapping ψ of a plane onto a cylinder given by setting z equal to x and θ equal to y, where x, y are Cartesian coordinates in the plane, and z, θ are cylindrical coordinates on the cylinder. This mapping is many-to-one, because the points (x, y), $(x, y \pm 2\pi)$, $(x, y \pm 4\pi)$, etc. are all mapped onto the same point of the cylinder. The preimage of a neighborhood, which, like $\mathfrak{U}$ in Figure 24.2 does not encircle the cylinder, consists of an infinite succession of neighborhoods $\mathfrak{U}'_j$ in the plane, obtained by displacing any one of them horizontally through distances $\pm 2\pi$, $\pm 4\pi$, etc., as in Figure 24.3; each $\mathfrak{U}'_j$ is homeomorphic to $\mathfrak{U}$; hence, $\mathfrak{U}$ is a good neighborhood. On the other hand, a neighborhood like $\mathfrak{V}$ in Figure 24.2, which is a band encircling the cylinder, is not a good neighborhood, because its preimage is an infinite strip in the plane and is mapped many-to-one onto $\mathfrak{V}$ under ψ.

EXERCISE

Discuss the various possible coverings of a torus by a plane, by a cylinder, and by another torus.

U'_{-1} U'_{o} U'_{1} etc.

Figure 24.3

If $\mathfrak{M}$ and $\mathfrak{N}$ are C^k manifolds, then the mapping ψ is required to be of class C^k; that is, if ψ maps P onto $Q = \psi(P)$, if $\mathfrak{B}$ is a good neighborhood of Q, if $\mathfrak{U}$ is the component of $\psi^{-1}(\mathfrak{B})$ containing P, and if $x^1, \ldots, x^n$ are coordinates of P in $\mathfrak{U}$, while $y^1, \ldots, y^n$ are coordinates of Q in $\mathfrak{B}$, then, as P varies, the x^i are functions of class C^k of the y^i, and conversely. (In nearly all cases of interest, $\mathfrak{M}$ and $\mathfrak{N}$ are analytic manifolds, and these functions are analytic.)

If the covering ψ is a one-to-one mapping, so that each of $\mathfrak{M}$ and $\mathfrak{N}$ is a covering of the other, then ψ is called a (C^k) *homeomorphism* and the manifolds are called *homeomorphic*; they are topologically indistinguishable. In the case $k = \infty$, a homeomorphism is sometimes called a *diffeomorphism*.

24.2 Principles of Lifting

If the covering manifold $\mathfrak{M}$ is connected, then $\mathfrak{N}$ is necessarily also connected. The converse is, of course, not true; however, only connected manifolds will be considered, from now on. If $\mathfrak{M}$ is *simply* connected (like the line and the plane the forgoing examples), then, according to the theorem below, $\mathfrak{M}$ represents the most that can be achieved, in a sense, by way of covering the given manifold $\mathfrak{N}$ with a connected manifold. Two lemmas are needed.

Lemma 1 (First Principle of Lifting). *Suppose that the manifold $\mathfrak{M}_1$ covers the manifold $\mathfrak{M}_0$ by the projection ψ. Let B_0 be an arbitrary base point of $\mathfrak{M}_0$. Of all the points P in $\mathfrak{M}_1$ such that $\psi(P) = B_0$, choose one and call it the base point B_1 of $\mathfrak{M}_1$. (See Figure 24.5 in the next section, where, however, a third manifold has been added, in connection with the theorem below.) Let $\mathscr{C}_0$ be a curve in $\mathfrak{M}_0$, described by a continuous function $P_0(t)$ $(0 \leq t \leq 1)$ starting from the base point B_0. Then, there is a unique curve $\mathscr{C}_1 : P_1(t)$ in $\mathfrak{M}_1$ such that $P_1(0) = B_1$, and $\psi(P_1(t)) = P_0(t)$. One says that the curve $\mathscr{C}_0$ has been* lifted *up to $\mathfrak{M}_1$ from $\mathfrak{M}_0$.*

PROOF. As noted under the definition of a covering, a neighborhood $\mathfrak{U}$ in $\mathfrak{M}_0$ is called a *good* neighborhood if each component of $\psi^{-1}(\mathfrak{U})$ is homeomorphic to $\mathfrak{U}$ under ψ. If I is any interval contained in $[0, 1]$, the symbol $P_0(I)$ denotes the segment of the curve $\mathscr{C}_0$ given by

$$P_0(I) = \{P_0(t) : t \in I\}.$$

A partition of $[0, 1]$ into closed intervals $[0, t_1], [t_1, t_2], \ldots, [t_{N-1}, 1]$ is called a *good* partition if each segment $P_0([t_j, t_{j+1}])$ of the curve $\mathscr{C}_0$ lies in a good neighborhood. A good partition exists, because each point of $\mathscr{C}_0$ is in a good neighborhood; hence each t in $[0, 1]$ is in an open interval I such that the curve segment $P_0(I)$ is in a good neighborhood. These open intervals cover $[0, 1]$; by the Heine-Borel theorem, a finite number of them cover $[0, 1]$; if these are arranged in order of increasing t, then t_1 can be chosen in the intersection of the first and second of these neighborhoods, t_2 in the intersection of the second and third, etc. See Figure 24.4. Thus, a good partition is obtained. Call $\mathfrak{U}_j$ the good neighborhood in which the curve segment $P_0([t_j, t_{j+1}])$ lies. For each $j = 0, 1, \ldots, N - 1$, a segment $P_1([t_j, t_{j+1}])$ of

Figure 24.4

the curve $\mathscr{C}_1 = \{P_1(t): 0 \leq t \leq 1\}$ in the upper manifold $\mathfrak{M}_1$ is now defined, inductively, as follows: First let $\mathfrak{B}_0$ be the component of $\psi^{-1}(\mathfrak{U}_0)$ that contains the base point B_1 of $\mathfrak{M}_1$, and define $P_1([0, t_1])$ to be $\hat{\psi}^{-1}(P_0[0, t_1])$, where $\hat{\psi}$ is the restriction of ψ to $\mathfrak{B}_0$; $\hat{\psi}$ is a homeomorphism of $\mathfrak{B}_0$ onto $\mathfrak{U}_0$; hence $P_1([0, t_1])$, thus defined, is a curve segment in $\mathfrak{M}_1$. Now suppose $P_1([t_j, t_{j+1}])$ has been defined; since $P_0(t_{j+1})$ lies in $\mathfrak{U}_{j+1}$ as well as in $\mathfrak{U}_j$, $\mathfrak{B}_{j+1}$ can be taken as the component of $\psi^{-1}(\mathfrak{U}_{j+1})$ that contains the endpoint $P_1(t_{j+1})$ of the previously defined segment of $\mathscr{C}_1$; then the segment $P_1([t_{j+1}, t_{j+2}])$ is defined as $\hat{\psi}^{-1}(P_0([t_{j+1}, t_{j+2}]))$, where now $\hat{\psi}$ is the restriction of ψ to $\mathfrak{B}_{j+1}$. In this way, the curve $\mathscr{C}_1$ in $\mathfrak{M}_1$ is constructed. It is uniquely determined by the curve $\mathscr{C}_0$ in the lower manifold and the choice of the base point B_1 in the upper one; in particular, it is independent of the choice of the good partition of $[0, 1]$, because any two partitions have a common refinement, and $\mathscr{C}_1$ is obviously not altered by refining the partition used (i.e., by adding further points of subdivision of $[0, 1]$). Each of the curves $\mathscr{C}_0$ and $\mathscr{C}_1$ uniquely determines the other.

Lemma 2   (Second Principle of Lifting).   *Under the conditions of Lemma 1, if $P_0(t, s)$ is a continuous function of two variables in $\mathfrak{M}_0$, defined in the square $0 \leq t, s = 1$, if $P_0(t, s_0)$ coincides with the base piont B_0, for some t_0 and s_0, then there is a unique continuous function $P_1(t, s)$ in $\mathfrak{M}_1$ such that (1) $\psi(P_1(t, s)) = P_0(t, s)$ and (2) $P_1(t_0, s_0)$ is the base point B_1 of $\mathfrak{M}_1$.*

The proof is similar to that of Lemma 1. The Heine–Borel theorem implies the existence of a finite collection of open sets in the t, s plane, which covers the square $0 \leq t, s \leq 1$, and such that each set determines a good neighborhood. Let these sets be arranged in a sequence such that each one contains points in common with at least one of the preceding sets. From here on the argument is as before.

Corollary.   *If two curves $\mathscr{C}_0$ and $\mathscr{C}'_0$ in the lower manifold, both running from B_0 to some point A_0, are homotopic, i.e., if one of them can be deformed in $\mathfrak{M}_0$ continuously into the other, keeping the endpoints fixed, then the curves that result from lifting them up to $\mathfrak{M}_1$ also have a common endpoint A_1 and are homotopic in $\mathfrak{M}_1$.*

SKETCH OF THE PROOF.   Let $P_0(t, s)$ in the lemma be such that, for each fixed s in $[0, 1]$, $P_0(t, s)$ traces a curve from B_0 to A_0, as t increases from 0 to 1, and such that for $s = 0$ this curve is $\mathscr{C}_0$ while for $s = 1$ it is $\mathscr{C}'_0$; then use a continuity argument, based on a good neighborhood of A_0, to show that the terminal point of the lifted curve, $P_1(1, s)$ cannot jump from one sheet of $\mathfrak{M}_1$ to another, as s varies.

24.3 Universal Covering Manifold

In Section 24.5, it will be proved that, if $\mathfrak{M}_0$ is any manifold, then there is a simply connected manifold that covers $\mathfrak{M}_0$. It is called the *universal covering manifold* of $\mathfrak{M}_0$, because of the theorem below, which says (1) that the universal covering manifold of a given $\mathfrak{M}_0$ is unique (up to homeomorphism), and (2) that it also covers any other manifold that covers $\mathfrak{M}_0$. The proof of existence is postponed until Section 24.5 because it is a little more abstruse than the proof of the present theorem.

Theorem. *If $\mathfrak{M}_1$ and $\mathfrak{M}_2$ are connected covering manifolds of $\mathfrak{M}_0$, and if $\mathfrak{M}_2$ is simply connected, then $\mathfrak{M}_2$ is a covering manifold of $\mathfrak{M}_1$. If $\mathfrak{M}_1$ is also simply connected, then $\mathfrak{M}_2$ and $\mathfrak{M}_1$ are homeomorphic, i.e., topologically indistinguishable.*

PROOF. Let ψ_{10} and ψ_{20} denote projections of $\mathfrak{M}_2$ and $\mathfrak{M}_1$, respectively, onto $\mathfrak{M}_0$. A projection ψ_{21} from $\mathfrak{M}_2$ to $\mathfrak{M}_1$ will be so constructed that $\psi_{10}\psi_{21} = \psi_{20}$. Choose base points B_2, B_1, and B_0 in the three manifolds such that $\psi_{20}(B_2) = \psi_{10}(B_1) = B_0$ (see Figure 24-5). To describe ψ_{21}, we must specify a point $\psi_{21}(Q_2)$ in $\mathfrak{M}_1$ for each point Q_2 in $\mathfrak{M}_2$; we do it as follows: Let $\mathscr{C}_2$ be a curve $P_2(t)$ in $\mathfrak{M}_2$ from B_2 to Q_2, namely, a curve such that $P_2(0) = B_2$, $P_2(1) = Q_2$. Then the projection of $\mathscr{C}_2$ onto $\mathfrak{M}_0$ is a curve $P_0(t) = \psi_{20}(P_2(t))$ from B_0 to Q_0. (*Comment*: $\mathscr{C}_0$ may be self-intersecting, even if $\mathscr{C}_2$ is not. For example, in Figure 24.6, $\mathscr{C}_2$ is a curve in a plane, and $\mathscr{C}_0$ is the result of wrapping the plane around a cylinder.) The mapping ψ_{10}^{-1} is, of course, generally multivalued, but according to the lifting principle of Lemma 1, there is a unique curve $\mathscr{C}_1 : P = P_1(t)$ in $\mathfrak{M}_1$, starting from the basepoint B_1 and such that $\psi_{10}(\mathscr{C}_1) = \mathscr{C}_0$. It is now asserted that the terminal point $Q_1 = P_1(1)$ of $\mathscr{C}_1$ is uniquely determined by the terminal point Q_2 of $\mathscr{C}_2$ in the upper manifold $\mathfrak{M}_2$. To prove this, let $\mathscr{C}_2'$ be any other curve in $\mathfrak{M}_2$ from B_2 to Q_2. Since $\mathfrak{M}_2$ is simply connected, $\mathscr{C}_2$ and $\mathscr{C}_2'$ are homotopic—either can be deformed into the other by a continuous deformation in $\mathfrak{M}_2$. Therefore, their images $\mathscr{C}_0$ and $\mathscr{C}_0'$ in $\mathfrak{M}_0$ are homotopic, and, by the corollary to Lemma 2, the curves $\mathscr{C}_1$ and $\mathscr{C}_1'$ that result from lifting $\mathscr{C}_0$ and $\mathscr{C}_0'$ up to $\mathfrak{M}_1$ have a common terminal point Q_1, which is now defined to be $\psi_{21}(Q_2)$. Since Q_2 was arbitrary, ψ_{21} is defined in all $\mathfrak{M}_2$, and

$$\psi_{10}(\psi_{21}(Q_2)) = \psi_{10}(Q_1) = Q_0 = \psi_{20}(Q_2).$$

EXERCISE

Complete the proof by showing (a) that ψ_{21} is an onto mapping, (b) that any Q_1 in $\mathfrak{M}_1$ has a neighborhood $\mathfrak{B}$ such that each component of $\psi_{21}^{-1}(\mathfrak{B})$ is homeomorphic to $\mathfrak{B}$ under ψ_{21}, and (c) that if $\mathfrak{M}_1$ is also simply connected then Q_1 uniquely determines Q_2, so that ψ_{21} is one-to-one.

Note. The statement that two manifolds are homeomorphic says nothing about how they might look if they are embedded in some Euclidean space of higher dimension. A circle in the plane, as a *one-dimensional manifold*, is homeomorphic to a simple knot in space; a simple loop of paper is not homeomorphic to a Möbius band, but it is homeomorphic to a loop of paper that has two half twists (one end was twisted through a full turn before being

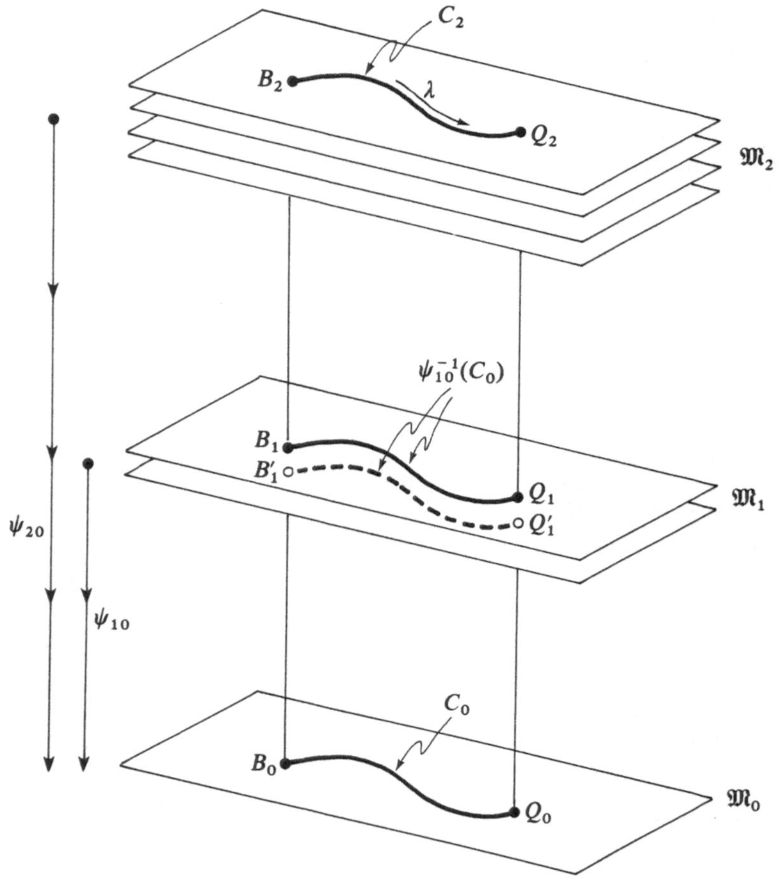

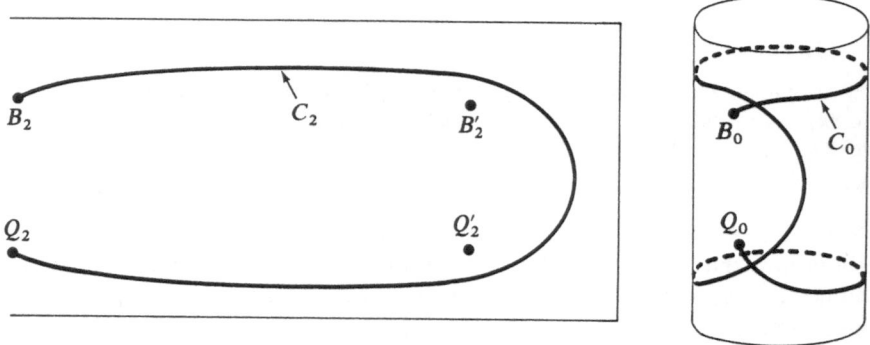

Figure 24.5 Covering manifolds—see text.

Figure 24.6

glued to the other end) or any even number of half twists, and a Möbius band is homeomorphic to such a loop with any odd number of half twists. This can be seen by considering the method of identification of edges, discussed in Section 23.1, for constructing these manifolds.

Because of this theorem, a simply connected covering manifold of $\mathfrak{N}$ is called the *universal covering manifold* of $\mathfrak{N}$. In the Section 24.5, below, it is shown that any manifold $\mathfrak{N}$ has a universal covering manifold. This is important in the theory of Lie groups and for the cosmological interpretation of Einstein manifolds. Since the construction of the universal covering manifold is somewhat abstruse, a few comments on constructions in general are in order.

24.4 Comments on the Construction of Mathematical Models

The extensive use of construction by mathematicians sometimes seems unnatural and artificial at first sight. To a physicist, for whom a real number is, for example, an instantaneous coordinate $x = x(t)$ of a moving point, it seems wholly inappropriate to regard this number as an infinite equivalence class of Cauchy sequences of rational numbers (especially when it is recalled that each rational number is an infinite equivalence class of ordered pairs of integers, etc.) However, once the construction has been completed and the properties of the resulting system have been found, one can then completely forget the details of the construction, and one can regard a real number as a thing just as atomic and indivisible as a "point" was in Euclidean geometry. The construction of mathematical models is now beginning to play a role in quantum theory and relativity, and for the same reason, basically, for which it plays a role in mathematics; when a structure is defined by writing down a set of axioms, one has to prove that the axioms are not mutually contradictory, and the best way to do that is to exhibit a model of the structure based on simpler axioms that have already been accepted. One of the earliest instances in mathematics dealt with complex numbers. To some mathematicians it seemed unnatural and dangerous to assert the existence of a number i such that $i^2 = -1$, and there was much argument about whether such a number "could exist." The question was settled (by Gauss in 1831 and independently by Hamilton in 1837) by considering ordered pairs (a, b) of real numbers and by defining the arithmetic operations on those pairs by

$$(a, b) + (c, d) = (a + c, b + d),$$

$$(a, b)(c, d) = (ac - bd, bc + ad),$$

whereupon the system of all ordered pairs acquired exactly the same properties as the system of numbers $a + bi$ would possess, if the number i existed. One then felt free to assert the existence of the complex number system, and it became irrelevant whether one wrote (a, b) or $a + bi$.

In the development of the relativistic wave equation for the electron, Dirac encountered the need for four quantities α_1, α_2, α_3, α_4 which should multiply according to the rules

$$\alpha_j \alpha_k + \alpha_k \alpha_j = 2\delta_{jk} \qquad (j, k = 1, \ldots, 4). \tag{24.4-1}$$

Instead of simply postulating the existence of the α_j and taking the equations (24.4-1) as axioms, Dirac exhibited four 4×4 matrices which multiply precisely according to (24.4-1) (when the right member is interpreted as $2\delta_{jk}$ times the unit matrix).

The theory of Lie algebras starts with abstract definitions and axioms. After an exceedingly long and complicated development, containing many difficult lemmas and theorems, one arrives at a classification of the so-called simple complex Lie algebras, in which the possible algebras have been narrowed down to nine types or classes. Two questions then arise: (1) Were the original axioms free from contradictions? (2) Is it possible that further work would narrow down the possibilities still further and eliminate some of the nine types? Both questions have been answered by constructing a mathematical model of every algebra in each of the nine classes, without using any axioms other than those of ordinary arithmetic. Some of the constructions are rather complicated and arbitrary, but they serve the purpose. See Hausner and Schwartz 1968.

As a further example, the question of the possible "existence" of certain systems of non-Euclidean geometry (i.e., based on axioms different from those of Euclid) was settled by constructing models, without using any new axioms. In these models, one constructed certain things that were arbitrarily called "points," one defined what was meant by the "distance" along a curve between two points, one determined a "straight line" in terms of the shortest curve between two points, etc., and in the end, one found that these "objects" satisfy all the axioms of Euclid except that through a point Q not on a straight line L there are many different straight lines L', L'', etc., parallel to L (or with a different construction that there are none at all). In this way it was proved that Euclid's parallel axiom could be modified without producing contradictions. See Chapters 26–28.

Distributions themselves are constructions. Dirac postulated the existence of something, called $\delta(x - x_0)$, which was supposed to behave like a function in most respects and to have certain specified properties. The functional $\langle \delta, \varphi \rangle \overset{\text{def}}{=} \varphi(x_0)$, when suitably interpreted, has precisely the required properties.

The mathematical model of the universal covering manifold $\mathfrak{M}$ of a given manifold $\mathfrak{N}$ is constructed in the next section by the method used in general relativity; see Note in Section 23.4. The space $\mathfrak{M}$ is not known in advance as a topological space or even as a collection of points. Instead, one has a collection of charts and specifies how they are to be pieced together to form $\mathfrak{M}$. According to the Whitney embedding theorem, which will not be proved here, an n-dimensional manifold such as $\mathfrak{M}$ (abstract or otherwise) is homeomorphic to an n-dimensional surface in some Euclidean space E^N of higher dimension; this surface thus provides an alternative mathematical model of $\mathfrak{M}$, once the first model has been constructed.

24.5 Construction of the Universal Covering

We are given a connected manifold $\mathfrak{N}$ of class C^k, and we wish to construct a simply connected manifold $\mathfrak{M}$ that covers $\mathfrak{N}$. We assume that the charts $K, L, \ldots$ of $\mathfrak{N}$ are simply connected. We choose a base point B_0 in $\mathfrak{N}$, and for any chart K we denote by $\alpha, \beta, \ldots$ the homotopy classes of paths from B_0 to K (the terminal point of those paths can be taken as any point of K, because K is simply connected). We construct duplicates of the chart K, called $K_\alpha, K_\beta, \ldots$, one for each such homotopy class. See Figure 24.7. Then the charts $K_\alpha, K_\beta, \ldots, L_\zeta, L_\eta, \ldots, \ldots$ are to be pieced together to form the manifold $\mathfrak{M}$, and we now specify how those charts overlap. Given any two of them, say K_α and L_η, we specify that there is no overlap unless K and L overlap in $\mathfrak{N}$. If K and L overlap, we can assume that the paths of α and η have a common terminal point in the overlap of K and L; then we specify that there is no overlap of K_α and L_η unless the paths of α are homotopic to those of η, and in that case we write $\alpha \sim \eta$ and we specify that K_α and L_η have the same overlap as K and L. In more detail, let

$$K = \{\mathfrak{U}, \varphi, N\}, \qquad L = \{\mathfrak{U}', \varphi', N'\}. \qquad (24.5\text{-}1)$$

Then, for each α, K_α is a copy of K distinguished from K and from other copies only by carrying "α" along as a distinguishing index. We write

$$K_\alpha = \{\mathfrak{U}_\alpha, \varphi, N\}, \qquad L_\eta = \{\mathfrak{U}'_\eta, \varphi', N'\}, \qquad (24.5\text{-}2)$$

where $\mathfrak{U}_\alpha$ and $\mathfrak{U}'_\eta$ are parts of the manifold $\mathfrak{M}$ under construction, and are defined as follows: We let each point $\mathbf{x}$ in the region N of the coordinate space

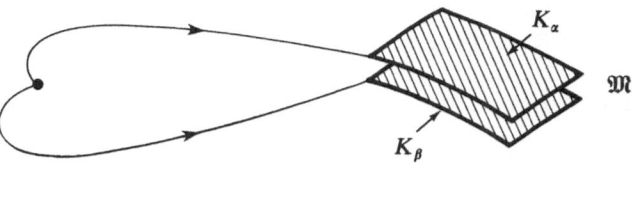

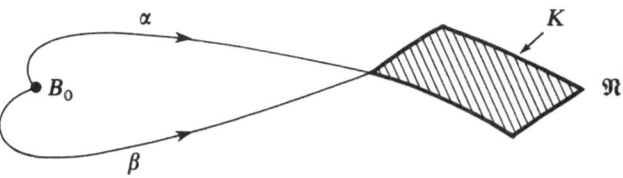

Figure 24.7 Construction of the universal covering manifold.

$\mathbb{R}^n$ determine a point p of $\mathfrak{U}_\alpha$ with coordinates $\varphi^j(p) \overset{\text{def}}{=} x^j$, where the x^j are the components of $\mathbf{x}$. $\mathfrak{M}$ is made up of the points determined in this way by all the charts $K_\alpha, K_\beta, \ldots, L_\zeta, L_\eta, \ldots, \ldots$; they are all distinct points of $\mathfrak{M}$ except for the identifications to be made as we specify the overlap of the charts.

The overlap of K and L in $\mathfrak{N}$ is described, according to (24.5-1), by

$$x'^j = x'^j(x^1, \ldots, x^n), \tag{24.5-3}$$

which gives a one-to-one mapping from part of N onto a part of N' and hence gives two coordinate systems in the region $\mathfrak{U} \cap \mathfrak{U}'$ of $\mathfrak{N}$. If $\alpha \sim \eta$, as above, we specify that the overlap of K_α and L_η is given by the same equations (24.5-3), and we identify the point of $\mathfrak{U}_\alpha$ having given coordinates $x^1, \ldots, x^n$ with the point of $\mathfrak{U}'_\eta$ having corresponding coordinates $x'^1, \ldots, x'^n$ determined by those equations.

Clearly, this procedure determines a manifold $\mathfrak{M}$ of the same differentiability class C^k as $\mathfrak{N}$. The projection ψ of $\mathfrak{M}$ onto $\mathfrak{N}$ is easily defined by projecting each K_α onto the corresponding K: Each point of $\mathfrak{U}_\alpha$ in $\mathfrak{M}$ is projected onto the point of $\mathfrak{U}$ in $\mathfrak{N}$ having the same coordinates $x^1, \ldots, x^n$. This projection ψ is of class C^k because in these coordinates it is just the identity mapping.

Lastly, to show that $\mathfrak{M}$ is simply connected, we first choose a base point A_0 in $\mathfrak{M}$ as one of the points that lie over the base point B_0 of $\mathfrak{N}$, as follows: B_0 lies in some chart in $\mathfrak{N}$, say L. Then the paths of the homotopy classes $\zeta, \eta, \ldots$ can be taken as closed paths beginning and ending at B_0. One of the classes, say η, is the class of nullhomotopic paths—paths that can be shrunk continuously in $\mathfrak{N}$ to the point B_0. Let A_0 be the point of L_η that lies over B_0, i.e., has the same coordinates $x^1, \ldots, x^n$ in L_η that B_0 has in L.

Given any chart K_α in $\mathfrak{M}$, we choose one of the paths in the class α in $\mathfrak{N}$ and lift it, according to the first principle of lifting in Section 24.2, up to $\mathfrak{M}$ as a unique path, which we call α', from the new base point A_0 to a point in K_α. Then, whenever charts K_α and L_ζ overlap, we have $\alpha \sim \zeta$; hence by the second principle of lifting, the paths α' and ζ' are homotopic in $\mathfrak{M}$, or, more precisely, they become homotopic if they are so chosen as to have a common terminal point in the intersection of K_α and L_ζ.

Now let $\mathscr{C}: P(\lambda), 0 \le \lambda \le 1$, be any path in $\mathfrak{M}$; we wish to show that it is homotopic to any other path that also goes from $P(0)$ to $P(1)$, i.e., that $\mathfrak{M}$ is simply connected. Each point $P(a)$ of $\mathscr{C}$ lies in some chart, i.e., $P(\lambda)$ lies in that chart for λ in some interval $(a - \varepsilon, a + \varepsilon)$. By the Heine–Borel theorem,

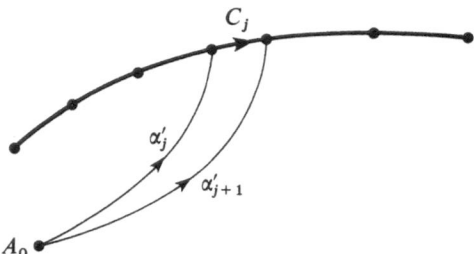

Figure 24.8

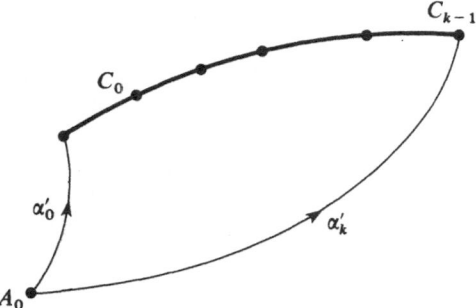

Figure 24.9

then, we can find a finite subdivision of $[0, 1]$ as

$$0 = \lambda_0 < \lambda_1 < \cdots < \lambda_k = 1$$

such that $P(\lambda)$ lies in some chart, say L_{α_j}, for λ in $[\lambda_j, \lambda_{j+1}]$, for each $j = 0, 1, \ldots, k - 1$. We call $\mathscr{C}_j$ the part of $\mathscr{C}$ for $\lambda_j \leq \lambda \leq \lambda_{j+1}$; $\mathscr{C}_j$ lies in L_{α_j}. For each j, we let α'_j be a path in $\mathfrak{M}$, as in the preceding paragraph, from the base point A_0 to $P(\lambda_j)$. Since $P(\lambda_j)$ and $P(\lambda_{j+1})$ both lie in L_{α_j}, it follows that $\alpha'_j \circ \mathscr{C}_j$ is homotopic to α'_{j+1}. See Figure 24.8. Therefore,

$$\alpha'_0 \circ \mathscr{C}_0 \circ \mathscr{C}_1 \circ \cdots \circ \mathscr{C}_{k-1} \sim \alpha'_k.$$

Since $\mathscr{C} = \mathscr{C}_0 \circ \mathscr{C}_1 \circ \cdots \circ \mathscr{C}_{k-1}$, we have

$$\mathscr{C} \circ \alpha'^{-1}_0 \circ \alpha'_k.$$

See Figure 24.9. The right member here depends only on the initial and terminal points $P(0)$ and $P(1)$ of $\mathscr{C}$; hence any other path from $P(0)$ to $P(1)$ is homotopic to $\mathscr{C}$, as required.

Summary. (Main Theorem).   *Every (connected) n-dimensional manifold $\mathfrak{N}$ has a universal covering manifold $\mathfrak{M}$ (n-dimensional), i.e., a simply connected covering manifold. $\mathfrak{M}$ covers any manifold that covers $\mathfrak{N}$, and the only simply connected manifolds that cover $\mathfrak{M}$ are homeomorphic to $\mathfrak{M}$. A model of $\mathfrak{M}$ has been constructed by the procedure given above.*

24.6 Manifolds Covered by a Given Manifold

A problem inverse to that of finding the universal covering manifold $\mathfrak{M}$ of a given manifold $\mathfrak{N}$ is now considered: Given a manifold $\mathfrak{M}$, construct the manifolds $\mathfrak{N}$ that can be covered by $\mathfrak{M}$. The procedure used is the identification of sets of points in $\mathfrak{M}$. A few examples are given first.

Let $\mathfrak{M}$ be the x, y plane, and let it cover the unit cylinder Z by being wrapped tightly about Z, the axis of Z lying in the y direction. We know, of course, that the plane then covers the cylinder, but we show how to establish that fact *a priori*. We know that, for given (x, y), all of the points $(x + 2\pi l, y)$, $l = 0$,

$\pm 1, \pm 2, \ldots$, of the plane coincide with a single point of the cylinder. There-fore, a manifold $\mathfrak{N}$ homeomorphic to the cylinder Z can be constructed by defining each set $\{(x + 2\pi l, y) : l = 0, \pm 1, \ldots\{ \stackrel{\text{def}}{=} \psi((x, y))$ to be a "point" of $\mathfrak{N}$ and by defining charts in $\mathfrak{N}$ in an obvious way. The mapping $(x, y) \rightarrow \psi((x, y))$ is then a projection of $\mathfrak{M}$ onto $\mathfrak{N}$. One says that all the points $(x + 2\pi l, y)$ of each set have been *identified* (i.e., made identical). Note that the factor 2π is irrelevant, because only the topological properties of $\mathfrak{N}$ are in-volved. Identification of the points $(x + n, y)$ or, more generally, of the points $(x + an, y)$, where a is any nonzero real number, would have the same effect.

Similarly, if, for each x, y in $\mathfrak{M}$, all the points of the form $(x + l, y + m)$, where l and m run independently over $0, \pm 1, \pm 2, \ldots$, are identified, then the resulting manifold $\mathfrak{N}$ is the torus (more properly, is homeomorphic to the torus).

Let $\mathfrak{M}$ be the infinite strip $-1 < x < 1, -\infty < y < \infty$. For given x, y in $\mathfrak{M}$, let the points $((-1)^l x, y + l), l = 0, \pm 1, \pm 2, \ldots$, be identified. The resulting manifold $\mathfrak{N}$ is the Möbius band.

To generalize from these examples, let $\mathfrak{M}$ be any (connected) manifold. Suppose that σ is a homeomorphism (of class C^k, if $\mathfrak{M}$ is a C^k manifold) of $\mathfrak{M}$ onto itself. Denote by σ^l the lth iterate of σ, i.e.,

$$\sigma^l(P) = \underbrace{\sigma(\sigma(\ldots \sigma(P) \ldots))}_{l \text{ repetitions}},$$

and denote by σ^{-l} the lth iterate of the inverse mapping σ^{-1}. For any point P in $\mathfrak{M}$, consider the point set

$$\psi(P) \stackrel{\text{def}}{=} \{\sigma^l(P) : l = 0, \pm 1, \ldots\}. \tag{24.6-1}$$

[In the first example, σ is the displacement $(x, y) \rightarrow (x + 2\pi, y)$ in the plane.] Suppose further that σ is such that the point set $\psi(P)$ is discrete in $\mathfrak{M}$, for every P; that is, suppose that there is a neighborhood of P that contains none of the other points $\sigma^l(P)$ with $l \neq 0$. Then, since σ is a homeomorphism, each $\sigma^k(P)$ has a neighborhood that contains no $\sigma^l(P)$ with $l \neq k$.

Under these assumptions, a manifold $\mathfrak{N}$ whose "points" are the sets $\psi(P)$ is now constructed:

$$\mathfrak{N} = \{\psi(P) : P \in \mathfrak{M}\}.$$

To define charts in $\mathfrak{N}$, let $\{\mathfrak{U}, \varphi, N\}$ be a chart in $\mathfrak{M}$. Assume that $\mathfrak{U}$ is small enough so that for no P in $\mathfrak{U}$ is $\sigma(P)$ also in $\mathfrak{U}$. (If this is not the case, replace the chart by a suitable subchart.) A chart $\{\tilde{\mathfrak{U}}, \tilde{\varphi}, \tilde{N}\}$ is then defined in $\mathfrak{N}$ as follows:

$$\tilde{\mathfrak{U}} = \{\psi(P) : P \text{ in } \mathfrak{U}\}.$$

$$\text{For } P \text{ in } \mathfrak{U}, \quad \tilde{\varphi}(\psi(P)) \stackrel{\text{def}}{=} \varphi(P),$$

hence,

$$\tilde{N} = N.$$

It is left as a quite obvious exercise to show that (a) $\tilde{\varphi}$ is one-to-one, (b) charts defined in this way are pairwise compatible and cover $\mathfrak{N}$, (c) the mapping

$P \to \psi(P)$ is onto $\mathfrak{N}$, and (d) any neighborhood $\tilde{\mathfrak{U}}$ of the kind defined above is a good neighborhood of each of its points, because the components of $\psi^{-1}(\tilde{\mathfrak{U}})$ are the sets $\sigma^l(\mathfrak{U}), l = 0, \pm 1, \ldots$. The conclusion is that ψ is a covering of $\mathfrak{N}$ by $\mathfrak{M}$.

It will now be shown that every covering of a manifold $\mathfrak{N}$ by a manifold $\mathfrak{M}$ is associated with a group of homeomorphisms in $\mathfrak{M}$ of the kind described above.

Let $\mathfrak{M}$ and $\mathfrak{N}$ be connected n-dimensional manifolds, and assume that $\mathfrak{M}$ covers $\mathfrak{N}$ by a projection ψ (which is assumed not to be one-to-one, so that the covering is not trivial). Let B_1 and B_0 be basepoints in $\mathfrak{M}$ and $\mathfrak{N}$, where B_1 lies over B_0, i.e., $\psi(B_1) = B_0$. We wish to show that, corresponding to each other point B_1' of $\mathfrak{M}$ that lies over B_0, there is a homeomorphism σ of $\mathfrak{M}$ that carries B_1 into B_1'.

The proof is quite simple if $\mathfrak{M}$ is simply connected, and that case will be discussed first. Let $\mathscr{C}_1$ be a curve in $\mathfrak{M}$ from B_1 to B_1', and let $\mathscr{C}_0$ be its image in $\mathfrak{N}$; $\mathscr{C}_0$ is a closed curve beginning and ending at B_0; $\mathscr{C}_1$ and $\mathscr{C}_0$ will be kept fixed during the discussion, and by means of them a homeomorphism σ of $\mathfrak{M}$ onto $\mathfrak{N}$ will be constructed, under which B_1 goes into B_1'. Let

$$P_1 : P_1(t) \qquad (0 \le t \le 1), \qquad P_1(0) = B_1$$

be a curve in $\mathfrak{M}$ going from B_1 to any point $P_1(1) = A_1$; the image of A_1 under σ will be constructed. Let P_0 be the image of the curve P_1 in $\mathfrak{N}$ with terminal point $A_0 = \psi(A_1)$. Let P_1' be the result of lifting the curve $C_0 \circ P_0$ up to $\mathfrak{M}$. Then $P_1'(1)$ is a point, which, like A_1, lies over A_0. A mapping in $\mathfrak{M}$ is defined by

$$\sigma : A_1 \to A_1' = P_1'(1).$$

It is well defined, because if the same construction starts with any other curve Q_1 in $\mathfrak{M}$ from B_1 to A_1, then Q_1 is homotopic to P_1 (because $\mathfrak{M}$ is simply connected); hence $C_0 \circ Q_0$ is homotopic to $\mathscr{C}_0 \circ P_0$, and, hence, by the corollary to the second principle of lifting, P_1' is homotopic to Q_1', and the same point A_1' is obtained as the image of A_1 under σ. Furthermore, σ is one-to-one, because C_0^{-1} exists. It is clear that σ is continuous, because a small displacement of A_1 can be achieved by a small change of the curve P_1, hence by a small change of the curve P_1', thus producing a small displacement of A_1'. Therefore, σ is a homeomorphism of $\mathfrak{M}$.

If $\mathfrak{M}$ is not simply connected, we can no longer assert that the curves P_1 and Q_1 from B_1 to A_1 in $\mathfrak{M}$ are homotopic. Let the terminal point $Q_1'(1)$ of Q_1' be called A_1''; we wish to show that $A_1'' = A_1'$, so that the mapping is still well defined. The curves $P_1^{-1} \circ P_1'$ and $Q_1^{-1} \circ Q_1'$ in $\mathfrak{M}$, which go from A_1 to A_1' and from A_1 to A_1'', respectively, are mapped under ψ onto the curves

$$P_0^{-1} \circ C_0 \circ P_0 \quad \text{and} \quad Q_0^{-1} \circ C_0 \circ Q_0,$$

But these are closed curves in $\mathfrak{N}$ beginning and ending at A_0, and they are both homotopic to C_0, and it follows from the corollary to the second principle of lifting that $P_1^{-1} \circ P_1'$ and $Q_1^{-1} \circ Q_1'$ are homotopic, and hence that

σ is well defined. The remainder of the discussion is the same as in the preceding case.

If B_1' is taken in succession as each of the points in $\mathfrak{M}$ lying over B_0 (including B_1 itself) a group of homeomorphisms of $\mathfrak{M}$ is obtained. The action of this group on $\mathfrak{M}$ is such that, for any A_1, the images $\{\sigma(A_1)\colon \text{all } \sigma\}$ are discrete. To see that, let $\mathfrak{U}$ be a good neighborhood of $A_0 = \psi(A_1)$. The points $\sigma(A_1)$ all lie over A_0, but since ψ is one-to-one in each component $\mathfrak{U}'$ of $\psi^{-1}(\mathfrak{U})$, there can be at most one point lying over A_0 in each such component.

The problem of finding all manifolds covered by a given manifold is thereby reduced to the problem of finding all homeomorphisms σ of the kind described above. This idea has applications to the theory of general relativity; see Chapter 30.

CHAPTER 25

Lie Groups

Lie Group G; linear Lie group; tangent vector; Lie algebra Λ of G; Lie product; Jacobi identity; abstract Lie algebra; structure constants; local isomorphism of SU(2) and SO(3); exponential mapping of Λ into G; logarithmic (or normal) coordinates in G; adjoint representations of Lie algebras and simply connected Lie groups; the Campbell–Baker–Hausdorff formula; translation of charts; ideals; simple Lie algebra; local and global homomorphisms of groups; homomorphism theory; center of a group; center of an algebra; covering group; direct and semidirect sums of Lie algebras; classifications of the simple Lie algebra.

Prerequisites: Chapters 18, 19, 23, 24; Sections 21.1–21.4.

The subject of this chapter is the advanced theory of continuous groups, often called, inaccurately, the theory of Lie groups. Most of the groups themselves play a role in physics and mathematics at a more elementary level. Among them are the rotation and rigid motion groups, the Lorentz and Poincaré groups, and the unitary and symplectic groups. What is new is the study of the groups and the structure they comprise from a deeper analytic, algebraic, and topological point of view. The key to the study is the theory of Lie algebras and of the interaction between the groups and their algebras. That inter-action has played a role in quantum mechanics from the beginning, in that the elements of the Lie algebras have appeared as operators derived from the symmetries of a system. In the last 25 years much of the terminology and certain specific groups, such as the groups derived from the Lie algebra G_2, have appeared in particle physics. So far, the applications to particle physics has been mostly heuristic, but it seems likely that as the physical theory becomes fully developed, the details of the mathematical structure will be of greater importance. Most presentations of the theory are quite recondite and hence rather difficult for the nonspecialist. I have attempted to present the subject in the most elementary way possible consistent with describing the complete structure. For instance, a vector field on the group manifold is defined as consisting of components subject to a transformation law, just as elsewhere in physics, rather than as an abstract mapping (derivation) in an algebra of C^∞ functions.

129

25.1 Definition and Statement of Objectives

A Lie group is a group G whose elements $g, h, \ldots$ can be regarded as the points of a manifold, in such a way that the group-theoretical properties of the elements vary continuously in the manifold. That is, if g, h, and gh are in charts $\{\mathfrak{U}, \varphi, N\}$, $\{\mathfrak{U}', \varphi', N'\}$, and $\{\mathfrak{U}'', \varphi'', N''\}$ (these charts are not necessarily distinct), then the coordinates of gh must be continuous functions of the coordinates of g and h, i.e., the components of $\varphi''(gh)$ must depend continuously on the components of $\varphi(g)$ and $\varphi'(h)$; also, the coordinates of g^{-1} must be continuous functions of the coordinates of g. If G is a group of matrices, like SU(2) or SO(3), and if the manifold is defined as in Section 19.5, then the continuous dependence just described is automatic, because the matrix elements of AB, and of A^{-1} (for A nonsingular), depend continuously on the matrix elements of A and B, and of A, respectively. If G is an abstract group, the continuous dependence must be postulated.

The formal definition can be given in many ways, because a very few basic properties imply many other properties. For example, the group is often postulated to be a C^∞ manifold, but Hilbert conjectured in 1900 that it was only necessary to require it to be a C^0 manifold, and then it would automatically be a C^∞ manifold; this conjecture was verified in 1952 by Gleason and independently by Montgommery and Zippin. They showed that the manifold of every Lie group is actually a real *analytic* manifold. Furthermore, it is only necessary to postulate one coordinate chart on the group, namely one located in a small neighborhood of the identity element and in which gh and g^{-1} are continuous; the rest of the manifold structure then follows from the group axioms. The definition given in this book postulates only what is necessary for deriving the remaining properties conveniently by elementary methods.

The familiar Lie groups, including all the ones (to the best of my knowledge) that have ever appeared in applications, are *linear* Lie groups, i.e., are isomorphic to groups of linear transformations in a finite-dimensional space, or, equivalently, groups of matrices. This applies even to the ones that originate as groups of nonlinear transformations. For example, the group of Möbius transformations

$$z \to \frac{\alpha z + \beta}{\gamma z + \delta} \qquad (\alpha\delta - \gamma\beta \neq 0)$$

in the complex plane is isomorphic to the restricted Lorentz group, which is linear. Furthermore, much of the theory is simpler for groups of matrices than for abstract Lie groups; for example, the exponential of a matrix, $\exp M$, is familiar and elementary, whereas the corresponding construction for the Lie algebra of an abstract Lie group requires the machinery developed in the next five sections. For physical applications, therefore, a theory of the linear groups alone would seem to be desirable. However, the abstract formulation appears necessary for a complete theory. Even if one starts with matrices, the theory leads to groups that are not, at least in any obvious way, groups of matrices, namely quotient groups G/H and semidirect products. Every *compact* Lie group is linear, but the proof depends on quite advanced developments in the theory. See Chevalley 1946. Two nonlinear Lie groups are described in the Appendix to this chapter. The abstract theory is presented below, but the specialization to matrices is mentioned at various points; see Exercises 1–7 in Section 25.14.

Let G be a group. Suppose that, in the space whose points are the elements of G, there is defined an n-dimensional coordinate chart $\{\mathfrak{U}, \varphi, N\}$ such that $\mathfrak{U}$ contains the identity element 1 of the group. (The symbol "1" is used because "e" is needed for exponentiation). It is assumed, for convenience, that φ maps 1 onto the origin of $\mathbb{R}^n$: $\varphi(1) = 0$. A subset $\mathfrak{U}_0$ of $\mathfrak{U}$ is called *open* (as in Chapter 23) if $\varphi(\mathfrak{U}_0)$ is an open subset of N in $\mathbb{R}^n$.

We assume that products and inverses of group elements are continuous in this chart insofar as their coordinates are defined. It then follows that we can define a smaller chart, with special properties, as follows: Let g and h be in $\mathfrak{U}$. If g and h are close enough to 1, that is, if $\varphi(g)$ and $\varphi(h)$ are close enough to the origin in $\mathbb{R}^n$, then gh, g^{-1}, and h^{-1} are also close to 1. In particular, if $g = h = 1$, then gh, g^{-1}, and h^{-1} are $= 1$ and their coordinates are defined and are all zero. Hence, by continuity there is a neighborhood $\mathfrak{U}_1$ of 1 such that if g and h are in $\mathfrak{U}_1$, the coordinates of gh, g, and h are defined and are in the open set N of $\mathbb{R}^n$. It is convenient to consider an even smaller neighborhood $\mathfrak{U}_0 = \mathfrak{U}_1 \cap \mathfrak{U}_1^{-1}$, where $\mathfrak{U}_1^{-1} \stackrel{\text{def}}{=} \{g^{-1}: g \in \mathfrak{U}_1\}$, and to call $N_0 = \varphi(\mathfrak{U}_0) \subset N$. Then, if g and h are in $\mathfrak{U}_0$, gh is in $\mathfrak{U}$, while g^{-1} and h^{-1} are in $\mathfrak{U}_0$. A vector-valued function $\mathbf{m}(\mathbf{x}_1, \mathbf{x}_2)$ is therefore defined, for all $\mathbf{x}_1$ and $\mathbf{x}_2$ in N_0, by

$$\mathbf{m}(\varphi(g),\ \varphi(h)) = \varphi(gh) \in N;$$

similarly, $\mathbf{l}(\mathbf{x})$ is defined by

$$\mathbf{l}(\varphi(g)) = \varphi(g^{-1}) \in N_0.$$

The group G, together with the n-dimensional chart $\{\mathfrak{U}, \varphi, N\}$, will be called an *$n$-dimensional Lie group* if the functions $\mathbf{m}(\cdot, \cdot)$ and $\mathbf{l}(\cdot)$ are defined in an open set N_0, as described above, and are of class C^4. Later, further charts will be obtained from $\{\mathfrak{U}, \varphi, N\}$, by use of the group operations, in such a way as to make G into a manifold.

All the groups described in Chapter 19 are Lie groups when coordinate charts are suitably defined in them.

(Some authors require the manifold of a Lie group to be connected; for reasons given in Section 25.11, that requirement is irrelevant.)

For example, let G be the rotation group SO(3) with the intrinsic coordinate $\theta_x, \theta_y, \theta_z$ discussed in Section 19.6. Then $\mathfrak{U}$ can be taken as the set of all group elements for which $\|\boldsymbol{\theta}\| < \pi$ (i.e., all for which $\|\boldsymbol{\theta}\| \neq \pi$), and $\mathfrak{U}_0$ as the set for which $\|\boldsymbol{\theta}\| < \pi/2$. Hence N is the interior of the ball K in $\mathbb{R}^3$ described in section, and N_0 is the open ball of half the radius of K. The same coordinates can be used for O(3); in that case the entire second component of the manifold is outside $\mathfrak{U}$.

To derive the properties of Lie groups from the above definitions, the Lie algebra $\Lambda = \Lambda(G)$ of a Lie group G is constructed; Λ is an n-dimensional linear space of elements $\lambda, \mu, \ldots$, in which a multiplicative operation $[\lambda, \mu]$, the so-called Lie product, is defined. The structure of Λ is completely determined by the properties of G in any arbitrarily small neighborhood of 1; on the other hand, Λ completely determines *many* of the properties of G. Then, the so-called exponential mapping from Λ into G is constructed; it generalizes the mapping $M \to e^M$ for matrices. In some neighborhood of the origin of Λ, the mapping is one-to-one, and the components of an element λ serve, via the inverse mapping, as the so-called logarithmic coordinates in G. Other coordinate charts are later obtained from this one by translations in G and are related to it analytically. An explicit formula (the CBH formula— see Section 25.10) then gives $\mathbf{m}(\lambda, \mu)$ in terms of λ and μ and shows that the dependence of gh on g and h is analytic in these coordinates. The formula depends only on the structure of the Lie algebra, which shows that, in a neighborhood of 1 where the logarithmic coordinates are defined, the structure of the group depends entirely on its infinitesimal group elements.

The study of Lie groups combines analysis, algebra, and topology in almost equal proportions. The application of powerful techniques of linear algebra yields a complete classification of Lie algebras, from which follows a classification of Lie groups. That may seem surprising in view of the origin of Lie groups often as groups of nonlinear transformations—see Eisenhart 1933.

25.2 The Expansions of $\mathbf{m}(\cdot, \cdot)$ and $\mathbf{l}(\cdot)$

Since the functions $\mathbf{m}(\mathbf{x}, \mathbf{y})$ and $\mathbf{l}(\mathbf{x})$ are of class C^4, they can be expanded in Taylor's series in the components x^i and y^i of $\mathbf{x}$ and $\mathbf{y}$, through 3rd order terms, with remainder terms of order 4. The group relation $a1 = 1a = a$ for arbitrary a shows that

$$\mathbf{m}(\mathbf{x}, 0) \equiv \mathbf{m}(0, \mathbf{x}) \equiv \mathbf{x} \qquad (25.2\text{-}1)$$

[it is recalled that $\varphi(1) = 0$]. Therefore, in the expansion of $\mathbf{m}(\mathbf{x}, \mathbf{y})$ about the origin, the constant term vanishes, the linear part is $\mathbf{x} + \mathbf{y}$, and the quadratic part contains terms like $x^j y^k$, but not terms like $x^j x^k$ or $y^j y^k$; that is,

$$m^i(\mathbf{x}, \mathbf{y}) = x^i + y^i + a^i_{jk} x^j y^k + b^i_{jkl} x^j x^k y^l + c^i_{jkl} x^j y^k y^l + \cdots, \qquad (25.2\text{-}2)$$

where the a's, b's, and c's are coefficients and where the summation convention has been used.

The associativity axiom of group theory imposes on $\mathbf{m}(\cdot, \cdot)$ the restriction

$$\mathbf{m}(\mathbf{m}(\mathbf{x}, \mathbf{y}), \mathbf{z}) = \mathbf{m}(\mathbf{x}, \mathbf{m}(\mathbf{y}, \mathbf{z})). \qquad (25.2\text{-}3)$$

If only the linear and quadratic terms of the expansion of $\mathbf{m}(\cdot, \cdot)$ are included then (25.2-3) is satisfied automatically; nevertheless, associativity imposes certain restrictions on the coefficients a^i_{jk} of the quadratic part, which can be seen only when the cubic terms are also included. Substitution of (25.2-2) into (25.2-3) gives, after considerable cancellation,

$$a^i_{jk}a^j_{lm}x^ly^mz^k + b^i_{jkl}(x^jy^k + x^ky^j)z^l = a^i_{jk}a^k_{lm}x^jy^lz^m + c^i_{jkl}x^j(y^kz^l + y^lz^k).$$

$$(25.2\text{-}4)$$

The b's and c's will now be eliminated from this relation. Since j, k, l, and m are summation indices, they can be renamed in each term in such a manner so that the factors $x^ky^lz^m$ appear throughout; then, since the equation is an identity in $\mathbf{x}$, $\mathbf{y}$, $\mathbf{z}$, the net coefficient of $x^ky^lz^m$ must vanish; this gives

$$a^i_{jm}a^j_{kl} - a^i_{kj}a^j_{lm} = c^i_{klm} + c^i_{kml} - b^i_{klm} - b^i_{lkm}.$$

This expression is now summed over the even permutations of k, l, m, and from the sum is subtracted the result of summing over the odd permutations; the right member gives zero and the left gives a sum that can be written as

$$0 = \sum_{\substack{\text{(even permutations of } k,\, l,\, m)}} (a^i_{jm} - a^i_{mj})(a^j_{kl} - a^j_{lk}). \qquad (25.2\text{-}5)$$

There are no further restrictions imposed on the coefficients a^i_{jk} by the group axioms, for the above restrictions just suffice to determine a Lie algebra, and it will appear eventually that every Lie algebra is the Lie algebra of some Lie group. (That is a deep result; see Hausner and Schwartz, Section III,7.)

From the equation

$$\mathbf{m}(\mathbf{l}(\mathbf{x}), \mathbf{x}) \equiv \mathbf{m}(\mathbf{x}, \mathbf{l}(\mathbf{x})) \equiv 0,$$

which expresses the group relation that $a^{-1}a = aa^{-1} = 1$ for all a, it is found that the expansion of $\mathbf{l}(x)$, through quadratic terms, is

$$l^i(\mathbf{x}) = -x^i + a^i_{jk}x^jx^k + \cdots. \qquad (25.2\text{-}6)$$

25.3 The Lie Algebra of a Lie Group

The Lie algebra Λ of a Lie group G is based on the so-called infinitesimal elements of G, that is, on the tangent vectors of smooth curves emanating from the identity element 1. Such a curve is given by a function $g(t)$, defined for some interval $0 \le t \le \varepsilon$ ($\varepsilon > 0$), such that $g(0) = 1$ and such that the corresponding curve $\mathbf{x}(t) = \boldsymbol{\varphi}(g(t))$ in the parameter space $\mathbb{R}^n$ has a tangent at each point (including $t = 0$, which is in fact the only point that matters). If

$$\mathbf{x}(t) = \boldsymbol{\varphi}(g(t)) = \boldsymbol{\lambda}t + \cdots, \qquad (25.3\text{-}1)$$

then the components λ^i of $\boldsymbol{\lambda}$ transform as the components of a contravariant vector at the point $\mathbf{x} = 0$ of the manifold, when the coordinates are changed (see Section 26.1). Namely, since

$$\boldsymbol{\lambda} = \frac{d}{dt}\, \boldsymbol{\varphi}(g(t))|_{t=0},$$

it is seen that if new coordinates

$$x'^i = x'^i(x^1, \ldots, x^n) \qquad (i = 1, \ldots, n)$$

are introduced, then

$$\lambda'^i = \frac{\partial x'^i}{\partial x^j}\bigg|_{x=0} \lambda^j.$$

This vector is called the *tangent vector at 1 to the curve g(t)*.

[According to the above definition, a *curve* is determined by the para-metrization as well as by the set of group elements in it; $g(t)$, $g(2t)$, and $g(2^t - 1)$ are different curves emanating from 1, and they have different tangent vectors, although those vectors all have the same direction.]

The set of all tangent vectors at 1 in an *n*-dimensional group G constitutes an *n*-dimensional vector space over the real field $\mathbb{R}$, for if $\lambda t + \cdots$ and $\mu t + \cdots$ are the coordinates of two smooth curves, then $(a\lambda + b\mu)t + \cdots$, where a and b are real numbers, is the coordinate of another smooth curve. This vector space will be made into an algebra, called the *Lie algebra* $\Lambda = \Lambda(G)$ *of the group G*, by defining a multiplication based on the group multiplication in G.

If $g(t)$ and $h(t)$ are smooth curves emanating from 1 in G, and if the function $k(t)$ is defined as the commutator of $g(t)$ and $h(t)$, that is, if

$$k(t) = g(t)h(t)g(t)^{-1}h(t)^{-1},$$

and if the coordinates of $g(t)$ and $h(t)$ are

$$\varphi(g(t)) = \lambda t + \cdots,$$

$$\varphi(h(t)) = \mu t + \cdots,$$

then a direct calculation shows that the coordinate of $k(t)$ is

$$\varphi(k(t)) = \nu t^2 + \cdots, \tag{25.3-2}$$

where

$$\begin{aligned} \nu^i &= a^i_{jk}(\lambda^j\mu^k - \lambda^k\mu^j) \\ &= (a^i_{jk} - a^i_{kj})\lambda^j\mu^k. \end{aligned} \tag{25.3-3}$$

Equation (25.3-2) shows that the function $\tilde{k}(t) \stackrel{\text{def}}{=} k(\sqrt{t})$ is a curve in G emanating from 1 and that ν is its tangent vector; ν is called the *Lie product* of λ and μ and is denoted by the *Lie bracket expression*

$$\nu = [\lambda, \mu]. \tag{25.3-4}$$

[One can also establish directly from (25.3-3) that the ν^i transform as the components of a vector, when coordinates are changed, after the rather complicated transformation laws of the coefficients a^i_{jk} have been worked out.]

From (25.3-3) it follows that the Lie product is linear in each factor and is antisymmetric: $[\mu, \lambda] = -[\lambda, \mu]$; from the identity (25.2-5), which was deduced from the associativity in G, it follows that the Lie product also satisfies the *Jacobi identity*

$$[\lambda, [\mu, \nu]] + [\mu, [\nu, \lambda]] + [\nu, [\lambda, \mu]] = 0. \tag{25.3-5}$$

Am example of a Lie algebra is the algebra of vectors in $\mathbb{R}^3$, where the Lie product is defined to be the vector product $[\lambda, \mu] = \lambda \times \mu$, in the notation of Gibbs. The Jacobi identity can be verified either by writing (25.3-5) in components or by use of the identity $\lambda \times (\mu \times \nu) = (\lambda \cdot \nu)\mu - (\lambda \cdot \mu)\nu$. Lie algebras of matrices are discussed in Section 25.5, below.

25.4 Abstract Lie Algebras

A finite-dimensional vector space, over $\mathbb{R}$ (or $\mathbb{C}$) as the field of scalars, in which a multiplication $[\lambda, \mu]$ is defined that is linear in each factor, is antisymmetric, and satisfies the Jacobi identity (25.3-5), is called a *real* (respectively *complex*) Lie algebra. The Lie algebra derived from a group is real.

A Lie algebra is not only generally noncommutative, but also non-associative; that is, generally, $[\lambda, [\mu, \nu]] \neq [[\lambda, \mu], \nu]$; it has no unity element because $[\lambda, \lambda] = 0$, for any λ, by the antisymmetry of $[\lambda, \mu]$.

An n-dimensional Lie algebra can be fully specified by choosing a basis $\varepsilon_1, \ldots, \varepsilon_n$ in it (a set of n linearly independent vectors) and then specifying the n^3 *constants of structure* C^i_{jk} (they are clearly not all independent), defined by

$$[\varepsilon_j, \varepsilon_k] = C^i_{jk}\varepsilon_i. \tag{25.4-1}$$

25.5 The Lie Algebras of Linear Groups

If G is a group of $m \times m$ matrices $L, M, \ldots$, that is, a subgroup of $GL(m, \mathbb{R})$ or $GL(m, \mathbb{C})$, then its Lie algebra can be realized as a Lie algebra of $m \times m$ matrices. A curve emanating from 1 has the form $A(t) = I + tL + \cdots$; matrices L obtainable in this way constitute a vector space Λ of dimension $\leq 2m^2$. To find the Lie product in Λ, let $B(t) = I + tM + \cdots$ be another curve emanating from 1, and define

$$K(t) = A(t)B(t)A(t)^{-1}B(t)^{-1}$$
$$= (I + tL + \cdots)(I + tM + \cdots)(I - tL + \cdots)(I - tM + \cdots), \tag{25.5-1}$$

as was done in Section 25.3 for the function $k(t)$ in the manifold of an abstract group. When these expressions are multiplied out, the linear terms cancel. Quadratic terms can arise from the nonconstant terms of at most two factors, the constant term I being taken for the other factors. All quadratic terms arising from $A(t)A(t)^{-1}$ and from $B(t)B(t)^{-1}$ cancel exactly, so the only remaining quadratic terms come from combining a linear term from $A(t)$ or $A(t)^{-1}$ with a linear term from $B(t)$ or $B(t)^{-1}$. From this it is seen that

$$K(t) = I + t^2(LM - ML) + \cdots. \tag{25.5-2}$$

According to the definition of Lie product [see equations (25.3-2) and (25.3-4)], then,

$$[L, M] = LM - ML. \tag{25.5-3}$$

This illustrates the general rule that any associative algebra can be made into a Lie algebra by defining $[\lambda, \mu]$ as $\lambda\mu - \mu\lambda$, where $\lambda\mu$ and $\mu\lambda$ denote products in the original associative algebra.

Matrices like L and M above were called infinitesimal group elements in Chapter 20; they are given by $L = (d/dt)A(t)|_{t=0}$, etc., and it was shown there that for SO(3), which is 3-dimensional, they can be taken as

$$\varepsilon_1 = \begin{pmatrix} 0 & 0 & 0 \\ 0 & 0 & -1 \\ 0 & 1 & 0 \end{pmatrix}, \quad \varepsilon_2 = \begin{pmatrix} 0 & 0 & 1 \\ 0 & 0 & 0 \\ -1 & 0 & 0 \end{pmatrix}, \quad \varepsilon_3 = \begin{pmatrix} 0 & -1 & 0 \\ 1 & 0 & 0 \\ 0 & 0 & 0 \end{pmatrix},$$

and that these matrices satisfy the relations

$$[\varepsilon_i, \varepsilon_j] = \varepsilon_k \qquad (ijk = 123, 231, \text{ or } 312).$$

In Chapter 21 it was shown that for SU(2), which is also 3-dimensional, the infinitesimal group elements can be taken as

$$\eta_1 = \frac{1}{2}\begin{pmatrix} 0 & -i \\ -i & 0 \end{pmatrix}, \quad \eta_2 = \frac{1}{2}\begin{pmatrix} 0 & -1 \\ 1 & 0 \end{pmatrix}, \quad \eta_3 = \frac{1}{2}\begin{pmatrix} -i & 0 \\ 0 & i \end{pmatrix},$$

and that these matrices satisfy the same relations as the ε_i, namely,

$$[\eta_i, \eta_j] = \eta_k \qquad (ijk = 123, 231, \text{ or } 312).$$

According to (25.4-1), these relations determine the structure of the corresponding Lie algebra completely. Therefore, if Λ_3 and Λ_2 denote the Lie algebras of SO(3) and SU(2), respectively, then the linear mapping of Λ_3 onto Λ_2 induced by $\varepsilon_i \to \eta_i$ ($i = 1, 2, 3$) is an isomorphism. Regarded as abstract Lie algebras, Λ_2 and Λ_3 are identical, whereas the corresponding groups are not. As will be seen later, the isomorphism of the algebras induces a one-to-one mapping of the groups only in a neighborhood of 1. This mapping is an isomorphism as far as it goes, but, when extended globally, it becomes the 2-to-1 homomorphism of SU(2) onto SO(3) discussed in Section 19.7.

Note that Λ_2 and Λ_3 are *real* Lie algebra. Although the matrices η_1, η_2, η_3 are complex, Λ_2 consists of the linear combinations of these matrices with real coefficients.

25.6 The Exponential Mapping; Logarithmic Coordinates

We seek a curve $g(t)$ in the group G such that

$$g(t + s) = g(t)g(s); \tag{25.6-1}$$

the points on this curve constitute a one-dimensional Abelian subgroup;

$g(0)$ is the identity 1 of G. The corresponding curve $x(t) = \varphi(g(t))$ passes through the origin of the coordinate space $\mathbb{R}^n$ and satisfies the equation

$$x^i(t + s) = m^i(\mathbf{x}(t), \mathbf{x}(s)). \tag{25.6-2}$$

Taking d/ds and then setting $s = 0$ in this equation gives

$$\dot{x}^i(t) = q_j^i(\mathbf{x}(t), 0)\lambda^j, \tag{25.6-3}$$

where $q_j^i(\cdot, \cdot)$ is defined by

$$q_j^i(\mathbf{x}, \mathbf{y}) = \frac{\partial}{\partial y^j} m^i(\mathbf{x}, \mathbf{y}) \tag{25.6-4}$$

and is of class C^3, and where λ^j $(j = 1, \ldots, n)$ are the components of the tangent vector λ to $\mathbf{x}(t)$ at $t = 0$. The system of ordinary differential equations (25.6-3), together with the initial condition $\mathbf{x}(0) = 0$, has a unique solution $\mathbf{x}(t)$ in some neighborhood of $t = 0$, for given λ. From the form of (25.6-3) it is clear that the solution depends on λ and t only through the combination $t\lambda$; hence we write the solution as $\mathbf{x}(t, \lambda) = \mathbf{X}(t\lambda)$. The corresponding curve $g(t)$ in G is denoted by $\exp(t\lambda)$ or $e^{t\lambda}$, because of (25.6-1), which now takes the form $e^{(t + s)\lambda} = e^{t\lambda}e^{s\lambda}$; hence $\mathbf{X}(t\lambda) = \varphi(e^{t\lambda})$. This extends the definition of the exponential function abstractly to a mapping from a Lie algebra Λ to a Lie group G, but in case the elements of G and Λ are matrices, the definition agrees with the usual one. The function $\mathbf{X}(\cdot)$ is of class C^3.

So far, the equation (25.6-1) has been used only for s in the immediate neighborhood of 0, but it will now be shown that this functional equation or, what amounts to the same thing, (25.6-2) is satisfied for all t and all s.

Theorem.   *The solution $\mathbf{x}(t)$ of (25.6-3) satisfies (25.6-2) for all t and s such that $\mathbf{x}(t)$, $\mathbf{x}(s)$, and $\mathbf{x}(t + s)$ are defined.*

Note 1. This does not follow in any trivial way from the mere forms of the equations (25.6-2, 3) because, as will be seen, the associative law of group multiplication has to be used in the proof.

Note 2. Once the functional equation $g(t + s) = g(t)g(s)$ has been established for t, s, and $t + s$ in an interval $(-T, T)$, the equation itself is then used to define $g(t)$ for t in $(-2T, 2T)$, then in $(-4T, 4T)$, etc. In consequence, $g(t)$ is uniquely defined for all t and satisfies the functional equation for all t and s—details to be supplied by the reader.

Note 3. If G is a group of matrices, so that the elements λ of Λ are also matrices, then the corresponding equation

$$e^{\lambda(t + s)} = e^{\lambda t}e^{\lambda s} \tag{25.6-5}$$

is usually established as follows: The matrix

$$\mu(s) = (e^{\lambda t})^{-1}e^{\lambda(t + s)},$$

as a function of s, satisfies the same differential equation and the same initial condition as the function $e^{\lambda s}$, namely

$$\frac{d}{ds}\,\mu(s) = \mu(s)\lambda,$$

$$\mu(0) = I;$$

since the solution of this initial value problem is unique, it follows that

$$(e^{\lambda t})^{-1}e^{\lambda(t+s)} = e^{\lambda s},$$

which is equivalent to (25.6-5). This argument is used as a model for the proof, given below, for the abstract case.

PROOF OF THE THEOREM.   It will be shown that the function

$$y(s) = m(l(x(t)), x(t+s)), \tag{25.6-6}$$

which is the coordinate of the group element $g(t)^{-1}g(t+s)$, satisfies the same differential equation (25.6-3) and the same initial condition as $x(s)$, which is the coordinate of the group element $g(s)$; since the solution of the initial-value problem is unique, it follows that $g(s) = g(t)^{-1}g(t+s)$; hence, $g(t+s) = g(t)g(s)$, as required. That $y(s)$ satisfies the initial condition $y(0) = 0$ follows from $g(t)^{-1}g(t) = 1$. Differentiation of (25.6-6) with respect to s and substitution of an expression like (25.6-3) for $\dot{x}^j(t+s)$ gives

$$\dot{y}^i(s) = q^i_j(l(x(t)), x(t+s))q^j_k(x(t+s), 0)\lambda^k. \tag{25.6-7}$$

It will now be shown that the total coefficient of λ^k in the right member of this equation is equal to $q^i_k(y(s), 0)$, so that $y(s)$ satisfies the same equation (25.6-3) as $x(s)$. For that purpose, the associativity of group multiplication is used in the following form:

$$\begin{aligned} m^i(y(s), z) &= m^i(m(l(x(t)), x(t+s)), z) \\ &= m^i(l(x(t)), m(x(t+s), z)), \end{aligned} \tag{25.6-8}$$

which is equivalent to saying that $[g(t)^{-1}g(t+s)]h = g(t)^{-1}[g(t+s)h]$. Differentiation with respect to z_k gives

$$q^i_k(y(s), z) = q^i_j(l(x(t)), m(x(t+s), z))q^j_k(x(t+s), z);$$

therefore

$$q^i_k(y(s), 0) = q^i_j(l(x(t)), x(t+s))q^j_k(x(t+s), 0),$$

as claimed, so that $y(s)$ and $x(s)$ satisfy the same equation, and the theorem is proved.

For $t = 1$, we have

$$X(\lambda) = \varphi(e^\lambda).$$

Furthermore, at $\lambda = 0$, the Jacobian

$$\det\left(\frac{\partial X^i}{\partial \lambda^j}\right) = \det(q^i_j(\lambda, 0))$$

is different from zero, because $q^i_j(0, 0) = \delta^i_j$. Therefore, the function $X(\lambda)$ has an inverse in some neighborhood of the origin, and the components of λ

can be taken as new coordinates, called *logarithmic* or *normal* coordinates of the group element $g = e^\lambda$ in some neighborhood of 1 in G. One also writes $\lambda = \log g$. The (generally many-to-one) mapping $\lambda \to e^\lambda$ of Λ into G is called the *exponential* mapping. The components of a given vector λ depend, of course, on the choice of the original coordinate system $\{\mathfrak{U}, \varphi, N\}$, but they transform as the components of a vector, so that each element of Λ (a vector) is mapped into a unique group element.

25.7 An Auxiliary Lemma on Inner Automorphisms; the Mappings Ad_μ

The lemma proved below is needed as part of the analytic machinery for proving the Campbell–Baker–Hausdorff theorem, but has independent interest. As in Chapter 21, a homomorphism of a group G (abstract or not) onto a group of nonsingular linear transformations of a vector space V is called a *representation* of G on V. If the homomorphism is one-to-one (i.e., an isomorphism), the representation is called *faithful*. If every Lie group had a faithful representation on a finite-dimensional space, the whole theory would reduce to the manipulation of matrices. Although that is not the case, any simply connected group has a particular representation (generally unfaithful), the so-called adjoint representation, described at the end of this section.

Note. The reader may wish to skip the proofs in this section and the next three. However, the definitions and the statements of the lemmas and theorems will be needed later.

If e^μ is a fixed group element, the mapping $g \to e^\mu g e^{-\mu}$ is an inner automorphism of G (see Section 18.10); it induces a linear mapping of Λ into itself, to be discussed below.

First, if μ is any fixd element of Λ, the linear transformation of Λ into itself defined by $\lambda \to [\mu, \lambda]$ is denoted by Ad_μ. Relative to a basis in Λ, this transformation is represented by an $n \times n$ matrix. [This matrix is not to be confused with and is generally of different order than the matrices of which Λ is composed, in case G is a linear group. If the elements of Λ are $m \times m$ matrices, then Ad_μ can be represented by an $m^2 \times m^2$ matrix.] Ad_μ^2 denotes the transformation

$$\lambda \to \mathrm{Ad}_\mu(\mathrm{Ad}_\mu \lambda) = \mathrm{Ad}_\mu[\mu, \lambda] = [\mu, [\mu, \lambda]];$$

similarly, $\mathrm{Ad}_\mu \mathrm{Ad}_\nu$, Ad_μ^k, and $\exp\{\mathrm{Ad}_\mu\}$ denote the transformations

$$\lambda \to [\mu, [\nu, \lambda]],$$

$$\lambda \to [\mu, [\mu, \dots [\mu, \lambda] \dots]],$$

$$\lambda \to \left(I + \mathrm{Ad}_\mu + \frac{1}{2!} \mathrm{Ad}_\mu^2 + \cdots\right)\lambda,$$

where I denotes the identity transformation in Λ.

Lemma. *Let e^μ be a fixed group element; for each λ in Λ let $g(t)$ be a smooth curve, with $g(0) = 1$, whose tangent vector at 1 is λ, and let λ' be the tangent vector at 1 to the curve $e^\mu g(t) e^{-\mu}$; then, the mapping $\lambda \to \lambda'$ is a linear transformation in Λ given explicitly by*

$$\lambda \to \lambda' = e^{\mathrm{Ad}_\mu}\lambda. \tag{25.7-1}$$

PROOF. For each fixed s in the interval $[0, 1]$, the group automorphism $g(t) \to e^{\mu s} g(t) e^{-\mu s}$ induces a mapping $\lambda \to \lambda(s)$, in the manner described in the lemma, and it will be proved that $\lambda(s)$ satisfies the same differential equation in s as $e^{s\,\mathrm{Ad}_\mu}\lambda$. The logarithmic coordinate of the group element $g(t, s) = e^{\mu s} g(t) e^{-\mu s}$ is

$$\mathbf{x}(t, s) = \log g(t, s) = t\lambda(s) + \cdots. \tag{25.7-2}$$

To find the s derivative at $s = s_0$, we write

$$g(t, s_0 + s) = e^{\mu s} g(t, s_0) e^{-\mu s}, \tag{25.7-3}$$

$$\mathbf{x}(t, s_0 + s) = \mathbf{m}(\mathbf{m}(s\mu, \mathbf{x}(t, s_0)), -s\mu), \tag{25.7-4}$$

where, as in the preceding sections, $\mathbf{m}(\cdot, \cdot)$ gives the coordinate (here, the logarithmic coordinate) of the product of two group elements in terms of the coordinates of the factors. We recall the definition (25.6-4) of $q_j^i(\mathbf{x}, \mathbf{y})$ and define similarly

$$p_j^i(\mathbf{x}, \mathbf{y}) = \frac{\partial}{\partial x_j} m^i(\mathbf{x}, \mathbf{y}). \tag{25.7-5}$$

Differentiating (25.7-4) with respect to t shows that the tangent vector at $t = 0$ to the curve $\mathbf{x}(t, s_0 + s)$ is given by

$$\lambda^i(s_0 + s) = p_j^i(s\mu, -s\mu) q_k^j(s\mu, 0)\lambda^k(s_0). \tag{25.7-6}$$

From the expansion (25.2-2) of $m^i(\mathbf{x}, \mathbf{y})$, the corresponding expansions of p_j^i and q_k^j are, to first order,

$$p_j^i = \delta_j^i + a_{jl}^i y^l + \cdots,$$

$$q_k^j = \delta_k^j + a_{lk}^j x^l + \cdots.$$

(These expansions and the quantities p_j^i and q_k^j now all refer to the logarithmic coordinates.) From (25.7-6), then,

$$\frac{d}{ds} \lambda^i(s + s_0)|_{s=0} = (a_{lk}^i - a_{kl}^i)\mu^l \lambda^k(s_0);$$

that is,

$$\frac{d}{ds} \lambda(s) = [\mu, \lambda(s)] = \mathrm{Ad}_\mu \lambda(s)$$

[see (25.3-3, 4)]; in terms of a basis in Λ, this is a system of n first-order differential equations with constant coefficients, and the solution is

$$\lambda(s) = e^{s\,\mathrm{Ad}_\mu}\lambda(0). \tag{25.7-7}$$

In particular, $\lambda' = \lambda(1) = e^{\mathrm{Ad}_\mu}\lambda$, as was to be proved.

Show that, for fixed μ, the linear mapping $\lambda \to e^{\mathrm{Ad}_\mu}\lambda$ is an automorphism of Λ, that is, (1) that it is one-to-one and (2) that

$$e^{\mathrm{Ad}_\mu}[\lambda, v] = [e^{\mathrm{Ad}_\mu}\lambda, e^{\mathrm{Ad}_\mu}v]. \qquad (25.7\text{-}8)$$

[Note that Ad_μ itself is not even a homomorphism of Λ.]

One says that e^{Ad_μ} is the *inner automorphism* of Λ induced by the inner automorphism $g \to e^\mu g e^{-\mu}$ of G.

Comment. If e^{μ_1} and e^{μ_2} are any two group elements, and if $e^{\mu_3} = e^{\mu_1}e^{\mu_2}$, then the automorphism

$$g \to e^{\mu_3}ge^{-\mu_3} = e^{\mu_1}e^{\mu_2}ge^{-\mu_2}e^{-\mu_1}$$

of G induces the automorphism

$$e^{\mathrm{Ad}_{\mu_3}} = e^{\mathrm{Ad}_{\mu_1}} e^{\mathrm{Ad}_{\mu_2}}$$

of Λ. Therefore, in a neighborhood of 1 such that logarithmic coordinates are defined, the association of e^{Ad_μ} with the group element e^μ is (locally) a homomorphism of G into a group of linear transformations in the vector space Λ. It will be seen later that if G is simply connected, this association can be extended to a homomorphism of all G, i.e., to the *adjoint representation* of G.

The group of linear transformations in Λ generated by transformations of the form e^{Ad_μ} is called the *inner-automorphism group* of Λ and is denoted by Int(Λ). Each of its elements is a finite product $e^{\mathrm{Ad}_{\mu_1}} e^{\mathrm{Ad}_{\mu_2}} \ldots e^{\mathrm{Ad}_{\mu_j}}$ and is called an *inner automorphism* of Λ (it cannot necessarily be written as e^{Ad_σ} for some $\sigma \in \Lambda$). If G is simply connected, this element of Int(Λ) is the image, under the homomorphism referred to above (the adjoint representation) of the element $g = e^{\mu_1}e^{\mu_2}\ldots e^{\mu_j}$ of G.

25.8 Auxiliary Lemmas on Formal Derivatives

It is customary to denote by $dg(t)/dt$ or $\dot{g}(t)$ the tangent vector at t to a curve $g(t)$ in the group manifold. This is a purely formal notation, unless G is a group of matrices, for in general no meaning has been attached to a "difference quotient" $(g(t_1) - g(t_2))/(t_1 - t_2)$. Nevertheless, the notation can be manipulated according to many of the rules of differentiation, often with considerable savings in the lengths of formulas and of their derivations.

If h and $g(t)$ are respectively a single group element and a curve in G, and if $\dot{g}(t_0)$ is the tangent vector to $g(t)$ at t_0, then $h\dot{g}(t_0)$ is defined to be the tangent vector to the curve $hg(t)$ at t_0. The one-to-one mapping of G onto itself, given by $g \to hg$ for fixed h, is called a *left translation* in G. It induces a one-to-one linear mapping of the space of tangent vectors at a point $g_0 = g(t_0)$ onto the space of tangent vectors at hg_0. Similarly, $\dot{g}(t_0)h$ is defined as the tangent vector to the curve $g(t)h$; hence a *right translation* $g \to gh$ in G induces a similar mapping of the space of tangent vectors at g_0 into that at $g_0 h$. According to these definitions,

$$\frac{d}{dt} hg(t) = h \frac{dg(t)}{dt}, \qquad \frac{d}{dt} g(t)h = \frac{dg(t)}{dt} h. \tag{25.8-1}$$

From the associativity in G it follows that if h_1 and h_2 are group elements and λ is a tangent vector to some curve, then $(h_1 h_2)\lambda = h_1(h_2 \lambda)$ and $(h_1 \lambda)h_2 = h_1(\lambda h_2)$; in consequence, all parentheses can be omitted. In such a product, any number of factors can be group elements, but only one can be a tangent vector, and the product is then a tangent vector. Generally, λ, $h\lambda$, λh, etc. are vectors at different points of G and cannot be compared, since, if λ_1 and λ_2 are vectors at different points, no meaning has even been given to an equation $\lambda_1 = \lambda_2$. However, if λ is in Λ, hence is a vector at 1, then λ and $h\lambda h^{-1}$ are vectors at the *same* point of G; in particular, the mapping $\lambda \to e^{\mu}\lambda e^{-\mu}$ is the inner automorphism $e^{\mathrm{Ad}\mu}$ of Λ discussed in the preceding section.

By working with a particular coordinate system, one can easily establish the following relations:

$$\frac{d}{dt} (g(t)h(t)) = g(t) \frac{dh(t)}{dt} + \frac{dg(t)}{dt} h(t), \tag{25.8-2}$$

$$\frac{d}{dt} (g(t)^{-1}) = -g(t)^{-1} \frac{dg(t)}{dt} g(t)^{-1}, \tag{25.8-3}$$

$$\frac{d}{dt} e^{t\lambda} = \lambda e^{t\lambda} = e^{t\lambda}\lambda. \tag{25.8-4}$$

Higher derivatives would generally take us out of the space of tangent vectors into other (probably less interesting) spaces. However if $g = g(s, t)$ is a smooth two-parameter family of elements of G, then the quantities defined by

$$\alpha = \alpha(s, t) = g^{-1} \frac{\partial g}{\partial s},$$
$$\tag{25.8-5}$$
$$\beta = \beta(s, t) = g^{-1} \frac{\partial g}{\partial t}$$

are tangent vectors through the point $g(s, t)^{-1}g(s, t) = 1$ for all s and t, i.e., are always in Λ and can be further differentiated.

Lemma. *With* α *and* β *as in* (25.8-5),

$$\frac{\partial \alpha}{\partial t} - \frac{\partial \beta}{\partial s} = [\alpha, \beta]. \tag{25.8-6}$$

[**Note.** If G is a linear group, so that $\partial^2 g/\partial t\partial s$ is well defined as a matrix (in addition to g, $\partial g/\partial t$, and $\partial g/\partial s$), then the lemma follows immediately from (25.8-5), after further differentiation, and (25.8-3); the terms in $\partial^2 g/\partial t\partial s$ cancel.]

PROOF.   To verify (25.8-6) for $s = s_0$, $t = t_0$, write $g(s, t) = g(s_0, t_0)\tilde{g}(s, t)$; then α and β can also be written as

$$\alpha(s, t) = \tilde{g}^{-1} \frac{\partial \tilde{g}}{\partial s},$$

$$\beta(s, t) = \tilde{g}^{-1} \frac{\partial \tilde{g}}{\partial t}.$$

Since $\tilde{g}(s_0, t_0) = 1$, the expansions of the coordinates of $\tilde{g}$ and $\tilde{g}^{-1}$ in powers of $s - s_0 = s_1$ and $t - t_0 = t_1$ start with the linear terms $[\varphi(1) = 0$ is assumed$]$:

$$x^i(s, t) \stackrel{\text{def}}{=} \varphi^i(\tilde{g}(s, t)) = \lambda^i s_1 + \mu^i t_1 + A^i s_1^2 + B^i s_1 t_1 + C^i t_1^2 + \cdots; \qquad (25.8\text{-}7)$$

hence,

$$y^i(s, t) \stackrel{\text{def}}{=} \varphi^i(\tilde{g}(s, t)^{-1}) = -\lambda^i s_1 - \mu^i t_1 - A^i s_1^2 - B^i s_1 t_1 - C^i t_1^2$$
$$+ a^i_{jk}(\lambda^j s_1 + \mu^j t_1)(\lambda^k s_1 + \mu^k t_1) + \cdots. \qquad (25.8\text{-}8)$$

Since $\alpha(s, t)$ is the tangent vector to the curve obtained from $\tilde{g}(s, t)^{-1}\tilde{g}(s', t)$ by varying s', for given s and t, and then setting $s' = s$, and similarly for $\beta(s, t)$, it follows that

$$\alpha^i(s, t) = \frac{\partial}{\partial s'} m^i(\mathbf{y}(s, t), \mathbf{x}(s't))|_{s'=s},$$

$$\beta^i(s, t) = \frac{\partial}{\partial t'} m^i(\mathbf{y}(s, t), \mathbf{x}(s, t))|_{t'=t};$$

a direct calculation from (25.8-7) and (25.8-8) then gives

$$\left(\frac{\partial \alpha^i}{\partial t} - \frac{\partial \beta^i}{\partial s} \right)_{s=s_0, t=t_0} = a^i_{jk}(\alpha^j \beta^k - \alpha^k \beta^j)_{s=s_0, t=t_0},$$

which is the desired result (25.8-6), by virtue of the definition (25.3-3, 4) of the Lie product.

25.9 An Auxiliary Lemma on the Differentiation of Exponentials

Lemma.   *If* $\lambda = \lambda(t)$ *is a smooth curve in* Λ, *and if* λ' *or* $\lambda'(t)$ *denotes* $d\lambda/dt$ *(which is also a curve in* Λ), *then*

$$e^{-\lambda} \frac{d}{dt} e^{\lambda} = f(\text{Ad}_\lambda)\lambda', \qquad (25.9\text{-}1)$$

where

$$f(z) = \frac{1 - e^{-z}}{z} = 1 - \frac{1}{2!} z + \frac{1}{3!} z^2 - \cdots. \qquad (25.9\text{-}2)$$

Comment 1. On the left side of (25.9-1), de^λ/dt is a tangent vector at the point e^λ of the group G; multiplication of this vector on the left by $e^{-\lambda}$ transforms it into a tangent vector at the point 1 of G, i.e., into an element of Λ.

Comment 2. On the right side of (25.9-1), since the transformation Ad_λ can be represented by an $n \times n$ matrix, and since the series for $f(z)$ converges absolutely for all z, it follows that $f(Ad_\lambda)$ is a well-defined linear transformation in Λ. [In particular, if λ and λ' commute, i.e., if $Ad_\lambda \lambda' = [\lambda, \lambda'] = 0$, then the right member of (25.9-1) is just λ' itself.]

Comment 3. If Λ is a Lie algebra of matrices, the lemma can in principle be obtained by multiplication of the power series for $e^{-\lambda(t)}$ with the power series obtained by term-by-term differentiation of the series for $e^{\lambda(t)}$, taking noncommutativity of the matrices λ and λ' into account by writing $[\lambda, \lambda'] = \lambda\lambda' - \lambda'\lambda$. It is easy to verify in this way that the first two or three terms of the expansion are as claimed in the lemma.

PROOF OF THE LEMMA.   Define

$$\alpha(s, t) = e^{-s\lambda(t)} \frac{\partial}{\partial s} e^{s\lambda(t)} = \lambda(t),$$

$$\beta(s, t) = e^{-s\lambda(t)} \frac{\partial}{\partial t} e^{s\lambda(t)},$$

so that $\beta(0, t) \equiv 0$, and use the lemma of Sections 25.8:

$$\frac{d}{dt} \lambda(t) - \frac{\partial}{\partial s} \beta(s, t) = [\lambda(t), \beta(s, t)].$$

For fixed t, this is a differential equation of the form

$$\lambda' = \frac{d}{ds} \beta(s) = A\beta(s),$$

where A is the matrix of the transformation $Ad_{\lambda(t)}$; the solution that satisfies the initial condition $\beta(0) = 0$ is

$$\beta(s) = \frac{1 - e^{-As}}{A} \lambda'.$$

[*Note. A* is a singular matrix, because $Ad_\lambda \lambda = 0$. The expression $(1 - e^{As})/A$ denotes the matrix obtained by substituting A for z in the entire function $(1 - e^{zs})/z$.] Therefore,

$$\beta(1, t) = e^{-\lambda} \frac{d}{dt} e^\lambda = f(Ad_\lambda)\lambda',$$

as was to be proved.

25.10 The Campbell–Baker–Hausdorff (CBH) Formula

If λ and μ are commutative elements of Λ (i.e., if $[\lambda, \mu] = 0$) or commutative matrices, then $e^\lambda e^\mu = e^{\lambda + \mu}$. In general, we seek a σ such that $e^\lambda e^\mu = e^\sigma$. The CBH formula gives σ explicitly in terms of λ and μ and Lie brackets containing λ and μ, for λ and μ in some neighborhood of the origin of Λ.

Theorem (CBH).   *Let $\psi(z) = z \log z/(z - 1) = 1 + w/1 \cdot 2 - w^2/2 \cdot 3 + w^3/3 \cdot 4 - \cdots$, for $|w| < 1$, where $z = 1 + w$; then, for λ and μ in some neighborhood of the origin in Λ,*

$$\sigma \stackrel{\text{def}}{=} \log(e^\lambda e^\mu) = \lambda + \int_0^1 \psi(e^{\mathrm{Ad}_\lambda} e^{t\,\mathrm{Ad}_\mu})\mu \, dt. \tag{25.10-1}$$

Comment 1. The argument of $\psi(\cdot)$ in the formula can be represented by an $n \times n$ matrix $M = I + W$, which can be made arbitrarily close to the unit matrix by taking λ and μ sufficiently small. [Ad_0 is the matrix 0, e^{Ad_0} is the matrix I.] The power series for $\psi(1 + w)$ converges absolutely for $|w| < 1$; hence, the series for $\psi(I + W)$ converges elementwise if each element w_{ij} of W satisfies $|w_{ij}| < 1/n$, i.e., if λ and μ are restricted to some neighborhood $\hat{N}$ of the origin in Λ.

Comment 2. The vectors λ, μ, and σ are the coordinates (logarithmic) of the group elements e^λ, e^μ, and e^σ; hence, $\sigma = \mathbf{m}(\lambda, \mu)$. The CBH formula is therefore a formula for $\mathbf{m}(\cdot, \cdot)$ in logarithmic coordinates, and the structure of Λ completely determines multiplication in the group in some neighborhood of 1.

Comment 3. All the expansions implied in (25.10-1) can be carried out; the first few terms are

$$\log(e^\lambda e^\mu) = \lambda + \mu + \tfrac{1}{2}[\lambda, \mu] + \tfrac{1}{12}[\lambda, [\lambda, \mu]] + \tfrac{1}{12}[\mu, [\mu, \lambda]] + \cdots . \tag{25.10-2}$$

PROOF OF THE THEOREM.   The function $\log(e^\lambda e^{t\mu})$ is denoted by $\sigma(t)$; a differential equation for $\sigma(t)$ will be found, whose solution gives the CBH formula. First, since

$$\frac{d}{dt} e^{\sigma(t)} = e^\lambda e^{t\mu}\mu = e^{\sigma(t)}\mu,$$

the lemma of the preceding section gives

$$\mu = e^{-\sigma(t)} \frac{d}{dt} e^{\sigma(t)} = f(\mathrm{Ad}_{\sigma(t)})\sigma'(t), \tag{25.10-3}$$

where $f(z) = (1 - e^{-z})/z$. If $\chi(z)$ is defined as $z/(1 - e^{-z}) = (f(z))^{-1}$, then $\chi(M)f(M) = I$ for any matrix M, so that (25.10-3) can be solved for $\sigma'(t)$:

$$\sigma'(t) = \chi(\mathrm{Ad}_{\sigma(t)})\mu. \tag{25.10-4}$$

From the definitions of $\chi(\cdot)$ and $\psi(\cdot)$ it follows that $\chi(z) = \psi(e^z)$; hence

$$\sigma'(t) = \psi(e^{\mathrm{Ad}_{\sigma(t)}})\mu.$$

This differential equation can be simplified by use of the lemma of Section 25.7, which says, in the notation of Section 25.8, that the mapping e^{Ad_σ} is the mapping $v \to e^\sigma v e^{-\sigma}$; specifically,

$$e^{\mathrm{Ad}_{\sigma(t)}}v = e^{\sigma(t)}v e^{-\sigma(t)} = e^\lambda e^{t\mu}v e^{-t\mu}e^{-\lambda}$$
$$= e^{\mathrm{Ad}_\lambda} e^{t\,\mathrm{Ad}_\mu}v.$$

Now, the unknown appears only on the left, and the CBH formula follows directly by integrating from $t = 0$ to $t = 1$ and by using the conditions

$$\sigma(0) = \log e^{\lambda} = \lambda, \qquad \sigma(1) = \log e^{\lambda}e^{\mu} = \sigma.$$

Since Ad_{λ} is the transformation $\mathbf{v} \to [\lambda, \mathbf{v}]$, the matrix elements of Ad_{λ} are linear functions of the components of λ. Therefore, the matrix elements of the transformations $\exp\{\text{Ad}_{\lambda}\}$ and $\exp\{t\,\text{Ad}_{\mu}\}$ are analytic functions of the components of λ and μ. Since $\psi(z)$ is analytic, for $|z - 1| < 1$, it follows that, for λ and μ restricted to the neighborhood $\hat{N}$ of the origin in Λ referred to in Comment 1, the components of $\sigma = \log e^{\lambda}e^{\mu}$ are analytic functions of the components of λ and μ; by analytic continuation, they are analytic for all λ and μ such that the logarithm is defined.

When logarithmic coordinates λ^i are used, $\log e^{\lambda}e^{\mu}$ is simply the multiplication function denoted by $\mathbf{m}(\lambda, \mu)$ previously. Although this function was only assumed to be of class C^4, it is now seen to be analytic, when logarithmic coordinates are used. In these coordinates, the inversion function $\mathbf{I}(\cdot)$ is given by $\mathbf{I}(\lambda) = -\lambda$, and hence is also analytic.

EXERCISE

Express the matrix elements of Ad_{λ} in terms of the components λ^i of λ, for a given basis $\varepsilon_1, \ldots, \varepsilon_n$ in Λ, and the corresponding structure constants C^i_{jk}.

25.11 Translations of Charts; Compatibility; G as an Analytic Manifold

In this section, a basic chart is chosen, with logarithmic coordinates in it, and other charts are obtained from it by translations in the group. To simplify the work, the basic chart is chosen as a rather small one, so as to have various convenient properties.

First, let $\mathfrak{U}$ be a neighborhood of 1 in the group G in which the mapping $e^{\lambda} \to \lambda$ into the Lie algebra is one-to-one, so that logarithmic coordinates can be used in $\mathfrak{U}$. Next, let $\mathfrak{B}$ be a sufficiently small subneighborhood in $\mathfrak{U}$ (an open subset of $\mathfrak{U}$), also containing 1, so that if g and h are in $\mathfrak{B}$, then gh is in $\mathfrak{U}$, *and* $\log gh$ is given by the CBH formula. Finally, let $\mathfrak{W}$ be a subneighborhood in $\mathfrak{B}$ such that if g and h are in $\mathfrak{W}$, then gh is in $\mathfrak{B}$, while g^{-1} and h^{-1} are in $\mathfrak{W}$ (this permits use of the CBH formula for triple products like $g_1 g_2^{-1} g_3$, etc.), and let N be the image of $\mathfrak{W}$ in Λ; i.e., $N = \{\log g : g \in \mathfrak{W}\}$. From now on, $\{\mathfrak{W}, \log, N\}$ will be regarded as the basic chart.

It is recalled that, for any fixed a in G, the one-to-one mapping of G onto itself given by $g \to ag$ is called a *left translation* by a and the mapping $g \to ga$ is a *right translation*. For any fixed a in G, a left-translated chart $\{a\mathfrak{W}, {}_a\varphi, N\}$ is defined as follows: First, the subset $a\mathfrak{W}$ of G is defined as

$$a\mathfrak{W} = \{ag_1 : g_1 \in \mathfrak{W}\};$$

then for each $g = ag_1$ in $a\mathfrak{W}$, $a\varphi(g)$ is defined as $\log g_1$. Note that the image of $a\mathfrak{W}$ under the mapping $g \to a\varphi(g)$ is the same open set N in the coordinate space Λ as the image of $\mathfrak{W}$ under the mapping $g \to \log g$. Right-translated charts $\{\mathfrak{W}a, \varphi_a, N\}$ are similar.

Theorem 1. *Any two charts obtained by translation of the basic chart $\{\mathfrak{W}, \log, N\}$ are compatible, in fact analytically compatible.*

Comment. If a lies in $\mathfrak{W}$, then the left (also right) translation by a is a homeomorphism in the basic chart, as far as it is defined, because the coordinates of ag are continuous (in fact, analytic) functions of the coordinates of g, by the CBH formula, and the coordinates of g are continuous functions of the coordinates of ag, because $g = a^{-1}(ag)$. Hence, any translated chart is compatible with the basic chart, and it will be shown that any two translated charts are also compatible with each other. G thus becomes a manifold, and Theorem 2 below then shows that the translations are homeomorphisms in all G.

PROOF OF THEOREM 1.   First consider charts obtained by left translation of the basic chart by a and b, respectively. If the intersection $a\mathfrak{W} \cap b\mathfrak{W}$ is empty, the charts are automatically compatible. Otherwise, it must be proved first that the intersection is mapped by $_a\varphi$ (also by $_b\varphi$) onto an open set in Λ (see Section 23.2). That is, if g is any point in the intersection, so that $g = ag_1 = bh_1$, where g_1 and h_1 are in $\mathfrak{W}$, then it must be shown that g is in some subset that is an *open* subset of each chart, according to the topology of that chart. Therefore, consider an element $g' = g\varepsilon$ that is close to g, so that ε is close to 1 (just how close it has to be will be seen in a moment). Then, $g' = ag_1\varepsilon = bh_1\varepsilon$. Now, $\mathfrak{W}$ is an open set; hence g_1 and h_1 are interior points. Since the coordinates of $g'_1 = g_1\varepsilon$ and $h'_1 = h_1\varepsilon$ depend continuously on the coordinates of ε, there is a neighborhood $\mathfrak{W}_0$ of 1 such that, for ε in $\mathfrak{W}_0$, g'_1 and h'_1 are in $\mathfrak{W}$. See Figure 25.1. The sets $g_1\mathfrak{W}_0 = \{g_1\varepsilon: \varepsilon \text{ in } \mathfrak{W}_0\}$ and $h_1\mathfrak{W}_0 = \{h_1\varepsilon: \varepsilon \text{ in } \mathfrak{W}_0\}$ are neighborhoods in $\mathfrak{W}$ of g_1 and h_1 respectively, i.e., *open* sets containing g_1 and h_1, respectively, because, according to the above comment, the left translations by g_1 and by h_1 are homeomorphisms in $\mathfrak{W}$. Therefore, since $g = ag_1 = bh_1$, $g\mathfrak{W}_0$ is an open set in the topology of either chart and is contained in both charts. Hence, the

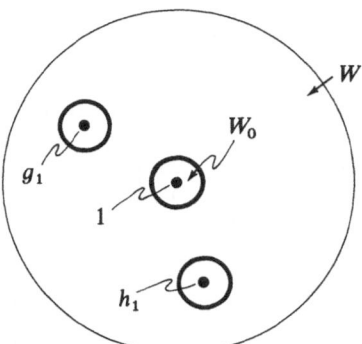

Figure 25.1

first condition for compatibility is satisfied. Second, it must be proved that the co-ordinates $_a\varphi(g) = \log g_1$ and $_b\varphi(g) = \log h_1$ are analytic functions of each other, for all g that can be written both as ag_1 and bh_1, with g_1 and h_1 in $\mathfrak{W}$. Since $b^{-1}a = h_1 g_1^{-1}$, and since h_1 and g_1 are in $\mathfrak{W}$, it follows that $b^{-1}a$ is in $\mathfrak{B}$ (even though a and b themselves may not even be in the neighborhood $\mathfrak{U}$ where logarithmic coordinates are defined). Hence, $\log h_1$, which is $= \log((b^{-1}a)g_1)$, depends analytically on $\log g_1$, according to the CBH formula. Similarly, $\log g_1$ depends analytically on $\log h_1$. Lastly, a similar argument shows the compatibility of two right-translated charts or of a right-translated and a left-translated one.

Theorem 2. *When a group G is covered by translated charts, as described above, the Hausdorff separation axiom is satisfied.*

PROOF. Let a and b be any two points of the group, and suppose that they are not separated, that is, that any neighborhood of a and any neighborhood of b have points in common. It is to be proved that then $a = b$. Let $\|\cdot\|$ be a norm in Λ, e.g., the Euclidean norm relative to some basis $\varepsilon_1, \ldots, \varepsilon_n$ in Λ. For any $\delta > 0$, let $\mathfrak{U}_\delta$ be the neighborhood of 1 consisting of those elements g in $\mathfrak{W}$ such that $\|\log g\| < \delta$. Then $a\mathfrak{U}_\delta$ and $b\mathfrak{U}_\delta$ are neighborhoods of a and b; let g be in $a\mathfrak{U}_\delta \cap b\mathfrak{U}_\delta$, so that $g = a\varepsilon = b\eta$, where ε and η are in $\mathfrak{U}_\delta$. Then $b^{-1}a = \eta\varepsilon^{-1}$, but since $\log \eta\varepsilon^{-1}$ is given by the CBH formula in terms of $\log \eta$ and $\log \varepsilon^{-1} = -\log \varepsilon$, both of which can be made arbitrarily small by suitable choice of δ, it is seen that $\log \eta\varepsilon^{-1}$, which is $= \log b^{-1}a$, is $= 0$; hence $b^{-1}a = 1$, as was to be proved.

Therefore, G, when covered by the translated charts, as will be assumed from now on, is real analytic manifold. We assume, as in Chapter 23, that G can be covered by a countable collection of these charts.

Lemma. *A left or right translation is an analytic mapping in all G.*

PROOF. The left translation $g \to ag$ maps the chart $\{b\mathfrak{W}, _b\varphi, N\}$ (b is arbitrary) onto the chart $\{ab\mathfrak{W}, _{ab}\varphi, N\}$; the corresponding coordinate mapping is the identity, for if $g = bh$, then

$$_b\varphi(g) = _{ab}\varphi(ag) = \log h.$$

The analyticity of the same mapping in right-translated charts follows from the compatibility of charts.

Theorem 3. *Product and inverse are analytic in all G.*

PROOF (For Product). Let a and b be arbitrary elements of G. Let $_a\varphi$, $_b\varphi$, and $_{ab}\varphi$ denote coordinates in charts obtained from the basic one by left translation by a, b, and ab. It will be shown that, for g and h close to 1, the $_{ab}\varphi$ coordinate of $(ag)(bh)$ depends analytically on the $_a\varphi$ coordinate of ag and the $_b\varphi$ coordinate of bh. Then, since all translated charts are analytically compatible, the same conclusion holds for any three charts in which a, b, and ab are respectively located. Now

$$_{ab}\varphi(agbh) = _{ab}\varphi(abb^{-1}gbh) = \log(b^{-1}gbh);$$

hence the last member of this equation must be shown to depend analytically on $\log g$ and $\log h$. The element $b^{-1}gb$ is obtained from g by a left translation by b^{-1}

followed (or preceded) by a right translation by b, and is $=1$ for $g = 1$; hence $\log(b^{-1}gb)$ exists for g near 1 and depends analytically on $\log g$, by the lemma above. By the CBH formula, $\log(b^{-1}gbh)$ depends analytically on $\log(b^{-1}gb)$ and $\log h$, and the conclusion follows.

The proof of the analyticity of the inverse is left as an exercise.

Theorem 4.   *The principal component of G, i.e., the component containing 1, is generated by any neighborhood $\mathfrak{W}_0$ of 1. That is, any g in that component can be written as a finite product $g = g_1 g_2 \ldots$, where each g_i is in $\mathfrak{W}_0$. (See Section 23.6 for "component.")*

PROOF.   Let G_0 be the subgroup of G generated by $\mathfrak{W}_0 \cap \mathfrak{W}_0^{-1}$, where $\mathfrak{W}_0^{-1}$ is the set consisting of the inverses of the elements of $\mathfrak{W}_0$. The set $\mathfrak{W}_{00} = \mathfrak{W}_0 \cap \mathfrak{W}_0^{-1}$ is an open set containing 1. Clearly G_0 is connected, because the factors $g_1, g_2, \ldots$ in such a finite product can each, in turn, be smoothly changed into 1 and then removed from the product, so that as a result g is connected to 1. It will be shown that G_0 is both open and closed. Therefore, by Section 23.6, it is an entire component of G. [Then, since $\mathfrak{W}_{00} \subset \mathfrak{W}_0$, it is *a fortiori* clear that G_0 is generated by $\mathfrak{W}_0$.] If $g = g_1 g_2 \cdots g_k$ is any element of G_0, then the open set $g\mathfrak{W}_{00}$, which contains g, consists of elements of the form $g_1 g_2 \cdots g_k g_{k+1}$, where g_{k+1} is also in $\mathfrak{W}_{00}$; therefore $g\mathfrak{W}_{00}$ is in G_0, and $\mathfrak{W}_0$ is open. On the other hand, if g' is any limit point of G_0, then the set $g'\mathfrak{W}_{00} = g'\mathfrak{W}_{00}^{-1} = \{g'h^{-1} : h \in \mathfrak{W}_{00}\}$ is a neighborhood of g', and hence contains an element of the form $g_1 \cdots g_k$; therefore, $g_1 \cdots g_k = g'h^{-1}$ for some h in $\mathfrak{W}_{00}$; hence, $g' = g_1 \cdots g_k h$ is in G_0, and G_0 is closed, as was to be proved.

Remarks. Left translation is a homeomorphism in all G. (It is one-to-one by group theory and is continuous, indeed analytic, by the lemma above.) Therefore, each left coset aG_0 of G_0 in G is a component, and conversely. The same argument with right translations shows that each component is also a right coset; therefore G_0 is a normal subgroup. The number of components is countable, because we have specified that the manifold can be covered by a countable collection of charts. The theory puts no other restriction on the nature of the factor group, because if G_0 is a connected Lie group and K any abstract countable group, then the direct product $G_0 \times K$ can be regarded as a Lie group G having one component for each element in K, and then $G/G_0 \cong K$. For this reason, G/G_0 is uninteresting, and many authors include a requirement of connectedness in their definition of a Lie group. We prefer not to exclude groups like O(3) from consideration.

25.12 Lie Algebra Homomorphisms

The homomorphism theory of Lie algebras and Lie groups, described in this section and the next two, is closely analogous to the homomorphism theory for groups, described in Sections 18.5-18.8.

Roughly speaking, a homomorphism is a mapping (generally many-to-one) of a mathematical structure onto a less complicated structure of the same kind such that any relation that holds among elements of the first structure also holds among their images in the second structure. Some of these relations generally turn out to be trivial ones in the second structure, like $1 \circ 1 = 1$ or $0 + 0 = 0$, while the remaining relations may be regarded as showing some of the main features of the first structure, but with less fine detail. Just as for groups, such a mapping exists if and only if the first structure contains a particular kind of substructure (e.g., a normal subgroup), which can serve as the kernel of the mapping; the theory shows how to reconstruct the mapping in question, when the substructure is given, by first forming the so-called quotient or factor structure (e.g., factor group) and then constructing the so-called natural homomorphism of the first structure onto the quotient structure, which is then shown to be equivalent to the original homomorphism.

The execution of this program is straightforward for Lie algebras. For Lie group, each idea introduced has an immediate parallel in the corresponding Lie algebras, and this interplay between the groups and their algebras provides powerful techniques for the investigation of the groups.

If Λ and $\overline{\Lambda}$ are real Lie algebras, a *homomorphism* $\Lambda \to \overline{\Lambda}$ is a mapping ψ of Λ into $\overline{\Lambda}$ which preserves all the operations of the Lie algebra; that is, if λ, μ are in Λ and a, b are in $\mathbb{R}$, then $\psi(a\lambda + b\mu) = a\psi(\lambda) + b\psi(\mu)$, and $\psi([\lambda, \mu]) = [\psi(\lambda), \psi(\mu)]$. For complex Lie algebras, the definition is the same, except that $\mathbb{C}$ replaces $\mathbb{R}$.

Comment. It is not really necessary to assume in advance that $\overline{\Lambda}$ is a Lie algebra; it only needs to be a structure in which $\lambda + \mu$, $a\lambda$, and $[\lambda, \mu]$ are defined. However, the part of it onto which Λ is mapped by the homomorphism, i.e., the image $\psi(\Lambda)$, is then necessarily a Lie algebra.

If ψ is one-to-one and onto, then it is an *isomorphism*; in symbols, $\Lambda \cong \overline{\Lambda}$; if, further, it is a mapping of Λ onto itself $(\overline{\Lambda} = \Lambda)$, then it is an *automorphism.*

A linear subspace A of Λ is a *subalgebra* if $[\lambda, \mu] \in A$ for all λ, $\mu \in A$: if, further, $[\lambda, \mu] \in A$ for all $\lambda \in A$ and all $\mu \in \Lambda$, then A is an *ideal* of Λ. A Lie algebra Λ of more than one dimension is *simple* if it contains no ideals except $\{0\}$ and Λ. (The reason for excluding one-dimensional algebras, in this definition, will appear in Section 25.16.) It is easy to see that the *kernel* of a homomorphism ψ, namely the set $\{\lambda : \psi(\lambda) = 0\}$, is an ideal in Λ. Therefore, an alternative definition is that a Lie algebra of more than one dimension is *simple* if it cannot be mapped homomorphically onto any less complicated algebra except the trivial algebra $\{0\}$.

Ideals play almost exactly the same role for Lie algebras as normal subgroups play for groups.

If Λ_0 is a subspace of Λ, then the relation $\lambda \equiv \mu \pmod{\Lambda_0}$, defined as meaning that $\lambda - \mu \in \Lambda_0$, is an equivalence relation, which decomposes Λ into disjoint classes called *residue classes modulo* Λ_0. If, for any fixed $\lambda \in \Lambda$, the residue class $\{\lambda + \mu : \mu \in \Lambda_0\}$ is denoted by $\overline{\lambda}$, and if we define $a\overline{\lambda} = \overline{a\lambda}$ and $\overline{\lambda} + \overline{\mu} = \overline{\lambda + \mu}$, then the set of residue classes constitutes a linear vector space, called the *factor space* of Λ modulo Λ_0. We shall show that if Λ_0 is an ideal, the factor space can be interpreted as a Lie algebra.

Theorem 1. *Let Λ_0 be a subspace of a Lie algebra Λ. For each choice of λ_1 and μ_1 in Λ, the set $\{[\lambda, \mu]: \lambda - \lambda_1 \in \Lambda_0, \mu - \mu_1 \in \Lambda_0\}$ is contained in a single residue class (namely the class $[\bar{\lambda}_1, \bar{\mu}_1]$) if and only if Λ_0 is an ideal in Λ. In this case, if the Lie product of two residue classes $\bar{\lambda}_1$ and $\bar{\mu}_1$ is defined by the equation $[\bar{\lambda}_1, \bar{\mu}_1] = \overline{[\lambda_1, \mu_1]}$, then the factor space of Λ modulo Λ_0 is a Lie algebra, denoted by Λ/Λ_0, called the* factor algebra *of Λ modulo Λ_0. The mapping $\lambda \to \bar{\lambda}$ is a homomorphism called the* natural homomorphism *of Λ onto Λ/Λ_0, and is denoted by ψ_n.*

Comment. Consideration of the commutative case, where every subspace of Λ is an ideal, and where every Lie product $[\lambda, \mu]$ is $=0$, shows that the set of elements $[\lambda, \mu]$ referred to in the theorem may be only part of the residue class $[\bar{\lambda}_1, \bar{\mu}_1]$, which is $=\Lambda_0$.

PROOF OF THEOREM 1. (1) Suppose that Λ_0 is an ideal. Generally,

$$[\lambda, \mu] - [\lambda_1, \mu_1] = [\lambda - \lambda_1, \mu] + [\lambda_1, \mu - \mu_1];$$

hence, if $\lambda - \lambda_1$ and $\mu - \mu_1$ are in Λ_0, then both terms on the right are in Λ_0; hence, $[\lambda, \mu]$ and $[\lambda_1, \mu_1]$ are in the same residue class, as claimed. (2) Conversely, if, for arbitrary λ and μ, $[\lambda, \mu + \sigma]$ is always in the same residue class as $[\lambda, \mu]$, for any σ in Λ_0, then $[\lambda, \sigma] \in \Lambda_0$, and it follows that Λ_0 is an ideal. (3) The definition $[\bar{\lambda}_1, \bar{\lambda}_2] = \overline{[\lambda_1, \lambda_2]}$ of the product of two residue classes shows that the mapping $\lambda \to \bar{\lambda}$ is a homomorphism, and then it follows from the comment after the definition of homomorphism that the factor space is a Lie algebra.

Theorem 2. (Homomorphism Law). *If Λ_0 is the kernel of a homomorphism ψ of Λ onto $\bar{\Lambda}$, then Λ_0 is an ideal in Λ (as already noted), and*

$$\frac{\Lambda}{\Lambda_0} \cong \bar{\Lambda}.$$

PROOF. Let ψ_n denote the natural homomorphism of Λ onto Λ/Λ_0. Whenever $\psi_n(\lambda) = \psi_n(\lambda_1)$, λ and λ_1 are in the same residue class, i.e., $\lambda - \lambda_1$ is in Λ_0; hence $\psi(\lambda - \lambda_1) = 0$, by definition of Λ_0 as the kernel of ψ, hence $\psi(\lambda) = \psi(\lambda_1)$. Therefore, $\psi\psi_n^{-1}$ is a one-to-one mapping of the set Λ/Λ_0 of residue classes onto $\bar{\Lambda}$; the mapping will be called χ. It is left to the reader to complete the proof by verifying that χ is linear and satisfies the equation $\chi([\bar{\lambda}, \bar{\mu}]) = [\chi(\bar{\lambda}), \chi(\bar{\mu})]$, where $\bar{\lambda}$ and $\bar{\mu}$ are any residue classes in Λ mod Λ_0, i.e., any elements of Λ/Λ_0.

25.13 Lie Group Homomorphisms

In the case of Lie groups, a homomorphism must preserve not only all algebraic relations but also all local topological and analytic relations arising from the manifold structure.

If G and $\bar{G}$ are Lie groups, then a mapping Ψ of G into $\bar{G}$ is called *a Lie-group homomorphism* if:

1. It is a homomorphism in the sense of group theory: $\Psi(gh) = \Psi(g)\Psi(h)$; $\Psi(g^{-1}) = \Psi(g)^{-1}$.
2. It is a continuous mapping; that is, if φ and $\bar{\varphi}$ are any coordinate systems in G and $\bar{G}$, respectively, then each component of the vector

$$\mathbf{y} = \bar{\varphi}(\Psi(\varphi^{-1}(\mathbf{x}))) \qquad (25.13\text{-}1)$$

   is a continuous function of the components of $\mathbf{x}$, for all $\mathbf{x}$ such that the above expression is defined.

Remark 1. The mapping is generally many-to-one; in fact, G may have more dimensions than $\bar{G}$.

Remark 2. Typical of non-group-theory relations in G that are preserved in $\bar{G}$ is the convergence of a sequence $g_1, g_2, \ldots$ to a limit h, which means that in a chart containing h the coordinates of $g_1, g_2, \ldots$ converge to the coordinates of h; then, because of the continuity of the functions (25.13-1), the sequence $\Psi(g_1), \Psi(g_2), \ldots$ converges to $\Psi(h)$ in $\bar{G}$. Similarly the image of a curve in G under a homomorphism is a curve in $\bar{G}$.

Remark 3. The property of Ψ of being a continuous mapping is invariant under the addition or deletion of compatible charts in either or both of the manifolds G and $\bar{G}$, because if $\mathbf{x}'$ and $\mathbf{y}'$ are coordinates in any other charts, then the schema

$$\mathbf{x}' \leftrightarrow \mathbf{x} \rightarrow \mathbf{y} \leftrightarrow \mathbf{y}'$$

shows that $\mathbf{y}'$ is continuous in $\mathbf{x}'$ whenever it is defined by the composition of the three mappings indicated.

Remark 4. If Ψ is one-to-one and onto, and if Ψ^{-1} is also continuous, then Ψ is a *Lie-group isomorphism*.

Theorem 1. *Let Ψ be a Lie-group homomorphism (i.e., a continuous homomorphism) of G onto $\bar{G}$, and let Λ and $\bar{\Lambda}$ be the Lie algebras of G and $\bar{G}$. Then, when expressed in logarithmic coordinates, the mapping Ψ is locally a Lie algebra homomorphism of Λ onto $\bar{\Lambda}$.*

Remark. As an obvious corollary, since the mapping Ψ is linear in these coordinates, every Lie-group homomorphism is locally an analytic mapping; it is also globally analytic, because if $g = g_0 h$ for arbitrary g_0, then the element $\Psi(g) = \Psi(g_0)\Psi(h) = \Psi(g_0)\Psi(g_0^{-1}g)$ is analytic in g for h in some neighborhood of 1, since products and inverses are analytic in each group.

PROOF OF THEOREM 1.   For any λ sufficiently near the origin in Λ, we can define an element of $\bar{\Lambda}$ as

$$\bar{\lambda} = \log(\Psi(e^{\lambda})).$$

We must show that the mapping $\lambda \to \bar{\lambda}$ is linear and maps $[\lambda, \mu]$ onto $[\bar{\lambda}, \bar{\mu}]$. We show first that it maps $t\lambda$ onto $t\bar{\lambda}$ for real t, that is, that $\overline{t\lambda} = t\bar{\lambda}$. For fixed λ, the set $\Psi(e^{t\lambda})$, $t \in \mathbb{R}$, is a one-parameter subgroup of $\bar{G}$ which includes the group element $e^{\bar{\lambda}}$; hence for each t there is a real number $s = f(t)$ such that

$$\Psi(e^{t\lambda}) = e^{s\bar{\lambda}} = e^{f(t)\bar{\lambda}},$$

where $f(t)$ is continuous, because Ψ is continuous, and where $f(0) = 0$ and $f(1) = 1$. Since Ψ maps products onto products, we have

$$\begin{aligned} e^{f(t+s)\lambda} = \Psi(e^{(t+s)\lambda}) &= \Psi(e^{t\lambda}e^{s\lambda}) \\ &= \Psi(e^{t\lambda})\Psi(e^{s\lambda}) = e^{f(t)\lambda}e^{f(s)\lambda} = e^{[f(t)+f(s)]\lambda}. \end{aligned}$$

Hence $f(t + s) = f(t) + f(s)$, and the only continuous functions with this property are linear. Since also $f(0) = 0$, $f(1) = 1$, we see that $f(t) \equiv t$, as was to be shown. We now apply the CBH formula to both sides of the equation $\Psi(e^{s\lambda}e^{t\mu}) = e^{s\bar{\lambda}}e^{t\bar{\mu}}$, and we see that

$$\overline{s\lambda + t\mu + \tfrac{1}{2}st[\lambda, \mu] + \cdots} = s\bar{\lambda} + t\bar{\mu} + \tfrac{1}{2}st[\bar{\lambda}, \bar{\mu}] + \cdots \qquad (25.13\text{-}2)$$

for all s and t. We write $s = \varepsilon s'$, $t = \varepsilon t'$. Since $\varepsilon\bar{v} = \overline{\varepsilon v}$, we can cancel one factor ε on each side of the above equation. Then, as $\varepsilon \to 0$, the quadratic and higher terms vanish; hence

$$\overline{s'\lambda + t'\mu} = s'\bar{\lambda} + t'\bar{\mu},$$

and this establishes the full linearity of the mapping $\lambda \to \bar{\lambda}$. Then the linear terms can be dropped from both sides of (25.13-2), and by a similar argument we see that

$$\overline{[\lambda, \mu]} = [\bar{\lambda}, \bar{\mu}],$$

i.e., the mapping $\lambda \to \bar{\lambda}$ is a Lie-algebra homomorphism, as required.

This theorem does not have a global converse, but only a local one, and a definition is needed, before the converse can be stated. If $\mathfrak{U}$ is a neighborhood of the identity 1 of a Lie group G, then an analytic mapping Ψ of $\mathfrak{U}$ into a Lie group $\bar{G}$ such that $\Psi(gh) = \Psi(g)\Psi(h)$ whenever g, h, and gh are in $\mathfrak{U}$ is called a *local homomorphism* of G into $\bar{G}$. If the inverse mapping is also a local homomorphism, i.e., is unique and analytic in some neighborhood of 1 in $\bar{G}$, then Ψ is a *local isomorphism*. If, furthermore, $G = \bar{G}$, then Ψ is a *local automorphism* of G.

We come now to the converse of Theorem 1.

Theorem 2. *If Λ and $\bar{\Lambda}$ are the Lie algebras of G and $\bar{G}$, then any Lie-algebra homomorphism $\psi: \Lambda \to \bar{\Lambda}$ induces a local homomorphism $\Psi: G \to \bar{G}$ given by the exponential mapping, namely $\Psi(e^{\lambda}) = e^{\psi(\lambda)}$, for e^{λ} in a suitable neighborhood of 1 in G.*

PROOF. By the CBH formula,

$$\begin{aligned} \Psi(e^{\lambda}e^{\mu}) &= \Psi(e^{\lambda + \mu + 1/2[\lambda, \mu] + \cdots}) \\ &= e^{\psi(\lambda + \mu + 1/2[\lambda, \mu] + \cdots)}; \end{aligned}$$

since ψ is a Lie-algebra homomorphism, the last expression is equal to

$$e^{\psi(\lambda)\,+\,\psi(\mu)\,+\,1/2[\psi(\lambda),\,\psi(\mu)]\,+\,\cdots},$$

and since the CBH formula holds also in $\bar{G}$, the above is equal to $e^{\psi(\lambda)}e^{\psi(\mu)}$; that is, $\Psi(e^{\lambda}e^{\mu}) = \Psi(e^{\lambda})\Psi(e^{\mu})$, as was to be proved.

As a corollary, if $\psi: \Lambda \to \bar{\Lambda}$ is a Lie-algebra isomorphism, then Ψ is a local isomorphism of the Lie group; if, furthermore, $G = \bar{G}$ and $\Lambda = \bar{\Lambda}$, then Ψ is a local automorphism of G.

The next theorem indicates circumstances under which a local homomorphism can be extended to a global one. This question arose in Section 21.1, where it was shown that quantum mechanics can lead to local representations of groups like SO(3) and $\mathscr{L}_p$. It is recalled that the groups SO(3) and SU(2) are locally but not globally isomorphic, and that $\mathscr{L}_p$ and SL(2, C) are also locally but not globally isomorphic. As a second example, let G be the two-dimensional torus group, regarded as a group of matrices of the form

$$\begin{pmatrix} e^{i\alpha} & 0 \\ 0 & e^{i\beta} \end{pmatrix}.$$

Its Lie algebra Λ is the commutative Lie algebra of matrices of the form

$$\alpha\begin{pmatrix} i & 0 \\ 0 & 0 \end{pmatrix} + \beta\begin{pmatrix} 0 & 0 \\ 0 & i \end{pmatrix}, \quad \alpha, \beta \in \mathbb{R}.$$

Any linear transformation

$$\alpha \to a\alpha + b\beta, \qquad \beta \to c\alpha + d\beta,$$

where a, b, c, and d are real and $ad - bc \neq 0$, is a Lie algebra automorphism of Λ. The corresponding local automorphism of G is

$$\Psi: \begin{pmatrix} e^{i\alpha} & 0 \\ 0 & e^{i\beta} \end{pmatrix} \to \begin{pmatrix} e^{i(a\alpha+b\beta)} & 0 \\ 0 & e^{i(c\alpha+d\beta)} \end{pmatrix};$$

it is valid only for sufficiently small α and β, because, for example, the element

$$g = \begin{pmatrix} e^{i\pi} & 0 \\ 0 & 1 \end{pmatrix}$$

has the property that g^2 is the identity of G, and this property is not preserved under Ψ, except for special choices of a and b.

Theorem 3. *Let Ψ be a local homomorphism of a Lie group G into a Lie group $\bar{G}$, defined in a neighborhood $\mathfrak{U}$ of 1 in G. Then:*

(a) *If there is an extension $\hat{\Psi}$ of the mapping Ψ to all of G, which is a homomorphism in the sense of elementary group theory, then $\hat{\Psi}$ is continuous, hence analytic, throughout G—i.e., it is a Lie-group homomorphism.*

(b) *If G is connected, then there is at most one Lie-group homomorphism $\hat{\Psi}$ which is an extension of Ψ.*

(c) *If G is simply connected, there is exactly one such extension $\hat{\Psi}$.*

PROOF OF PART (a). For arbitrary fixed g_0, let $g = g_0 h$, where h is in the neighborhood $\mathfrak{U}$, so that the mapping $h \to \Psi(h)$ is analytic. Then, $\hat{\Psi}(g) = \hat{\Psi}(g_0)\Psi(h)$, but products and inverses are analytic throughout both groups (Section 25.11); hence h is analytic in g, and $\hat{\Psi}(g)$ is analytic in g.

PROOF OF PART (b). Any g in G can be written as $g_1 g_2 \ldots g_k$, where all the factors are in $\mathfrak{U}$; hence, if $\hat{\Psi}$ is any extension of the homomorphism Ψ, then $\hat{\Psi}(g) = \hat{\Psi}(g_1)\ldots$ $\hat{\Psi}(g_k) = \Psi(g_1)\ldots\Psi(g_k)$, which is completely determined by Ψ; hence any two such extensions must agree for every g.

PROOF OF PART (c). As on previous occasions, we let $\mathfrak{B}$ be a subneighborhood of $\mathfrak{U}$, containing 1, such that if g and h are in $\mathfrak{B}$, g^{-1} and h^{-1} are in $\mathfrak{B}$, while gh and hg are in $\mathfrak{U}$. The mapping $\hat{\Psi}$ will be constructed. Let h and k be elements of G joined by a smooth curve $\mathscr{C}$. Let $\mathscr{C}$ be partitioned into smaller segments by points (group elements) $g_0, g_1, \ldots, g_l$ so that $g_0 = h$ and $g_l = k$, while $g_i^{-1}g_{i+1}$ is always in $\mathfrak{B}$. We call

$$\bar{g} = \Psi(g_0^{-1}g_1)\Psi(g_1^{-1}g_2)\ldots\Psi(g_{l-1}^{-1}g_l);$$

$\bar{g}$ is an element of the group $\bar{G}$, and if Ψ were a global homomorphism $\bar{g}$ would be $= \Psi(h^{-1}k)$. From the above formula, it is seen that $\bar{g}$ is unaltered by a refinement of the partition of $\mathscr{C}$; hence, since any two partitions have a common refinement, $\bar{g}$ depends only on h and k and the curve $\mathscr{C}$, and we show next that it is independent of $\mathscr{C}$ for given h and k. Namely, let $\tilde{\mathscr{C}}$ be another curve from h to k, close to $\mathscr{C}$, partitioned by points $\tilde{g}_0\,(=g_0=h), \tilde{g}_1, \ldots, \tilde{g}_l\,(=g_l=k)$ such that $\tilde{g}_i$ is close to g_i for each i, and call

$$\tilde{\bar{g}} = \Psi(g_0^{-1}\tilde{g}_1)\Psi(\tilde{g}_1^{-1}\tilde{g}_2)\ldots\Psi(\tilde{g}_{l-1}^{-1}g_l).$$

Then $\tilde{\bar{g}}$ can be obtained from $\bar{g}$ by inserting suitable factors, for instance,

$$\Psi(g_1^{-1}\tilde{g}_1)\Psi(\tilde{g}_1^{-1}g_1)$$

(which cancel) between the first two factors of $\bar{g}$ and by then noticing that

$$\Psi(\tilde{g}_1^{-1}g_1)\Psi(g_1^{-1}g_2)\Psi(g_1^{-1}\tilde{g}_2) = \Psi(\tilde{g}_1^{-1}\tilde{g}_2).$$

In this way it is seen that $\tilde{\bar{g}} = \bar{g}$, that is, that $\bar{g}$ is unaltered by a continuous deformation of the curve $\mathscr{C}$, keeping the endpoints h and k fixed. Lastly, if G is simply connected any two curves from h to k are homotopic; hence $\bar{g}$ depends only on h and k, so that we can write $\bar{g} = \bar{g}(h, k)$. Furthermore,

$$\bar{g}(gh, gk) = \bar{g}(h, k) \qquad \text{for any } g \text{ in } G,$$

since the same is true of each factor $\Psi(g_i^{-1}g_{i+1})$. It follows that the mapping $\hat{\Psi}$ of G into $\bar{G}$ given by $\hat{\Psi}(g) = \bar{g}(1, g)$ is the required extension of the local homomorphism Ψ, because

$$\hat{\Psi}(g_1 g_2) = \bar{g}(1, g_1)\bar{g}(g_1, g_1 g_2) = \bar{g}(1, g_1)\bar{g}(1, g_2),$$

by addition of curves in G.

25.14 Law of Homomorphism for Lie Groups

The law of homomorphism for Lie groups is the same as that for abstract groups, except that it has additional topological aspects.

A subgroup of a Lie group G is called *closed* if its elements constitute a closed point set in the manifold of G.

Theorem 1.   *The kernel of a Lie group homomorphism* Ψ *of G onto $\bar{G}$, that is, the set*

$$G_0 = \{g \in G: \Psi(g) = 1 \ (identity \ of \ \bar{G})\}$$

is a closed normal subgroup of G.

It is a normal subgroup by elementary group theory, and it is closed, because of the continuity of the mapping Ψ: If $\{g_i\}$ is a sequence in G_0 that converges to h, then $\Psi(g_i)$ converges to $\Psi(h)$, but $\Psi(g_i) = 1$ for every i; hence $\Psi(h) = 1$, hence h is in G_0.

The next two theorems show that if G_0 is a closed subgroup of G (whether normal or not), then logarithmic coordinates can be so chosen in G that, when restricted to G_0, they are logarithmic coordinates in G_0, whereby G_0 becomes a Lie group in its own right, and its Lie algebra is a subalgebra of the Lie algebra of G.

If $\mathscr{C}: g(t)$ ($0 \le t \le \varepsilon$) is any smooth curve emanating from the identity ($g(0) = 1$) and lying in the subgroup G_0, then the tangent vector to $\mathscr{C}$ at 1 is called a *tangent vector to G_0 at 1.*

Theorem 2.   *Let G_0 be a closed subgroup of G. Then the set Λ_0 of all tangent vectors to G_0 at 1 is a subalgebra of the Lie algebra of G. If G_0 is a normal subgroup, then Λ_0 is an ideal.*

PROOF.   Suppose λ is in Λ_0. Then there is a smooth curve $g(t)$ that lies in G_0 and is such that

$$\log(g(t)) = \lambda t + \cdots,$$

where the dots denote terms of order t^2. For each positive integer m, $g(t/m)^m$ is in G_0, and, by the CBH formula,

$$\log\left(g\left(\frac{t}{m}\right)^m\right) = \lambda t + \cdots,$$

where now the dots denote terms of order t^2/m. By letting $m \to \infty$, we see, since G_0 is a closed set, λt is the coordinate of a point in G_0; hence, for λ in Λ_0, e^λ is in G_0. If λ_1 and λ_2 are both in Λ_0, then, by the CBH formula, the vector

$$\log(e^{\lambda_1 t}e^{\lambda_2 t}) = t(\lambda_1 + \lambda_2) + \tfrac{1}{2}t^2[\lambda_1, \lambda_2] + \cdots$$

is the coordinate of a point in G_0. By an argument similar to the one above, we see from the linear terms that $\lambda_1 + \lambda_2$ is in Λ_0, and more generally so is $t\lambda_1 + s\lambda_2$ for real t and s, so that Λ_0 is a subspace. In a like manner we see from the quadratic terms that $[\lambda_1, \lambda_2]$ is in Λ_0; hence Λ_0 is a subalgebra. Lastly, if G_0 is a normal subgroup, only one of λ_1 and λ_2 need be in Λ_0, and we see that Λ_0 is an ideal.

It is now clear how to define coordinates in G_0. Let $\varepsilon_1, \ldots, \varepsilon_n$ be a basic of the vector space Λ such that $\varepsilon_1, \ldots, \varepsilon_k$ ($k < n$) is a basis of Λ_0. For any λ, we write $\lambda = \lambda^1\varepsilon_1 + \cdots + \lambda^n\varepsilon_n$, and then $\lambda^1, \ldots, \lambda^k$ can be taken as coordinates in G_0 in a suitable neighborhood of 1. To find such a neighborhood, let N be a neighborhood of 0 in the Lie algebra Λ of G in which logarithmic coordinates can be used. Then the intersection $N \cap \Lambda_0$ is an open set in Λ_0, and hence contains a neighborhood N_0 (connected open set) of 0 in Λ_0. The set $\mathfrak{U}_0$ in the group G_0 given by

$$\mathfrak{U}_0 = \{g \in G_0 : \log g \in N_0\}$$

and the mapping $\varphi_0(g) = (\lambda^1, \ldots, \lambda^k)$ of $\mathfrak{U}_0$ into $\mathbb{R}^k$ define a chart $\{\mathfrak{U}_0, \varphi_0, N_0\}$ in G_0. It, and the other charts obtained from it by translation in G_0, as described in Section 25.11, are called *inherited* (*from G*). In this way we arrive at the following theorem.

Theorem 3.   *If G_0 is a closed subgroup of G, and if Λ and Λ_0 are as above, then G_0, with its inherited charts, is a Lie group, and Λ_0 is its Lie algebra.*

The next two theorems show that, if G_0 is a closed normal subgroup, then the factor group G/G_0 can be endowed with a manifold structure which makes it into a Lie group; the Lie algebra of this group is isomorphic to Λ/Λ_0; the natural homomorphism of G onto G/G_0 is analytic with respect to this manifold structure. The last theorem of the section is the homomorphism law itself.

It is easily seen that the last $n - k$ components $\lambda^{k+1}, \ldots, \lambda^n$ of λ with respect to the basis $\varepsilon_1, \ldots, \varepsilon_n$ described above can be taken as coordinates in G/G_0, for they are constant in each coset (of G_0 in G), or more precisely in the intersection of the coset and a suitable neighborhood $\mathfrak{B}$ of 1 in G in which logarithmic coordinates can be used. Namely, if $e^{\lambda'} = e^{\lambda} e^{\mu}$, where μ is in Λ_0, so that $e^{\lambda'}$ and e^{λ} are in the same coset, then, by the CBH formula,

$$\lambda' = \lambda + \mu + \tfrac{1}{2}[\lambda, \mu] + \cdots;$$

all the terms on the right, starting with μ, are in Λ_0, because Λ_0 is an ideal; hence λ' and λ have the same last $n - k$ components. It is also clear that these last components are different (i.e., at least one of them is different) in different cosets. In these coordinates, the product and inversion functions $m(\cdot, \cdot)$ and $l(\cdot)$ in G/G_0 are continuous (in fact analytic): For any product $e^{\lambda} e^{\lambda'} = e^{\lambda''}$ in G, all n components of λ'' depend continuously on all the components of λ and of λ'; hence in particular the same is true of the last $n - k$ components, which are the coordinates of the corresponding cosets. The natural homomorphism of G onto G/G_0, in which a group element e^{λ} is mapped onto the coset in which it lies, is continuous in these coordinates, for the mapping consists in simply ignoring the first k components of λ.

In this way the following theorem is established.

Theorem 4.   *Let G, G_0, Λ, Λ_0, $\varepsilon_1, \ldots, \varepsilon_n$, $\lambda^1, \ldots, \lambda^n$ be as above. Then $\lambda^{k+1}, \ldots, \lambda^n$ can be taken as coordinates $\bar{\varphi}(\bar{g})$ ($\bar{g}$ denotes a coset) in a subset $\bar{\mathfrak{U}}$ of G/G_0 (consisting of cosets that intersect the neighborhood $\mathfrak{B}$ in G), thus defining a chart which makes G/G_0 an $(n - k)$-dimensional Lie group, called the Lie factor group (it is also denoted by G/G_0) of G with respect to G_0. The natural homomorphism of G onto G/G_0 is continuous, and hence is a Lie group homomorphism (hence is analytic).*

The following example shows that the conclusions of the theorem need not hold if G_0 is not a *closed* subgroup: Let G be the 2-dimensional torus group consisting of matrices

$$\begin{pmatrix} e^{i\alpha} & 0 \\ 0 & e^{i\beta} \end{pmatrix} \qquad (\alpha, \beta \text{ real}).$$

Let G_0 be the subgroup of matrices

$$\begin{pmatrix} e^{it} & 0 \\ 0 & e^{i\theta t} \end{pmatrix} \qquad (t \text{ real}),$$

where θ is a fixed real irrational number. The manifold of G is a torus, and that of G_0 is a helical curve everywhere dense on the torus. G_0 is a normal subgroup, but is not closed. The Lie algebra Λ is a plane, and, under the mapping $e^\lambda \to \lambda$ of G onto Λ, G_0 maps onto a set of parallel straight lines dense in the plane. Λ_0 is that line of the set that passes through the origin. With respect to the basis $(\varepsilon_1, \varepsilon_2)$ referred to, where now ε_1 lies in Λ_0, the second coordinate λ^2 has different values on different lines of the set that constitutes G_0, and since any neighborhood of the origin intersects infinitely many of the lines, λ^2 is not constant in G_0 or in any coset.

The last two theorems are now stated without proof.

Theorem 5.   *The Lie algebra of G/G_0 is isomorphic to Λ/Λ_0.*

Theorem 6 (Law of Homomorphism).   *If G_0 is the kernel of a Lie group homomorphism of G onto $\bar{G}$, then*

$$\frac{G}{G_0} \cong \bar{G}.$$

In other words, if Ψ is the isomorphism (in the sense of elementary group theory) given in Section 18.8, then Ψ and its inverse Ψ^{-1} are analytic mappings.

The first seven of the exercises below deal with the somewhat elusive question, mentioned in Section 25.1, as to which Lie groups have faithful representations, and hence are linear groups, i.e., can be regarded as groups of matrices (or of the corresponding linear transformations), as is normally true of the groups that appear in applications. Every Lie algebra Λ has at least one representation, the so-called adjoint representation of Λ on itself (Exercise 2). By means of the exponential mapping, this gives a local representation of a Lie group G on its algebra Λ. This may or may not be extendable to a representation on all of G, and, if it can, it may or may not be faithful.

Exercise 8 deals with covering groups. SU(2) is the universal covering group of SO(3). The basic theorem of Section 24.3 on covering manifolds shows that there is no group not isomorphic to SU(2) that covers SU(2). Hence, as stated in Section 21.1, there are no multivalued representations of SO(3) except the 2-valued ones.

EXERCISES

The *center* C of a group G is the set of those group elements that commute with every group element, i.e.,

$$C = \{g \in G : gh = hg \; \forall h \text{ in } G\}.$$

Similarly, the center Z of a Lie algebra is the set of those elements that commute with every element in the algebra, i.e.,

$$Z = \{\lambda \in \Lambda : [\lambda, \mu] = 0 \; \forall \mu \text{ in } \Lambda\}.$$

1. Show that the center of a Lie group is a closed normal subgroup, and the center of a Lie algebra is an ideal.

2. Recall that, for any λ in Λ, Ad_λ is the linear transformation $\mu \to [\lambda, \mu]$ of Λ into itself. Show that if the Lie product of two such transformations is defined in the usual way

$$[\text{Ad}_\lambda, \text{Ad}_\kappa] = \text{Ad}_\lambda \text{Ad}_\kappa - \text{Ad}_\kappa \text{Ad}_\lambda,$$

whereupon the set $\{\text{Ad}_\lambda : \lambda \in \Lambda\}$ of all such transformations becomes a Lie algebra, then the mapping $\lambda \to \text{Ad}_\lambda$ is a homomorphism of Λ onto this new algebra. This homomorphism is called the *adjoint representation of* Λ (on itself).

3. Show that if Λ is *center-free*, which means that $Z = \{0\}$, then the adjoint representation is faithful (i.e., the above homomorphism is an isomorphism).

4. If G is simply connected, then the local homomorphism $e^\lambda \to e^{\text{Ad}_\lambda}$ of G onto a group of linear transformations in Λ, discussed in Section 25.7, can be extended to a homomorphism of all G, called the *adjoint representation* of G on Λ. Show that a necessary condition for this homomorphism to be an isomorphism is that G be *center-free*, which means that $C = \{1\}$; then the homomorphism is locally an isomorphism.

5. To illustrate Exercise 4, let G be SU(2) and take the 2 × 2 matrices T_1, T_2, T_3 defined by (21.2-4) as a basis of the Lie algebra Λ of SU(2). Then, the transformations Ad_λ are represented by 3 × 3 real matrices. Show that the matrices e^{Ad_λ} constitute the group SO(3) and that the homomorphism $e^\lambda \to e^{\text{Ad}_\lambda}$ is then the familiar 2-to-1 homomorphism of SU(2) onto SO(3). What are the centers of SU(2) and SO(3)?

6. Show that if C is the center of a group G, then the factor group G/C is not necessarily center-free by considering the finite group

$$G = \{\pm 1, \pm i, \pm j, \pm k\},$$

where i, j, k are the quaternion units, which satisfy the equations

$$i^2 = j^2 = k^2 = -1,$$

$$ij = -ji = k, \quad jk = -kj = i, \quad ki = -ik = j.$$

7. Show that the center Z of a Lie algebra Λ is an ideal, and that the factor algebra Λ/Z is center free.

8. Let G be a connected Lie group, let $\mathfrak{M}'$ be the universal covering manifold of the manifold $\mathfrak{M}$ of G, and let ψ be the projection of $\mathfrak{M}'$ onto $\mathfrak{M}$. $\mathfrak{M}'$ is made into a group G', called the *universal covering group* of G, by defining a multiplication in it, as follows: First, choose one of the points in $\mathfrak{M}'$ that lie over the identity 1 in G, and call it 1' [hence $\psi(1') = 1$; 1' will be the identity in G']. Then, let g' and h' be any two points of $\mathfrak{M}'$, and let $g'(s)$ and $h'(s)$ be curves in $\mathfrak{M}'$ connecting g' and h', respectively, with 1', so that

$g'(0) = h'(0) = 1'$, while $g'(1) = g'$ and $h'(1) = h'$. Let $g(s)$ and $h(s)$ be the projections of $g'(s)$ and $h'(s)$ down into $\mathfrak{M}$; i.e., $g(s) = \psi(g'(s))$ and $h(s) = \psi(h'(s))$. Then, $k(s)$, defined as $= g(s)h(s)$, is a curve in $\mathfrak{M}$ starting at 1. Let $k'(s)$ be the curve in $\mathfrak{M}'$ that results from lifting $k(s)$ up to $\mathfrak{M}'$ in such a way that $k'(0) = 1'$. (See Section 24.2). Then, the product $g'h'$ in $\mathfrak{M}'$ is defined to be $= k'(1)$. Show that this definition is consistent (i.e., independent of the choice of the curves; recall that $\mathfrak{M}'$ is simply connected) and that it makes G' into a Lie group. Show that the projection ψ is a Lie group homomorphism of G' onto G. Show that if g'_0 lies over 1 [i.e., if $\psi(g'_0) = 1$; i.e., if g'_0 is in the kernel of the homomorphism just referred to], then g'_0 commutes with every h' in G'. *Hint*: Choose the defining curves in $\mathfrak{M}'$ in such a way that $h'(s) = 1'$ for $0 \le s \le \frac{1}{2}$ while $g'_0(s) = g'_0$ for $\frac{1}{2} \le s \le 1$.

25.15   Direct and Semidirect Sums of Lie Algebras

The concepts introduced in this section are analogous to the direct and semidirect products of groups, defined in Section 22.9, in connection with the crystallographic space groups. Suppose that a Lie algebra can be decomposed as the direct sum (in the vector space sense) of two subspaces Λ_1 and Λ_2; i.e., any λ in Λ can be uniquely decomposed as $\lambda_1 + \lambda_2$, where λ_1 is in Λ_1 and λ_2 is in Λ_2. Suppose, further, that $[\lambda_1, \lambda_2] = 0$ for every λ_1 in Λ_1 and λ_2 in Λ_2. Then Λ_1 and Λ_2 are both ideals in Λ, and Λ is said to be their direct sum.

Now suppose that Λ is the direct sum (in the vector space sense) of Λ_0 and M, where Λ_0 is an ideal, while M is merely a subalgebra. Then, for λ_1 and λ_2 in Λ_0 and μ_1 and μ_2 in M,

$$[\lambda_1 + \mu_1, \lambda_2 + \mu_2] = [\lambda_1, \lambda_2] + [\mu_1, \lambda_2] + [\lambda_1, \mu_2] + [\mu_1, \mu_2].$$

The first three terms on the right are all in Λ_0 (because Λ_0 is an ideal) and can be rewritten as

$$[\lambda_1, \lambda_2] + \text{Ad}_{\mu_1} \lambda_2 - \text{Ad}_{\mu_2} \lambda_1,$$

while the fourth term, $[\mu_1, \mu_2]$, is in M.

The same result is obtained if we start with Lie algebras Λ_0 and M and then construct Λ, after first noting some of the properties of the linear transformations Ad_μ. Each of them transforms the ideal Λ_0 into itself, and the mapping

$$\mu \to \text{Ad}_\mu$$

is a representation of M on Λ_0, according to Exercise 2 of the preceding section, because, for μ and ν in M,

$$\text{Ad}_{[\mu, \nu]} = \text{Ad}_\mu \text{Ad}_\nu - \text{Ad}_\nu \text{Ad}_\mu.$$

For fixed μ, the transformation Ad_μ is a derivation, that is,

$$\text{Ad}_\mu[\lambda_1, \lambda_2] = [\text{Ad}_\mu \lambda_1, \lambda_2] + [\lambda_1, \text{Ad}_\mu \lambda_2].$$

[In any algebra, a *derivation* is a linear transformation ρ such that $\rho(x \circ y) = \rho(x) \circ y + x \circ \rho(y)$, where the circle denotes the multiplication in the algebra.]

Now let Λ_0 and M be any given Lie algebras, and let the mapping

$$\mu \to \rho(\mu)$$

be a representation of M by derivations in Λ_0. A Lie algebra called a *semidirect sum* of Λ_0 and M, and denoted $\Lambda_0 \oplus_\rho M$, is defined as follows: As a vector space, it is the direct sum of Λ_0 and M, so that its elements are ordered pairs $\{\lambda, \mu\}$, where λ is in Λ_0 and μ is in M, and a Lie product is defined in it as

$$[\{\lambda_1, \mu_1\}, \{\lambda_2, \mu_2\}] = \{[\lambda_1, \lambda_2] + \rho(\mu_1)\lambda_2 - \rho(\mu_2)\lambda_1, [\mu_1, \mu_2]\}.$$

This product is obviously linear in each factor and antisymmetric.

EXERCISE

1. Show that the product just defined satisfies the Jacobi identity.

If $\Lambda = \Lambda_0 \oplus_\rho M$, and if Λ_0 is identified with the set of elements of the form $\{\lambda, 0\}$ and M with the set of elements of the form $\{0, \mu\}$, then $\rho(\mu)$ is just the transformation Ad_μ in Λ, because

$$\mathrm{Ad}_{\{0,\mu\}}\{\lambda, 0\} = [\{0, \mu\}, \{\lambda, 0\}] = \{\rho(\mu)\lambda, 0\}.$$

EXERCISE

2. Let G_0 and H be closed subgroups of a Lie group G, where G_0 is normal. Assume that each g in G has a unique representation as $g_0 h$, where g_0 and h are in G_0 and H, respectively. Let Λ, Λ_0, and M be the Lie algebras of G, G_0, and H, and show that $\Lambda = \Lambda_0 \oplus_\rho M$, where, for any μ in M, $\rho(\mu) = \mathrm{Ad}_\mu$. *Hint*: For λ in Λ_0 and μ in M, let $\{\lambda, \mu\}$ denote $\log(e^\lambda e^\mu)$ and find the Lie product of two such curly bracket expressions by applying the expansion of the CBH formula to

$$e^{\{\lambda_1, \mu_1\}} e^{\{\lambda_2, \mu_2\}},$$

and to each factor separately.

The semidirect sum is a direct sum $\Lambda_0 \oplus M$ if ρ is the trivial homomorphism which maps every μ in M onto the zero transformation; i.e., $\rho(\mu)\lambda = 0$ for all λ. Λ_0 and M are both ideals in $\Lambda_0 \oplus M$.

A fundamental structure theorem, proved at a much later stage of the theory, says that any Lie algebra can be written as a repeated semidirect sum

$$(\ldots((\Lambda_0 \oplus_{\rho_1} \Lambda_1) \oplus_{\rho_2} \Lambda_2) \ldots \oplus_{\rho_k} \Lambda_k)$$

of Lie algebras, each of which is either 1-dimensional or simple; hence, a main objective of the theory is a classification of the simple algebras. The theorem holds for both real and complex Lie algebras; its proof, which is quite deep, is found in Hausner and Schwartz 1968.

25.16 Classification of the Simple Complex Lie Algebras

The relations of the various objects in the theory are indicated in the following schema:

$$
\begin{array}{ccc|ccc}
 & \text{real} & \text{complex} & \begin{array}{c}\text{simple}\\ \text{complex}\end{array} & \begin{array}{c}\text{simple}\\ \text{real}\end{array} & \\
\text{Lie} & \text{Lie} & \text{Lie} & \text{Lie} & \text{Lie} & \text{Lie} \\
\text{group} & \text{algebra} & \text{algebra} & \text{algebra} & \text{algebras} & \text{groups}
\end{array}
$$

Each Lie group determines a unique real Lie algebra, which in turn determines a unique complex Lie algebra, by a process called complexification, described below. The complex case is simpler than the real case, just as in elementary matrix theory, because the complex number system $\mathbb{C}$ is algebraically closed, while $\mathbb{R}$ is not. (It is recalled that a real matrix generally has complex eigenvalues and eigenvectors.) There exists a complete classification of the simple complex Lie algebras into four main series of algebras and five so-called exceptional algebras. The next step is to find all the simple real algebras whose complexification leads to a given complex algebra. This step is carried through in Hausner and Schwartz 1968, where the reader can find a complete classification of the simple real algebras. The result is considerably more elaborate than the classification of the complex algebras, but it is still two steps removed from a classification of the Lie groups; for this, one must first find all possible repeated indirect sums of 1-dimensional and simple algebras, as described at the end of the preceding section, and then find all (say connected) Lie groups that yield a given real Lie algebra.

We shall sketch the development very briefly through the classification of the simple complex algebras. For the algebraic details and the many lemmas needed for the proofs, the reader is referred to Hausner and Schwartz 1968.

As indicated in the preceding section, we are mainly interested in the simple algebras, but, in the analysis of them, certain nonsimple algebras appear, namely the semisimple, solvable, and nilpotent Lie algebras. To define those, we note first that if Λ_1 and Λ_2 are any ideals in a real or complex Lie algebra Λ, then $[\Lambda_1, \Lambda_2]$, defined as the subspace spanned by elements of the form $[\lambda_1, \lambda_2]$, where λ_1 is in Λ_1 and λ_2 is in Λ_2, namely, the subspace

$$[\Lambda_1, \Lambda_2] = \operatorname{span}\{[\lambda_1, \lambda_2] : \lambda_1 \in \Lambda_1, \lambda_2 \in \Lambda_2\}$$

is an ideal contained in both Λ_1 and Λ_2. We then define two descending sequences of ideals in Λ, namely,

$$\Lambda^1 = \Lambda \supset \Lambda^2 \supset \Lambda^3 \supset \cdots$$

and

$$\Lambda^{(1)} = \Lambda \supset \Lambda^{(2)} \supset \Lambda^{(3)} \supset \cdots,$$

inductively by

$$\Lambda^{k+1} = [\Lambda, \Lambda^k], \qquad \Lambda^{(k+1)} = [\Lambda^{(k)}, \Lambda^{(k)}].$$

Λ is said to be *solvable* if $\Lambda^{(k)} = 0$ for some k and *nilpotent* if $\Lambda^k = 0$ for some k. It is easy to see that a nilpotent algebra is solvable; in fact, $\Lambda^{(k)} \subset \Lambda^k$, for all k, by induction on k.

As in Section 25.12, a Lie algebra Λ of more than one dimension is *simple* if it contains no proper nonzero ideals. It is called *semisimple* if it contains no proper nonzero solvable ideals (in which case Λ itself cannot be solvable, so the word "proper" can be omitted in this last definition).

It turns out, at a considerably later stage of the development, that an algebra Λ is semisimple if and only if $\Lambda^2 = \Lambda$ (hence, we require that dim $\Lambda > 1$, for if dim $\Lambda = 1$, then $\Lambda^2 = 0$), also if and only if it can be written as a direct sum of ideals

$$\Lambda = \Lambda_1 \oplus \cdots \oplus \Lambda_k,$$

where each Λ_k is a simple algebra.

If Λ is a real Lie algebra, its *complexification* is defined as the complex Lie algebra $\hat{\Lambda}$ whose elements are formal sums $\lambda + i\mu$, where λ and μ are in Λ, and in which the linear combinations and Lie products are defined in the obvious way; in particular,

$$[\lambda_1 + i\mu_1, \lambda_2 + i\mu_2] = [\lambda_1, \lambda_2] + i[\mu_1, \lambda_2] + i[\lambda_1, \mu_2] - [\mu_1, \mu_2].$$

$\hat{\Lambda}$ is semisimple if and only if Λ is semisimple. If Λ is simple, then its complexification is either simple or is the direct sum of two identical (i.e., isomorphic) simple complex algebras.

Any real or complex Lie algebra Λ contains nilpotent subalgebras (they are, of course, not ideals, if Λ is semisimple); in particular it contains a so-called Cartan subalgebra M, defined below, which is nilpotent. To investigate the structure of Λ, one investigates the structure of M and the relationship between the elements of M and the other elements of Λ. That relationship is described by the operators Ad_μ, $\mu \in M$; Ad_μ transforms an element of Λ into another element of Λ, namely λ into $[\mu, \lambda]$. The mapping $\mu \to \mathrm{Ad}_\mu$ is a representation of M on the vector space Λ; hence the theory starts by considering general representations of solvable and nilpotent Lie algebras.

The study of a representation ρ of an abstract algebra Λ has the merit that while λ in Λ is an abstract object, $\rho(\lambda)$ is a linear transformation in a vector space, and standard linear algebra can be applied; for example, in the complex case, the transformation $\rho(\lambda)$ has at least one eigenvalue and one eigenvector. Also, the Lie product of $\rho(\lambda)$ and $\rho(\mu)$ is simply $\rho(\lambda)\rho(\mu) - \rho(\mu)\rho(\lambda)$.

If ρ is a representation of any Lie algebra Λ on a vector space V, we call $\mathbf{v}$ in V a *weight vector* of ρ if it is a simultaneous eigenvector of all the transformations $\rho(\lambda)$, $\lambda \in \Lambda$, i.e., if

$$\rho(\lambda)\mathbf{v} = \alpha(\lambda)\mathbf{v}, \qquad \forall \lambda \in \Lambda,$$

where $\alpha(\cdot)$ is a numerical-valued function, obviously linear, defined on Λ, called the corresponding *weight* of ρ. A vector $\mathbf{v}$ in V is a *generalized weight vector* of ρ corresponding to the weight $\alpha(\cdot)$ if, for some integer k,

$$(\rho(\lambda) - \alpha(\lambda)I)^k\mathbf{v} = 0, \qquad \forall \lambda \in \Lambda,$$

I being the identity transformation in V. The set of all generalized weight vectors for given $\alpha(\cdot)$ is called the corresponding *weight space* and is denoted by V_α. Thus, weight, weight vector, and weight space correspond to eigenvalue, eigenvector, and algebraic eigenspace, in the case of a single linear transformation. In the latter case, if V is a complex vector space, V is the direct sum of all the algebraic eigenspaces $V_1 \oplus \cdots \oplus V_k$ corresponding to the eigenvalues $z_1, \ldots, z_k$—this corresponds to the fact that any matrix can be put into Jordan normal form. Analogous results hold for weights and weight vectors if the Lie Algebra in question is solvable or nilpotent.

Theorem 1. *If ρ is a representation of a solvable complex Lie algebra M on a vector space V, then ρ has at least one weight vector $\mathbf{v}$ and corresponding weight $\alpha(\cdot)$.*

Under the further assumption that M is nilpotent, we have the following:

Theorem 2. *If ρ is a representation of a nilpotent complex Lie algebra M on a vector space V, then the weight spaces of ρ span V as a direct sum*

$$V = V_{\alpha_1} \oplus \cdots \oplus V_{\alpha_k},$$

where V_{α_j} is the weight space that corresponds to the weight $\alpha_j(\cdot), j = 1, \ldots, k.$

In each case the proof proceeds by induction on the dimension n of M; the ideal M^2 is $\neq M$, and if N is a subspace of M of dimension $n-1$ that contains M^2, then N is a subalgebra (in fact, an ideal) and is solvable in the case of Theorem 1 and is nilpotent in the case of Theorem 2; hence the inductive hypothesis can be applied to N. The induction starts for $n = 1$, in which case there is only one linear transformation $\rho(\lambda)$ involved up to a scalar multiplier, and the statements in the theorems reduce to the corresponding known facts of linear algebra. The algebraic work in the proofs is straightforward, but its quantity is enough to discourage all but the strong at heart.

Next, two important tools in the analysis of a general Lie algebra Λ are introduced: the *symmetric bilinear form* and the notion of a *Cartan subalgebra*. The first is the symmetric form (λ, μ) defined for all λ and μ in Λ by the equation

$$(\lambda, \mu) = \text{tr}(\text{Ad}_\mu \, \text{Ad}_\lambda),$$

where "tr" denotes trace. It is real- or complex-valued, according as Λ is a real or complex Lie algebra, but is not positive definite, except in a special case mentioned below. A basic theorem, *Cartan's Criterion*, says that a real or complex Lie algebra is semisimple if and only if the symmetric bilinear form is nonsingular, which means that for no $\lambda \neq 0$ is $(\lambda, \mu) = 0$ for all μ.

If Λ is any complex Lie algebra and M is a nilpotent subalgebra, we can apply Theorem 2 to the adjoint representation of M on Λ, so that the symbols $\rho(\mu)$ and V are replaced by Ad_μ and Λ. The weights, weight vectors, and weight spaces of this representation are then called *roots, root vectors*, and *root spaces* of M in Λ. If $\alpha(\,\cdot\,)$ is a root, the corresponding root space is denoted by Λ_α; it is a subspace of Λ. From the nilpotence of M it follows that the zero function, $\alpha(\lambda) = 0$ for all λ, is one of the roots, and the corresponding root space, called Λ_0, contains M. If the nilpotent subalgebra M can be so chosen that Λ_0 is $= M$, then M is a *Cartan subalgebra* of Λ. A basic theorem that every complex Lie algebra has a Cartan subalgebra.

It turns out that if Λ is a complex *semisimple* Lie algebra, then (a) the Cartan subalgebra M is commutative, (b) for each $\alpha \neq 0$, the root space Λ_α is one-dimensional, (c) if α is a root, $-\alpha$ is also a root, and (d) if λ and λ' are nonzero vectors in Λ_α and $\Lambda_{-\alpha}$, then $[\lambda, \lambda']$ is a nonzero vector in M, and $(\lambda, \lambda') \neq 0$. We number the nonzero roots $\pm\alpha_1, \pm\alpha_2, \ldots, \pm\alpha_k$; we choose vectors λ_i and λ_{-i} in Λ_{α_i} and $\Lambda_{-\alpha_i}$, so normalized that $(\lambda_i, \mu_{-i}) = 1$, and we call

$$\mu_i = [\lambda_i, \lambda_{-i}], \qquad i = 1, \ldots, k.$$

It can be shown that the vectors μ_i span M.

It follows from (a) and (b) in the preceding paragraph that for a semisimple algebra, only ordinary root vectors (i.e., no generalized ones) appear. For the root vectors λ_α ($\alpha \neq 0$), that follows from the one-dimensionality of Λ_α; and every vector v in $\Lambda_0 = M$ is a root vector, because $\mathrm{Ad}_\mu v = 0$ for all μ in M.

The Cartan subalgebra is not unique, but it can be shown that if M' is any other Cartan subalgebra in Λ, then M and M' have the same dimension, and there is an automorphism of Λ that carries M onto M'; hence, either of them can be used to investigate the structure of Λ.

It is found that the configuration of the vectors μ_i completely determines the Lie algebra. The description of this configuration is greatly simplified by the fortunate fact that if M_r denotes the real vector space consisting of linear combinations of the μ_i with real coefficients, then the natural bilinear from $(\,\cdot\,,\,\cdot\,)$ is real and positive definite in M_r; hence, M_r is a Euclidean space, if $(\cdot, \cdot)$ is taken as the scalar product. It can be shown that the real dimension of M_r is the same as the complex dimension of M, and we call it m. The (complex) dimension of Λ is then $m + 2k$. The length of a vector μ in M_r is $\|\mu\| = (\mu, \mu)^{1/2}$, and the angle between two such vectors is

$$\cos \angle\, \mu, v = \frac{(\mu, v)}{\|\mu\|\,\|v\|}.$$

The *star* in M_r consisting of the vectors μ_i, thought of as radiating out from the origin, has a rather high degree of symmetry and can be described in the following terms: (1) For any given simple algebra Λ, either all the μ_i have the same length, or there are just two lengths, some of the μ_i being short, and the others long. (2) The angle between any two of the vectors is an integer multiple of $30°$ or $45°$. (3) If the angle is $30°$ or $150°$, one vector is long and the other short; the ratio of the lengths is $\sqrt{3}$. If the angle is $45°$ or $135°$, the ratio of the lengths is $\sqrt{2}$. If the angle is $60°$ or $120°$, the two vectors have the same length. (4) The entire star is symmetric with respect to reflection in each hyperplane perpendicular to one of the μ_i. Every minimal star that satisfies these conditions determines a unique simple complex Lie algebra, and different stars determine different algebras. If Λ is merely semisimple and is a direct sum $\Lambda = \Lambda_1 \oplus \cdots \oplus \Lambda_k$ of simple algebras, then M_r is spanned by k mutually orthogonal subspaces, each containing the star of one of the simple algebras.

We now assume that Λ is simple. When M_r is one-dimensional, the star consists of two opposed vectors of equal length, and the algebra Λ, called A_1, has dimension $l = 3$. When M_r is two-dimensional, there are three possible stars, shown in Figure 25.2 together with the designations and dimensions l of the corresponding algebras, which are called A_2, B_2, and G_2. When M_r is three-dimensional, there are again three possible stars, corresponding to algebras called A_3, B_3, and C_3. The star of A_3 consists of six pairs of opposed vectors μ_i and μ_{-i}, all equal in length, and extending from the origin to the midpoints of the edges of a cube; the angles that occur are $60°$, $90°$, $120°$, and $180°$. In the star of the algebra B_3 there are six pairs of long vectors, arranged as for A_3, extending to the edges of a cube, and three pairs of mutually orthogonal short vectors making angles of $45°$ with the nearest long ones, the length ratio being $\sqrt{2}$; the short vectors extend to the midpoints of the faces of the cube referred to. The star of C_3 is the same as that of B_3, but with the long and short vectors interchanged, so that the star fits into a rhombic dodecahedron. The dimension number l is equal to 8, 10, 14, 15, 21, and 21, for the algebras A_2, B_2, G_2, A_3, B_3, and C_3, respectively.

It is of course meaningless to talk about the lengths and directions of the vectors λ_i, because they lie in the complex space Λ, for which $(\cdot, \cdot)$ is not even a Hermitian inner product. What is meaningful is to find the Lie products $[\lambda, \mu]$ for sufficiently many pairs λ, μ so as to determine the structure of Λ. That is best done by means of the models described in the next section.

To determine the possible stars, when M_r is of more than 3 dimensions, one makes use of a device due to the Soviet mathematician E. B. Dynkin. A *simple set* of vectors in the star is a certain set Π of just m of the vectors μ_i (m is always $< 2k$) such that all the vectors of the star can be obtained by repeated additions and subtractions, starting with the vectors of Π, and such that only one set of vectors satisfying the conditions (1)–(4), above, i.e.,

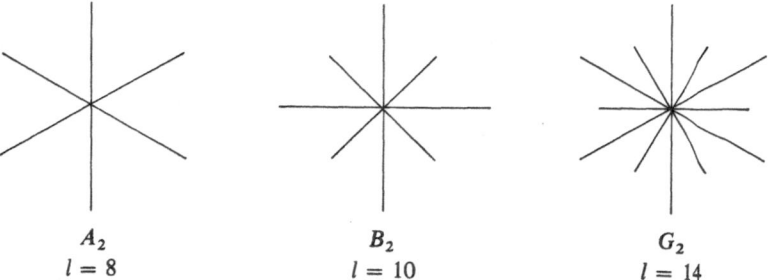

$$A_2 \qquad\qquad B_2 \qquad\qquad G_2$$
$$l = 8 \qquad\qquad l = 10 \qquad\qquad l = 14$$

Figure 25.2 The 2-dimensional stars of simple complex Lie algebras.

only one star, can be obtained in this way. It can be proved that it is always possible to choose a simple set of vectors. Furthermore, although the set Π is not in general unique, if Π' is another simple set, then there is an automorphism of Λ under which M is invariant and Π is carried into Π'; hence it is unimportant which simple set is used. The possible angles between any two vectors of Π are 90°, 120°, 135°, and 150°. A *Dynkin diagram* is a set of m points or small circles on a plane, one for each vector in Π. If the angle between two vectors in Π is 120°, 135°, or 150°, then the corresponding points of the diagram are joined by a single, double, or triple line respectively; if the angle is 90°, the points are not directly connected. If the angle is 135° or 150°, the point corresponding to the shorter vector is indicated by an asterisk. Then, a number of things can be proved about the diagrams of simple complex algebras; for example, a diagram can contain no loops, it is connected, it can contain at most one double or triple line, it can have at most one branching, and so on. In consequence of these rules, it is found that there can be just seven types of Dynkin diagrams, as follows (m is the number of points and is equal to the dimension of M, l is the dimension of Λ):

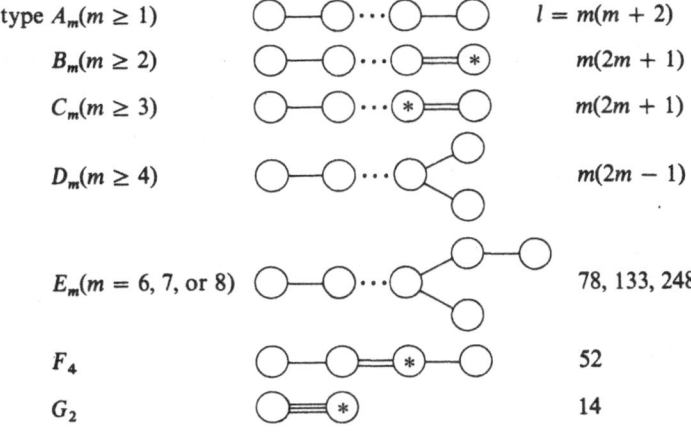

type $A_m(m \geq 1)$		$l = m(m + 2)$
$B_m(m \geq 2)$		$m(2m + 1)$
$C_m(m \geq 3)$		$m(2m + 1)$
$D_m(m \geq 4)$		$m(2m - 1)$
$E_m(m = 6, 7,$ or $8)$		78, 133, 248
F_4		52
G_2		14

Figure 25.3 Dynkin diagrams for the simple complex Lie algebras.

Types A_m, B_m, C_m, and D_m constitute the *regular series*, and the remaining five algebras are called *exceptional*.

25.17 Models of the Simple Complex Lie Algebras

The above classification results from imposing various conditions, arrived at by very lengthy algebraic considerations, which a simple complex Lie algebra must satisfy. To show that there are no further conditions, and hence that all the above algebras actually exist, models of them are constructed. The models of the regular series are algebras of matrices, defined below. We continue to denote the elements of the algebras by the symbols λ, μ, ..., even though other symbols might seem more appropriate for matrices.

1. A_m consists of all $(m + 1) \times (m + 1)$ complex matrices of trace zero. See exercises below. For B_m and D_m it is necessary to introduce the antidiagonal matrix

$$J = p \times p \text{ matrix} \begin{pmatrix} 0 \cdots 1 \\ \vdots \diagup \vdots \\ 1 \cdots 0 \end{pmatrix}.$$

2. B_m consists of all $(2m + 1) \times (2m + 1)$ complex matrices λ such that $\lambda J + J\lambda^T = 0$ $(p = 2m + 1)$.

3. D_m consists of all $2m \times 2m$ complex matrices λ such that $\lambda J + J\lambda^T = 0$ $(p = 2m)$.

    For C_m it is necessary to introduce the $2m \times 2m$ antidiagonal matrix

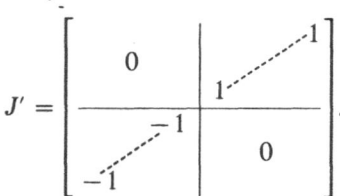

4. C_m consists of all $2m \times 2m$ complex matrices λ such that $\lambda J' + J'\lambda^T = 0$.

The following exercises concern the series A_m. The series B_m, C_m, and D_m are similar. The models of the exceptional algebras are more complicated and are given in Hausner and Schwartz.

EXERCISES

1. Let Λ' be the Lie algebra of $(m + 1) \times (m + 1)$ complex matrices λ, with $[\lambda, \mu] = \lambda\mu - \mu\lambda$. Compute the natural bilinear form

$$(\lambda, \mu) = \operatorname{tr}(\operatorname{Ad}_\lambda \operatorname{Ad}_\mu).$$

[Note that $\operatorname{Ad}_\lambda$ and Ad_μ are linear transformations in an $(m + 1)^2$ dimensional space, namely Λ'.] Show that

$$(\lambda, \mu) = 2((m + 1)\operatorname{tr}(\lambda\mu) - (\operatorname{tr} \lambda)(\operatorname{tr} \mu)).$$

Show that (λ, μ) is singular in Λ' but nonsingular in the subalgebra $\Lambda = A_m$ of matrices of trace zero, so that A_m is semisimple.

2. Let M denote the commutative subalgebra of the Lie algebra Λ of Exercise 1 consisting of diagonal matrices (with trace zero). Consider the adjoint representation of M on $\Lambda: \mu \to \operatorname{Ad}_\mu$, where

$$(\operatorname{Ad}_\mu \lambda)_{rs} = (\mu_{rr} - \mu_{ss})\lambda_{rs} \qquad (r, s = 1, \ldots, n).$$

Consider the roots and root vectors of this representation. Show that the root space Λ_0, which consists of all matrices λ such that $(\mu_{rr} - \mu_{ss})^k \lambda_{rs} = 0$ for some k, for all μ in M, consists also of the diagonal matrices; hence $\Lambda_0 = M$, hence M is a Cartan subalgebra. Show that the other roots $\alpha(\cdot)$ and corresponding root vectors λ_α are obtained by choosing fixed i and k and setting

$$\alpha(\mu) = \mu_{ii} - \mu_{kk},$$
$$\lambda_\alpha = \lambda(i, k),$$

where $\lambda(i, k)$ is the matrix

$$(\lambda(i, k))_{i'k'} = \frac{1}{\sqrt{2(m + 1)}} \delta_{ii'}\delta_{kk'}$$

and that the vector μ_α is the diagonal matrix given by

$$(\mu_\alpha)_{ii} = \frac{1}{2(m + 1)},$$

$$(\mu_\alpha)_{kk} = \frac{-1}{2(m + 1)},$$

$$\text{otherwise} \quad (\mu_\alpha)_{i'k'} = 0.$$

Show that a simple set of roots is the set

$$\alpha_i(\mu) = \mu_{i+1\,i+1} - \mu_{ii}, \qquad i = 1, \ldots, m.$$

Show that the angle between μ_{α_i} and $\mu_{\alpha_{i+1}}$ is 120° and that otherwise the angle between μ_{α_i} and μ_{α_j} is 90°, so that the Dynkin diagram of Λ is as given above for A_m, namely,

(m small circles).

For the classification and models of the simple real Lie algebras, which are needed for a classification of Lie groups, the reader is referred to Hausner and Schwartz. It is recalled that if Λ is a simple real Lie algebra, its complexification $\hat{\Lambda}$ is either simple or is the direct sum of two identical (i.e., isomorphic) simple complex algebras. Hence, to classify the simple real algebras, one must examine each simple complex algebra and then find all simple real algebras from which it can be obtained by complexification.

Given a simple complex algebra $\hat{\Lambda}$, one possible choice of Λ consists of the elements of $\hat{\Lambda}$ but regarded as a linear space over the real field $\mathbb{R}$, rather than $\mathbb{C}$, as the field of scalars, but that is not the only possibility. Other possibilities are found by considering the so-called conjugations in $\hat{\Lambda}$. A *conjugation* in a complex Lie algebra is an antilinear mapping C [that is, $C(a\lambda + b\mu) = \bar{a}C\lambda + \bar{b}C\mu$], which preserves Lie products (that is, $C[\lambda, \mu] = [C\lambda, C\mu]$), and whose square is the identity mapping [that is, $C(C\lambda) = \lambda$]. The set of all λ in $\hat{\Lambda}$ such that $C\lambda = \lambda$, with $\mathbb{R}$ as the field of scalars, is a simple real Lie algebra. A complete analysis of the conjugations in the simple complex Lie algebras, and the enumeration of the resulting simple real algebras, is given in Hausner and Schwartz. As an example, if $\hat{\Lambda}$ is the simple complex algebra A_1 of 2×2 matrices of trace zero, there are three corresponding simple real algebras, namely A_1 itself (with $\mathbb{R}$ as the field of scalars) and

$$RA_1 = \{2 \times 2 \text{ real matrices of trace zero}\}$$

and

$$QA_1 = \{2 \times 2 \text{ matrices of the form } iH, \text{ where } H \text{ is Hermitian and of trace zero}\}.$$

We mention in passing that some of the corresponding Lie groups are SL(2, $\mathbb{C}$), $\mathscr{L}_p$, SL(2, $\mathbb{R}$), SU(2), and SO(3).

Corresponding to each simple complex algebra A_m, for $m > 1$, there are $4 + [(m + 1)/2]$ simple real algebras, where [ ] denotes integer part.

Corresponding to the exceptional algebra G_2 there are three real algebras, called G_2 (over $\mathbb{R}$), $HG_2^{(3)}$, and $HG_3^{(1)}$.

25.18 Note on Lie Groups and Lie Algebras in Physics

The role of the rotation group SO(3) as a symmetry group in quantum mechanics was discussed in Chapter 22. In calculations, the corresponding Lie algebra usually appears, rather than the group itself. The Lie algebra of SO(3), which is of course the same as the Lie algebra of SU(2), the universal covering group of SO(3), may be realized either as the Lie algebra of 3×3 real skew symmetric matrices or that of 2×2 skew Hermitian matrices of trace zero, called QA_1 in the preceding section. It may also be realized as a Lie algebra of operators in the Hilbert space $\mathfrak{H}$ of states of a physical system. If $\mathfrak{H}$ is taken as the space $L^2(\mathbb{R}^3)$ of wave functions $\psi(\mathbf{x})$ of a spin-zero nonrelativistic particle, the mappings

$$g: \psi(\mathbf{x}) \rightarrow \psi(g^{-1}\mathbf{x}) \qquad [g \in SO(3)]$$

form a representation of SO(3) on $\mathfrak{H}$, as in Section 20.9. The so-called infinitesimal operators of this representation were discussed in that section and are

$$L_j = x^l \frac{\partial}{\partial x^k} - x^k \frac{\partial}{\partial x^l} \qquad (jkl = 123, 231, 312).$$

The corresponding self-adjoint operators $i\hbar L_j$ are the angular momentum observables. The linear combinations of the L_j with real coefficients give a realization of the Lie algebra Λ of SO(3).

In elementary particle physics, the relevant symmetry group is often not known, owing to the lack of a complete theory of elementary particles, but various Lie algebras Λ are often found to play a role, on empirical grounds.

Confusion can arise because the word "group" is often used for a Lie algebra. In particular, one finds references to the "group G_2." According to Section 25.16, G_2 is a Lie algebra, and in fact a complex Lie algebra. The corresponding group in the physical theory is presumably a group whose Lie algebra is one of the three real Lie algebras, mentioned in the preceding section, whose complexification gives G_2. The nearest that one can come to identifying a unique group in such a case is this: Of the real Lie algebras Λ whose complexification is a given algebra $\hat{\Lambda}$, just one is the Lie algebra of a compact group (or possibly of several compact groups), and of the Lie groups G having a given real algebra Λ, just one is simply connected. Hence, in particular, there is a unique compact simply connected Lie group associated with the algebra G_2. Since many of the symmetry groups of physics are neither compact nor simply connected, the identification of the group to be associated with a given Lie algebra in particle physics must presumably await further theoretical developments.

The algebra G_2 has also been used in the study of atoms having partially filled f shells; see Racah 1949. In that case the theory is sufficiently complete that presumably there is a clearly defined corresponding group, which, according to Racah (see also Behrends, Dreitlein, Fronsdal, and Lee 1962) is a subgroup of SO(7).

Appendix to Chapter 25—Two Nonlinear Lie Groups

In this appendix, we give two examples of Lie groups that are not linear, that is, have no faithful finite-dimensional representations, hence cannot be realized as groups of matrices.

For the first example, let G denote the so-called *Heisenberg group*

$$G = \left\{ \begin{pmatrix} 1 & x & z \\ 0 & 1 & y \\ 0 & 0 & 1 \end{pmatrix} : x, y, z \in \mathbb{R} \right\}.$$

If the above matrix is denoted by $g_{x,y,z}$, direct calculation shows that

$$
\begin{aligned}
g_{x,0,0}^{-1} &= g_{-x,0,0}, \\
g_{0,y,0}^{-1} &= g_{0,-y,0}, \qquad\qquad (25.\text{A-}1)\\
g_{x,0,0}\, g_{0,y,0}\, g_{x,0,0}^{-1}\, g_{0,y,0}^{-1} &= g_{0,0,xy}.
\end{aligned}
$$

It follows that if ρ is any representation of G, then $\rho(g_{0,0,z})$ is a unimodular matrix for every z, because $\det(\rho(g_{0,0,xy}))$ is equal to

$$\det \rho(g_{x,0,0})\det \rho(g_{0,y,0})\det \rho(g_{x,0,0})^{-1} \det \rho(g_{0,y,0})^{-1} = 1.$$

Now let G_0 denote the normal subgroup

$$G_0 = \left\{ \begin{pmatrix} 1 & 0 & n \\ 0 & 1 & 0 \\ 0 & 0 & 1 \end{pmatrix} : n \text{ integer} \right\}.$$

It will be shown that every finite-dimensional representation of the factor group G/G_0 is nonfaithful; hence G/G_0 is not a linear group. Let $\bar{g}_{x,y,z}$, where $0 \leq z < 1$, denote the element of G/G_0 (a coset in G) that contains $g_{x,y,z}$. That is, $\bar{g}_{x,y,z}$ is the infinite set

$$\bar{g}_{x,y,z} = \left\{ \begin{pmatrix} 1 & x & z+n \\ 0 & 1 & y \\ 0 & 0 & 1 \end{pmatrix} : n = 0, \pm 1, \pm 2, \ldots \right\}$$

of 3×3 matrices. It is easily seen that, in analogy with (25.A-1),

$$\bar{g}_{x,0,0}\, \bar{g}_{0,y,0}\, \bar{g}_{x,0,0}^{-1}\, \bar{g}_{0,y,0}^{-1} = \bar{g}_{0,0,z},$$

where $z \equiv xy \pmod 1$. Hence, as above, if ρ is any representation of G/G_0, then $\det \rho(\bar{g}) = 1$ for every $\bar{g}$ in the subgroup

$$H = \{\bar{g}_{0,0,z} : 0 \leq z < 1\} < \frac{G}{G_0}.$$

But H is isomorphic to SO(2), with $2\pi z$ playing the role of θ; hence H is compact and Abelian. According to the general theory of representations given in Sections 21.1–21.4, every representation of a compact group is equivalent to a unitary representation, and every unitary representation of an Abelian group is completely reducible as the direct sum of 1-dimensional representations. Hence, if ρ is any m-dimensional representation of G/G_0, then, relative to a suitable basis in V^m, the representation ρ of the subgroup $H \cong SO(2)$ has the diagonal form

$$\rho(\bar{g}_{0,0,z}) = \begin{pmatrix} e^{2\pi i n_1 z} & & (0) \\ & \ddots & \\ (0) & & e^{2\pi i n_m z} \end{pmatrix}.$$

Each of the 1-dimensional representations given by the diagonal elements of this matrix has determinant $= 1$ for all $\bar{g}$ in H; hence all the integers $n_1, \ldots, n_m$ are zero. That is, all elements of the subgroup H are presented by the $m \times m$ unit matrix; hence the representation ρ of G/G_0 is not faithful.

The second example is less elementary (at least, the present discussion of it is), because use is made of a moderately deep theorem on representations of Lie algebras. It will be shown that if $\tilde{G}$ is the universal covering group of $SL(2, \mathbb{R})$ (the group of 2×2 real matrices with unit determinant), then $\tilde{G}$ has no faithful finite-dimensional representation.

A canonical form for any matrix M in $SL(2, \mathbb{R})$ is now obtained. Let R be the rotation [an element of the subgroup $SO(2)$] that transforms the first column of M into a vector having components $a, 0$, where $a > 0$. Then RM is of the form

$$RM = \begin{pmatrix} a & b \\ 0 & c \end{pmatrix}, \qquad ac = 1.$$

We write $a = e^x$, $c = e^{-x}$. Then,

$$\begin{pmatrix} e^{-x} & 0 \\ 0 & e^x \end{pmatrix} RM = \begin{pmatrix} 1 & y \\ 0 & 1 \end{pmatrix},$$

for some real y. Hence, if the rotation R^{-1} is R_θ, we have

$$M = \begin{pmatrix} \cos\theta & -\sin\theta \\ \sin\theta & \cos\theta \end{pmatrix} \begin{pmatrix} e^x & 0 \\ 0 & e^{-x} \end{pmatrix} \begin{pmatrix} 1 & y \\ 0 & 1 \end{pmatrix}. \tag{25.A-2}$$

This is the desired canonical form. It follows that the manifold of $SL(2, \mathbb{R})$ is the Cartesian product of a circle and two lines, $C^1 \times \mathbb{R}^2$. Since the universal covering of C^1 is $\mathbb{R}$, the manifold of $\tilde{G}$ is $\mathbb{R}^3$.

The Lie algebra Λ of $SL(2, \mathbb{R})$ has as basis the matrices

$$L_1 = \left. \frac{\partial M}{\partial \theta} \right|_0 = \begin{pmatrix} 0 & -1 \\ 1 & 0 \end{pmatrix},$$

$$L_2 = \left. \frac{\partial M}{\partial x} \right|_0 = \begin{pmatrix} 1 & 0 \\ 0 & -1 \end{pmatrix},$$

$$L_3 = \left. \frac{\partial M}{\partial y} \right|_0 = \begin{pmatrix} 0 & 1 \\ 0 & 0 \end{pmatrix},$$

where, in each case, the subscript zero indicates that the matrix in question is evaluated at $\theta = x = y = 0$. A direct calculation shows that

$$[L_1, L_2] = 2L_1 + 4L_3,$$
$$[L_2, L_3] = 2L_3, \qquad\qquad (25.A-3)$$
$$[L_3, L_1] = L_2.$$

These equations can be solved for L_1, L_2, and L_3; i.e., the Lie products on the left also form a basis for Λ. It follows from the definition of the Lie product in terms of commutators in the group that the group is generated by commutators, i.e., by elements of the form $ghg^{-1}h^{-1}$. By the argument used in the first example, it then follows that if ρ is any representation, then $\det \rho(g) = 1$ for all g in SL(2, $\mathbb{R}$).

Since $\tilde{G}$ and SL(2, $\mathbb{R}$) are isomorphic in a neighborhood of the identity element, they have identical Lie algebras, and Λ is also (in the sense of isomorphism) the Lie algebra of $\tilde{G}$. Hence, if ρ is any representation of $\tilde{G}$, it follows that $\det \rho(\tilde{g}) = 1$ for all $\tilde{g}$ in $\tilde{G}$. That is, any representation of $\tilde{G}$ is unimodular.

It can be shown that the Lie algebra Λ is simple. Namely, if Λ has an ideal J that contains a nonzero vector $\lambda = aL_1 + bL_2 + cL_3$, then J also contains the three vectors $[L_j, \lambda]$ and the nine vectors $[L_k, [L_j, \lambda]]$. A direct calculation, starting with (25.A-3), shows that L_1, L_2, and L_3 can be expressed as suitable linear combinations of those 13 vectors (it is not necessary to go to higher Lie products); hence J coincides with Λ, i.e., Λ is a simple Lie algebra.

Let $g_{\theta, x, y}$ denote the element (25.A-2) of SL(2, $\mathbb{R}$), where $0 \le \theta \le 2\pi$. Then θ, x, y with θ unrestricted, can be taken as coordinates of elements $\tilde{g}_{\theta, x, y}$ in $\tilde{G}$ in such a way that in the covering of SL(2, $\mathbb{R}$) by $\tilde{G}$, the element $\tilde{g}_{\theta', x, y}$ of $\tilde{G}$ lies over the element $g_{\theta, x, y}$ of SL(2, $\mathbb{R}$) if $\theta' \equiv \theta \pmod{2\pi}$.

Now let ρ be a representation of $\tilde{G}$ on an m-dimensional vector space $V^m = \mathbb{C}^m$. It will be shown that ρ is nonfaithful. A representation of Λ, which will also be called simply ρ, is induced in the usual way:

$$\rho(L_1) = \frac{\partial}{\partial \theta} \rho(\tilde{g}_{\theta, x, y})|_{\theta = x = y = 0}, \quad \text{etc.}$$

According to Hausner and Schwartz, p. 143, Theorem 2, a representation of a (real or complex) *simple* Lie algebra Λ is completely reducible; i.e., V^m can be written as a direct sum $V^{k_1} \oplus V^{k_2} \oplus \cdots$ of invariant subspaces on each of which ρ is irreducible ($\Sigma k_j = m$). The group $\tilde{G}$ is generated by elements of the form e^λ ($\lambda \in \Lambda$), and $\rho(e^\lambda)$ is $= e^{\rho(\lambda)}$; hence the representation ρ of $\tilde{G}$ is also completely reducible. Specifically, each of the subspaces V^{k_j} of V^m is invariant under $\rho(\tilde{g})$, for all $\tilde{g}$ in $\tilde{G}$, and the restriction ρ_j of ρ to V^{k_j} is irreducible. Let $\tilde{g}_1$ denote the element $\tilde{g}_{2\pi, 0, 0}$; $\tilde{g}_1$ and all its powers lie over the identity element of SL(2, $\mathbb{R}$), hence commute with all $\tilde{g}$ in $\tilde{G}$; hence, by Schur's lemma, $\rho_j(\tilde{g}_1) = \lambda I$, where I is the $k_j \times k_j$ unit matrix. Since every representation is unimodular, $\det \rho_j(\tilde{g}) = 1$ for all $\tilde{g}$; hence $\lambda^{k_j} = 1$. This is true for each j; hence there is some power $\tilde{g}_1^l$ of $\tilde{g}_1$ (e.g., $l = \prod_{(j)} k_j$) such that $\rho(\tilde{g}_1^l) = \rho(\tilde{g}_1)^l = I$ ($= m \times m$ unit matrix). But $\tilde{g}_1^l$ is not the identity of $\tilde{G}$; hence ρ is nonfaithful.

CHAPTER 26

Metric and Geodesics on a Manifold

Scalar, vector, and tensor fields; Lie brackets; covariant and contravariant vectors; transformation laws; inner and outer multiplication; contraction; quotient law; derivations; metric tensor; definite and indefinite metric; Riemannian and pseudo-Riemannian manifolds; raising and lowering of indices; geodesics; Euler variational equation; natural, affine, or preferred parameter; Christoffel three-index symbols; spacelike, null, and timelike geodesics; initial-value and two-point problems of geodesics; Volterra integral equations; Picard iterations; Whitehead's theorem; continuation of geodesics; affinely connected manifolds; Riemannian and pseudo-Riemannian covering manifolds.

Prerequisite: Elementary theory of manifolds (Chapters 23 and 24).

A manifold, as defined in Chapter 23, is a thing characterized just by its local topology: It is a locally n-dimensional space in which the Hausdorff separation axiom holds. In this chapter and the next two, a manifold is made into a geometric structure by introducing further notions such as geodesics (a geodesic is the analogue of a straight line in Euclidean geometry), lengths, angles, and so on. The most fundamental notion is *geodesic*, which is derived, in the main geometries of interest to physics, either from a metric or from an affine connection; we start with a metric, because of its similarity with distance in familiar Euclidean geometry.

Roughly speaking, geometric properties contrast with the global topological properties of a given manifold in that the latter are expressed in terms of integer-valued or discrete quantities like the number of components, the fundamental group, and higher homotopy groups, whereas geometry involves continuous real quantities like lengths and angles and the natural parameter along a geodesic (explained in Sections 26.6, 26.7, and 26.12).

In Riemannian geometry, one starts with a metric differential form $ds^2 = \sum g_{jk}\, dx^j\, dx^k$ in a given coordinate system, from which geometric ideas are derived. A simple familiar example of a nonplane 2-dimensional geometry is the geometry on the unit 2-sphere, where $ds^2 = d\theta^2 + \sin^2\theta\, d\varphi^2$ in terms of the polar angles θ and φ. The geodesic (shortest curve) between two points is an arc of a great circle, and so on. Various theorems of spherical geometry follow; for example, the sum of the angles of a triangle (whose sides are geodesics) exceeds π by an amount equal to the area of the triangle.

Although a general Riemannian manifold can in principle be regarded as embedded in some Euclidean space E^N of higher dimension, just as the 2-sphere can be regarded as the surface of the unit ball in E^3, it is easier and more appropriate to use intrinsic coordinates for the study of the intrinsic geometric concepts determined by the metric.

If the metric form on a manifold $\mathfrak{M}$ is positive definite (see Section 26.4), then $\mathfrak{M}$ is called Riemannian; if it is indefinite, $\mathfrak{M}$ is called pseudo-Riemannian; the latter case appears in general relativity. As will be seen, there are rather profound differences between the two kinds of manifold. It will be assumed that $\mathfrak{M}$ is connected and of class C^k, where k is large enough to accommodate any derivatives that appear in the discussion—it suffices to take $k = 4$ in this chapter and the next two. In relativity, it cannot be assumed that $\mathfrak{M}$ is of class C^∞, because the metric is determined by the distribution of matter, which is not necessarily infinitely smooth, and because the Einstein field equations, since they are hyperbolic, can propagate discontinuities of various derivatives of the $g_{\mu\nu}$ in the form of gravitational waves.

26.1 Scalar and Vector Fields on a Manifold

If $x^1, \ldots, x^n$ are Cartesian coordinates in a Euclidean space, a vector field in the space is described by n functions $v^i(x^1, \ldots, x^n)$, $i = 1, \ldots, n$. If the Cartesian coordinates are transformed by rotating the axes, then the components $v^1, \ldots, v^n$ of the vectors are transformed by the same rotation matrix by which the coordinates $x^1, \ldots, x^n$ are transformed.

If curvilinear coordinates are introduced, it becomes necessary to distinguish two kinds of vectors, covariant and contravariant, which have different physical or mathematical origins. (Still further kinds, the so-called vector densities of various order, are sometimes introduced; they are often convenient, but not necessary.)

First, a function $f(P)$ on a manifold $\mathfrak{M}$ (real-valued, unless otherwise specified) is called a *scalar* (*field*) on $\mathfrak{M}$. When $f(P)$ is given for all points P in $\mathfrak{M}$, then, as in Section 23.5, a function $\hat{f}(x^1, \ldots, x^n) = \hat{f}(\mathbf{x})$ is associated with any given coordinate chart $\{\mathfrak{U}, \varphi, N\}$ by the equations

$$\hat{f}(\mathbf{x}) = f(P), \qquad \mathbf{x} = \varphi(P), \quad \text{for all } P \text{ in } \mathfrak{U}. \tag{26.1-1}$$

This collection of functions $\hat{f}(\cdots)$, one associated with each chart, can be regarded as determining the scalar field $f(P)$. If $\{\mathfrak{U}', \varphi', N'\}$ is a second coordinate chart, the relation between the corresponding functions $\hat{f}$ and $\hat{f}'$ in the overlap of the two charts is simply

$$\hat{f}(\mathbf{x}) = \hat{f}'(\mathbf{x}'), \tag{26.1-2}$$

where

$$\mathbf{x} = \varphi(P) \quad \text{and} \quad \mathbf{x}' = \varphi'(P). \tag{26.1-3}$$

The interpretation of (26.1-2) is that it becomes an identity in $x^1, \ldots, x^n$ if the x'^i in the right member are expressed in terms of the x^i, or an identity in $x'^1, \ldots, x'^n$ if the x^i in the left member are expressed in terms of the x'^i; the values of $\hat{f}$ and $\hat{f}'$ are the same at a given point P in $\mathfrak{M}$.

While a scalar field is a collection of functions $\hat{f}(\cdots)$, one associated with each chart, a vector field or a tensor field is a collection of *sets* of functions, one set associated with each chart; in the overlap of two charts, the two corresponding sets of functions are related by a transformation law which generalizes the law (26.1-2) for scalars.

As a first example of a vector field, consider the gradient of the scalar $f(P)$, which is assumed to be of class C^1. In the chart $\{\mathfrak{U}, \varphi, N\}$, this consists of the partial derivatives of the function (26.1-1), namely,

$$v_i(x^1, \ldots, x^n) = \frac{\partial}{\partial x^i} \, \hat{f}(x^1, \ldots, x^n), \tag{26.1-4}$$

or, more concisely,

$$v_i(\mathbf{x}) = \frac{\partial}{\partial x^i} \, \hat{f}(\mathbf{x}) \qquad (i = 1, \ldots, n). \tag{26.1-5}$$

If the relation established by the equations (26.1-3) in the overlap of two charts is described by writing

$$x^i = \hat{x}^i(x'^1, \ldots, x'^n), \tag{26.1-6}$$

$$x'^i = \hat{x}'^i(x^1, \ldots, x^n), \tag{26.1-7}$$

as in Section 23.2, and if, in the second chart, one writes correspondingly

$$v_i'(\mathbf{x}') = \frac{\partial}{\partial x'^i} \, \hat{f}(\mathbf{x}'), \tag{26.1-8}$$

then the relation between the two sets of functions $\{v_i\}$ and $\{v_i'\}$ in the overlap of the two coordinate systems is

$$v_i'(\mathbf{x}') = \frac{\partial \hat{x}^k}{\partial x'^i} (\mathbf{x}') v_k(\mathbf{x}). \tag{26.1-9}$$

(Here, the summation convention is used, according to which the right member is understood to be summed for $k = 1, 2, \ldots, n$; k is called a *dummy index*.) As in (26.1-1), the interpretation of (26.1-9) is that it is an identity if the x^i are expressed in terms of the x'^i, or the x'^i in terms of the x^i.

Henceforth, the circumflex will be dropped; then, equation (26.1-9) is abbreviated as

$$v_i' = \frac{\partial x^k}{\partial x'^i} v_k, \tag{26.1-10}$$

which is the transformation law for a covariant vector.

The symbol x^k may stand for a variable or for a function; in (26.1-10) it stands for a function, and one must look at the denominator of the partial derivative to learn what the independent variables are; if a prime appears, the independent variables are $x'^1, \ldots, x'^n$, while if a double prime appears, the independent variables are $x''^1, \ldots, x''^n$, etc. These conventions are standard in this subject.

A *covariant vector field* on $\mathfrak{M}$ is defined as a collection of sets $\{v_i\}$ of n functions each, one set associated with each chart in $\mathfrak{M}$, such that the transformation law (26.1-10) holds for any two such sets in the overlap of the corresponding charts.

Comments. (1) A vector field is not necessarily the gradient of a scalar, as was the case in the foregoing example. (2) The two charts may cover exactly the same portion $\mathfrak{U}$ of $\mathfrak{M}$, in which case the transformation law (26.1-10) refers to a "change of independent variables," in the ordinary sense.

The transformation law is *transitive*: If (26.1-10) is followed by another transformation from the coordinates x'^j to coordinates x''^j, then v''_j and v_j have the correct relation, because

$$v''_j = \frac{\partial x'^k}{\partial x''^j} v'_k = \frac{\partial x'^k}{\partial x''^j} \frac{\partial x^l}{\partial x'^k} v_l = \frac{\partial x^l}{\partial x''^j} v_l, \tag{26.1-11}$$

in the overlap of the three charts.

Next, to provide an example of a so-called contravariant vector field, we consider a fluid flow in $\mathfrak{M}$. If a fluid particle is at the point $P(t)$ at time t, then, in a suitable chart $\{\mathfrak{U}, \varphi, N\}$ its coordinates are $x^k(t)$, where $\mathbf{x}(t) = \varphi(P(t))$, and the quantities

$$v^k(t) = \frac{d}{dt} x^k(t), \qquad k = 1, \ldots, n,$$

are called its *generalized velocity components*. [They are the Cartesian velocity components of the representing point $\mathbf{x}(t)$ in the coordinate space $\mathbb{R}^n$.] If $x'^k(t)$ and $v'^k(t)$ are the corresponding coordinates and velocity components relative to a second chart $\{\mathfrak{U}', \varphi', N'\}$, then

$$v'^k(t) = \frac{\partial \hat{x}'^k}{\partial x^j} v^j(t).$$

If, to describe the flow of an entire fluid, not just one particle, the velocity components of the fluid particle which is at point $\mathbf{x}$ at some instant t are called $v^i(\mathbf{x})$, then the transformation law is

$$v'^k(\mathbf{x}') = \frac{\partial \hat{x}'^k}{\partial x^j} (\mathbf{x}) v^j(\mathbf{x}).$$

Like (26.1-1) and (26.1-9), this is an identity in the overlap of the charts if both sides are expressed in terms of the x^i or both in terms of the x'^i. Again, the circumflex will be dropped.

A *contravariant vector* field on $\mathfrak{M}$ is defined as a collection of sets $\{v^i\}$ of n functions each, one set associated with each chart, such that the transformation law

$$v'^k = \frac{\partial x'^k}{\partial x^j} v^j \tag{26.1-12}$$

holds for any two such sets in the overlap of the corresponding charts. Contrast this with (26.1-10) by noting where the prime appears in the derivative.

This transformation law is also transitive.

Note. In Riemannian and pseudo-Riemannian spaces (including Euclidean spaces), any vector can be represented in either covariant or contravariant form; the formulas for the raising and lowering of indices by the metric tensor g_{jk} are given in Section 26.5, below. In some circumstances, however, for example in Weyl's form of the unified field theory, distances and vector magnitudes are only relative, and there is no metric tensor, but only a so-called affine connection (see Section 26.12). In those circumstances, covariant and contravariant vectors are essentially different entities. Note finally that the coordinates $x^1, \ldots, x^n$ are *not* the components of a vector, because they do not transform according to the law (26.1-12), except under homogeneous linear transformations.

When compatible coordinate charts are added to or deleted from the manifold, as described in Chapter 23, it is understood that corresponding sets of vector components $\{v_j\}$ or $\{v^j\}$ are also added or deleted in accordance with the transformation laws (26.1-10, 12). Properties of the vector fields that are invariant under such additions and deletions are regarded as intrinsic properties. In this sense, the concept of a covariant or contravariant vector field is a coordinate-free concept.

EXERCISES

1. Let u^j and v^j be smooth contravariant vector fields, and call

$$w^j = u^k \frac{\partial}{\partial x^k} v^j - v^k \frac{\partial}{\partial x^k} u^j. \tag{26.1-13}$$

Show that the quantities $\{w^j\}_1^n$ transform according to the law (26.1-12), hence constitute another vector field. We write

$$\mathbf{w} = [\mathbf{u}, \mathbf{v}],$$

and we call $\mathbf{w}$ the *Lie bracket* of $\mathbf{u}$ and $\mathbf{v}$. Clearly $[\mathbf{v}, \mathbf{u}] = -[\mathbf{u}, \mathbf{v}]$. Show that if u^j, v^j, and w^j are any given smooth vector fields, then

$$[[\mathbf{u}, \mathbf{v}], \mathbf{w}] + [[\mathbf{v}, \mathbf{w}], \mathbf{u}] + [[\mathbf{w}, \mathbf{u}], \mathbf{v}] = 0 \qquad \text{(Jacobi identity)}.$$

It follows that if we start two or more C^∞ vector fields in a C^∞ manifold, and form all possible vector fields by repeated construction of linear combinations (with constant coefficients) and Lie brackets, the result is a Lie algebra (possibly infinite-dimensional).

2.   Let u^j and v^j be smooth contravariant vector fields whose Lie product is $=0$, so that

$$u^k \frac{\partial}{\partial x^k} v^j = v^k \frac{\partial}{\partial x^k} u^j.$$

It is to be shown that if a point Q in the manifold is obtained by starting at a point P and following a solution curve of the field u^j, i.e., of the system

$$\frac{d}{dt} x^j = u^j \qquad (j = 1, \ldots, n),$$

for a time t_1 and then following a solution curve of the field v^j for a time t_2, the same point Q can be reached by first following a solution curve of the field v^j for some time t_3 and then following one of the field u^j for some time t_4. To this end, let a surface $x^j(s, t)$ in the manifold be defined by the initial-value problems

$$\frac{\partial}{\partial s} x^j(s, 0) = u^j(\mathbf{x}(s, 0)),$$

$$x^j(0, 0) \quad \text{given} \quad (\text{point } P),$$

and

$$\frac{\partial}{\partial t} x^j(s, t) = v^j(\mathbf{x}(s, t)),$$

$$x^j(s, 0) \quad \text{given} \quad (\text{by the preceding problem}).$$

See Figure 26.1.

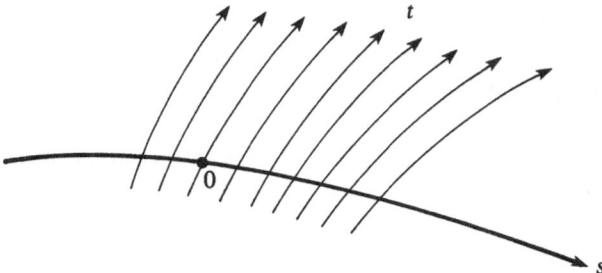

Figure 26.1

Show that then the equation

$$\frac{\partial}{\partial s} x^j(s, t) = u^j(\mathbf{x}(s, t))$$

holds also for $t \neq 0$, by showing that the two sides of this equation satisfy identical initial value problems with respect to t, namely, the same "ordinary" differential equations (s fixed)

$$\frac{\partial}{\partial t} \left(\frac{\partial}{\partial s} x^j(s, t) \right) = \frac{\partial v^j}{\partial x^k} (\mathbf{x}(s, t)) \left(\frac{\partial}{\partial s} x^k(s, t) \right),$$

that is,

$$\frac{\partial}{\partial t}(u^j(\mathbf{x}(s, t))) = \frac{\partial v^j}{\partial x^k}(\mathbf{x}(s, t))(u^k(\mathbf{x}(s, t)))$$

and the same initial condition, because

$$\frac{\partial}{\partial s} x^j(s, 0) = u^j(\mathbf{x}(s, 0)).$$

The surface $x^j(s, t)$ consists of points that can be reached from P by zig-zag paths, each segment of which is either a piece of solution curve of the field u^j or one of the field v^j.

26.2 Tensor Fields

Contravariant tensors T^{jk}, T^{jkl}, etc. of arbitrary rank (the *rank* is the number of indices) transform according to the law

$$T'^{jk\cdots} = \frac{\partial x'^j}{\partial x^r}\frac{\partial x'^k}{\partial x^s}\cdots T^{rs\cdots}. \tag{26.2-1}$$

The summation convention applies to all the repeated indices $r, s, \ldots$ on the right; a multiple sum results. Covariant tensors T_{jk}, T_{jkl}, etc. transform according to the law

$$T'_{jk\cdots} = \frac{\partial x^r}{\partial x'^j}\frac{\partial x^s}{\partial x'^k}\cdots T_{rs\cdots}, \tag{26.2-2}$$

and mixed tensors according to the law illustrated by the example

$$T'^{jk}{}_l{}^m = \frac{\partial x'^j}{\partial x^r}\frac{\partial x'^k}{\partial x^s}\frac{\partial x^t}{\partial x'^l}\frac{\partial x'^m}{\partial x^u} T^{rs}{}_t{}^u. \tag{26.2-3}$$

All these transformation laws are transitive. The last tensor is said to have *contravariant rank* 3 and *covariant rank* 1.

From the transformation laws, it is seen that if u^j, v^j are contravariant vectors and w_j, z_j are covariant vectors, then the quantities defined by

$$T^{jk} = u^j v^k, \qquad T_{jk} = w_j z_k, \qquad T^j{}_k = u^j w_k, \qquad T^{jk}{}_l = u^j v^k w_l,$$

etc. are tensors of the indicated kind. More generally, if

$$T^{j_1 j_2 \cdots}{}_{k_1 k_2 \cdots}{}^{l_1 l_2 \cdots}\cdots \quad \text{and} \quad S^{r_1 r_2 \cdots}{}_{s_1 s_2 \cdots}{}^{t_1 t_2 \cdots}\cdots$$

are any two tensors, then the products

$$(T^{j_1 j_2 \cdots}{}_{k_1 k_2 \cdots}{}^{l_1 l_2 \cdots}\cdots)(S^{r_1 r_2 \cdots}{}_{s_1 s_2 \cdots}{}^{t_1 t_2 \cdots}\cdots)$$

are the components of a tensor

$$R^{j_1 j_2 \cdots}{}_{k_1 k_2 \cdots}{}^{l_1 l_2 \cdots}\cdots{}^{r_1 r_2 \cdots}{}_{s_1 s_2 \cdots}{}^{t_1 t_2 \cdots}\cdots,$$

whose contravariant and covariant ranks are the sums of the corresponding ranks of T and S. The process just described is called *outer multiplication* of vectors and tensors.

If a tensor carries the same symbol for a superscript as for a subscript, then the summation convention applies to that symbol, and the result is a tensor of lower rank; e.g., given a tensor $R^j{}_{klm}$, a tensor R_{kl} can be defined as $R_{kl} = R^j{}_{klj}$. Similarly, from the tensor $S^{jk}{}_{lm}$ there can be formed the scalars $S^{jk}{}_{jk}$ and $S^{jk}{}_{kj}$. This process is called *contraction*. Outer multiplication followed by contraction is called *inner multiplication*; for example, if v^j and w_j are vectors, then $v^j w_j$ is a scalar.

An easily verified converse to the last result is the *quotient law* which says that if a collection of sets of quantities $\{v^i\}$ is given, one set associated with each coordinate chart that contains some point P_0, and if for *every* covariant vector $\{w_j\}$ defined at P_0 the quantity $v^j w_j$ is a scalar (an invariant under coordinate changes), then the sets $\{v^i\}$ define a contravariant vector (at P_0). The roles of the co- and contravariant vectors can be interchanged. More generally, if, for example, sets of n^3 quantities $\{T_{jk}{}^l\}$ are given and are such that the quantities

$$S_j = T_{jk}{}^l v^k w_l$$

transform according to the law (26.1-10) for a covariant vector, for arbitrary vectors $\{v^k\}$ and $\{w_l\}$ defined at P_0, then the sets $\{T_{jk}{}^l\}$ transform according to the law for a tensor of the indicated kind—covariant of rank 2 and contravariant of rank 1. One says more briefly that $T_{jk}{}^l$ *is such a tensor*.

Note on the Terminology Found in Some Books on Differential Geometry. An alternative definition of vector field, given below, is often used. First, if v^j is a contravariant vector field, then the linear operator $L \overset{\text{def}}{=} v^j \, \partial/\partial x^j$, when applied to a differentiable scalar f, gives the scalar $Lf = v^j(\partial f/\partial x^j)$. If the manifold and the functions f and v^j are of class C^∞, then L maps this class of function into itself. Furthermore, if f and g are any two such functions, then

$$Lfg = fLg + gLf; \tag{26.2-4}$$

a linear operator having this property is called a *derivation*, as in Chapter 25. Conversely, if L is any derivation in the algebra of C^∞ functions, then a corresponding vector field v^j can be constructed such that $L = v^j(\partial/\partial x^j)$, provided the operator L has a sufficiently local character. In order that L be so representable, it is clearly necessary that whenever scalars f and g agree in a neighborhood $\mathfrak{U}$ in $\mathfrak{M}$, then Lf and Lg must agree in $\mathfrak{U}$. Even more is true: If L can be so represented, if P is any point, and if the equation

$$\left.\frac{\partial f}{\partial x^j}\right|_P = \left.\frac{\partial g}{\partial x^j}\right|_P \qquad (j = 1, \dots, n) \tag{26.2-5}$$

holds in any chart (hence in every chart containing P), then Lf and Lg must agree at P. (It turns out that if L is any linear operator that satisfies this last condition, it is necessarily a derivation.) To construct the field v^j at any point P in a given chart, we define functions $f^{(j)}(\mathbf{x})$, $j = 1, \dots, n$, by the requirement that

$$f^{(j)}(\mathbf{x}) = x^j$$

in a neighborhood of P, and we call

$$v^j(\mathbf{x}) = Lf^{(j)}(\mathbf{x}).$$

This defines functions $v^j(\mathbf{x})$ throughout the chart. Then, if $f(\mathbf{x})$ is a C^∞ scalar field and P any point in the chart, we call

$$g(\mathbf{x}) = x^j \left(\frac{\partial f}{\partial x^j} \bigg|_P \right)$$

Clearly f and g satisfy (26.2-5); hence

$$Lf|_P = Lg|_P = v^j \frac{\partial f}{\partial x^j} \bigg|_P. \tag{26.2-6}$$

Since the left member here is a scalar, it is seen from the quotient law that the v^j transform as the components of a contravariant vector field under changes of coordinates, and it is seen also that L is a derivation. Hence there is a one-to-one correspondence between local derivations and vector fields. Because of this correspondence, some authors define a vector field as a *local derivation in the algebra of C^∞ functions on a manifold*. We shall continue to use the older definition, which has been traditional in most branches of physics.

26.3 Metric in Euclidean Space

If $X^1, \ldots, X^n$ are Cartesian coordinates in a Euclidean space and $x^1, \ldots, x^n$ are curvilinear coordinates, in terms of which the Cartesian coordinates are expressed as $X^i = X^i(x^1, \ldots, x^n), i = 1, \ldots, n$, then an important tensor g_{jk} is obtained by writing

$$g_{jk} = g_{jk}(x^1, \ldots, x^n) = \frac{\partial X^1}{\partial x^j} \frac{\partial X^1}{\partial x^k} + \cdots + \frac{\partial X^n}{\partial x^j} \frac{\partial X^n}{\partial x^k}. \tag{26.3-1}$$

It is clear that the same functions $g_{jk}(\ldots)$ are obtained, for the given coordinate system $x^1, \ldots, x^n$, if the X^j are replaced by other Cartesian coordinates, say $\tilde{X}^j$, obtained from the X^j by a rotation of the axes (or by a rigid motion, generally). Now, let x^i and $x^i + \Delta x^i$ ($i = 1, \ldots, n$) be the coordinates of two points P_1 and P_2, and let X^i and $X^i + \Delta X^i$ denote the Cartesian coordinates of the same two points, i.e.,

$$X^i = X^i(x^1, \ldots, x^n),$$
$$X^i + \Delta X^i = X^i(x^1 + \Delta x^1, \ldots, x^n + \Delta x^n). \tag{26.3-2}$$

The square of the distance from P_1 to P_2 is given by

$$(d(P_1, P_2))^2 = (\Delta X^1)^2 + \cdots + (\Delta X^n)^2; \tag{26.3-3}$$

if the Δx^i are regarded as small quantities, then, by expanding (26.3-2) in a Taylor's series, it is seen that

$$(d(P_1, P_2))^2 = g_{jk}(x^1, \ldots, x^n)\Delta x^j \, \Delta x^k + O(|\Delta x|^3), \tag{26.3-4}$$

where $|\Delta x|$ stands for $\max|\Delta x^j|$. This equation is usually paraphrased by writing

$$ds^2 = g_{jk}\, dx^j\, dx^k; \qquad (26.3\text{-}5)$$

ds is called the *line element*. From (26.3-1) it follows that if a transformation is made from $x^1, \ldots, x^n$ to new coordinates $x'^1, \ldots, x'^n$, then the functions $g_{jk}(\cdots)$ transform like the components of a second-rank covariant tensor, as the notation indicates. Furthermore, this tensor is symmetric: $g_{jk} = g_{kj}$. From (26.3-4) it follows that if the g_{jk} are regarded as the elements of a matrix, then this matrix is positive definite. A similar metric tensor will be assumed to exist in any Riemannian space, but will not generally be expressible in the form (26.3-1) because Cartesian coordinates do not exist if the space is non-Euclidean.

26.4 Riemannian and Pseudo-Riemannian Manifolds

A *Riemannian manifold* is an n-dimensional manifold $\mathfrak{M}$ on which a second-rank tensor g_{jk} is defined, which, in all $\mathfrak{M}$, is (1) *symmetric*, i.e., $g_{jk} = g_{kj}$, and (2) *positive definite*, i.e., $g_{jk}v^jv^k > 0$ for any nonvanishing vector $\{v^j\}$. The eigenvalues of the matrix (g_{jk}) are all positive. Note that if g_{jk} is symmetric, then

$$g'_{jk} = \frac{\partial x^r}{\partial x'^j} \frac{\partial x^s}{\partial x'^k} g_{rs} \qquad (26.4\text{-}1)$$

is also symmetric. Note also that the positive definiteness is compatible with this transformation law, because $g_{jk}v^jv^k$ is a scalar.

The determinant of the matrix (g_{jk}) is denoted by g or $g(x^1, \ldots, x^n)$; it is not a scalar, since its value at a given point in the manifold depends on the coordinate system.

The transformation law (26.4-1) can be written in matrix notation as

$$G' = JGJ^T, \qquad (26.4\text{-}2)$$

where G is the matrix (g_{jk}) and where J is the Jacobian matrix.

In a *pseudo-Riemannian manifold*, the matrix G is not required to be positive definite, but only nonsingular and symmetric. Each eigenvalue of G is then either >0 or <0, and the *signature* s of G is defined as the number of positive eigenvalues minus the number of negative ones. According to Sylvester's *law of inertia* of quadratic forms, the signature of the matrix $G' = JGJ^T$ is the same as that of G—a proof can be found in Bôcher 1922; hence, the signature s is independent of the choice of coordinate system, at each point. The eigenvalues of G are continuous functions of its elements g_{jk}, hence of the coordinates $x^1, \ldots, x^n$, and no eigenvalue is ever zero; hence no eigenvalue can ever switch its sign as the x^i vary. Therefore, since $\mathfrak{M}$ is connected, the signature s is constant throughout $\mathfrak{M}$. In general relativity, $\mathfrak{M}$ has dimension 4 and signature 2, so that G has three positive eigenvalues and one negative one throughout $\mathfrak{M}$.

Since det $G \neq 0$, G has an inverse; the elements of the inverse are denoted by $g^{jk} = g^{jk}(x^1, \ldots, x^n)$; these functions transform according to the law

$$g'^{jk} = \frac{\partial \hat{x}'^j}{\partial x^r} \frac{\partial \hat{x}'^k}{\partial x^s} g^{rs} \qquad (26.4\text{-}3)$$

[this equation can be obtained by taking matrix inverses in (26.4-2) and by noting that the Jacobian matrices of inverse transformations are inverse matrices]; hence g^{jk} is a contravariant tensor of rank 2.

If $\mathcal{S}$ is an n-dimensional hypersurface in a Euclidean space E^N, where $N > n$, then $\mathcal{S}$ may be regarded as a Riemannian manifold with the metric that $\mathcal{S}$ inherits from E^N. If $x^1, \ldots, x^n$ are intrinsic coordinates in a portion $\mathfrak{U}$ of $\mathcal{S}$, if $X^1, \ldots, X^N$ are Cartesian coordinates in E^N, and if, in $\mathfrak{U}$,

$$X^i = X^i(x^1, \ldots, x^n), \qquad i = 1, \ldots, N,$$

then, as in Section 26.2, the distance d (in E^N) between two points x^j, $x^j + \Delta x^j$ in $\mathcal{S}$ is given by

$$d^2 = \sum_{i=1}^N [X^i(x^1, \ldots, x^n) - X^i(x^1 + \Delta x^1, \ldots, x^n + \Delta x^n)]^2$$

$$= g_{jk} \, \Delta x^j \, \Delta x^k + O(|\Delta x|^3),$$

where

$$g_{jk} = \sum_{i=1}^N \frac{\partial X^i}{\partial x^j} \frac{\partial X^i}{\partial x^k}.$$

One says that $\mathcal{S}$ is *immersed* in E^N and that g_{jk} is the *inherited* metric tensor.

EXAMPLE

If $x^1 = \theta$ and $x^2 = \varphi$ are polar coordinates on the unit sphere, where $0 < \theta < \pi$, $-\pi < \varphi < \pi$, and if X, Y, Z are Cartesian coordinates in E^3, then

$$X = \sin x^1 \cos x^2, \qquad Y = \sin x^1 \sin x^2, \qquad Z = \cos x^1,$$

and

$$g_{11} = 1, \qquad g_{12} = g_{21} = 0, \qquad g_{22} = \sin^2 x^1;$$

this result is usually written as $ds^2 = d\theta^2 + \sin^2 \theta \, d\varphi^2$; it can also be obtained from the line element given by $ds^2 = dr^2 + r^2 \, d\theta^2 + r^2 \sin^2 \theta \, d\varphi^2$ for spherical polar coordinates in E^3 by setting $r = 1$ and $dr = 0$.

We state, without proof, that an n-dimensional Riemannian manifold can always be immersed in E^N, with $N = \frac{1}{2}n(n + 1)$, although a higher number of dimensions may be needed if the manifold is to be immersed as a hypersurface without self-intersections (see discussion of the Klein bottle in Section 23.1). For the proof, see L. P. Eisenhart, *Riemannian Geometry* (1966).

A pseudo-Riemannian manifold can be similarly immersed in a flat (generalized Minkowski) space of suitable dimension and signature. The converse to this statement, however, has a limitation: A smooth surface in such a flat space is a pseudo-Riemannian manifold only if it is nowhere parallel to the light cone.

EXERCISES

1. Consider the Riemannian manifold $\mathscr{S}$ determined by the single coordinate system $x^1 = \xi, x^2 = \eta$, where the metric tensor is given in matrix form as

$$(g_{jk}) = \frac{1}{(1 + \xi^2 + \eta^2)^2} \begin{pmatrix} 1 + \eta^2 & -\xi\eta \\ -\xi\eta & 1 + \xi^2 \end{pmatrix},$$

and where ξ and η are allowed to vary over the entire ξ, η plane. Show that $\mathscr{S}$ can be immersed in E^3 as a hemisphere.

2. Find similarly an immersion in E^3 of the manifold with metric tensor

$$(g_{jk}) = \begin{pmatrix} 1 + \eta^2 & \xi\eta \\ \xi\eta & 1 + \xi^2 \end{pmatrix}.$$

26.5 Raising and Lowering of Indices

In a Riemannian or pseudo-Riemannian manifold, if v^j is any contravariant vector, then the covariant vector $v_j = g_{jk}v^k$ obtained by inner multiplication with the metric tensor is regarded as simply another representation of v^j; one says that v_j has been obtained from v^j by *lowering the index*, and when the same symbol v is used for a contravariant and a covariant vector, it is understood that they are related in this way. Similarly, if w_j is any covariant vector, one says that the contravariant vector $w^j = g^{jk}w_k$ has been obtained by *raising the index*. If an index is first raised and then lowered (or conversely), then the original vector is obtained again, because (g_{jk}) and (g^{jk}) are inverse matrices, so that $g_{jk}g^{kl}w_l = w_j$. Any index on any tensor can be raised or lowered in the same way; e.g.,

$$S^{\ k}_{j\ l} = g^{km}S_{jml} = g_{lm}S^{\ km}_j, \quad \text{etc.;}$$

note that the horizontal ordering of the indices must be maintained, unless the tensor is symmetric. Clearly g^{jk} is simply the result of raising both indices of g_{jk}; the mixed form of the metric tensor is

$$g^j_k = g^{jl}g_{lk} = \begin{cases} 1 & \text{for } j = k \\ 0 & \text{for } j \neq k \end{cases} = \delta^j_k;$$

in this special case, it is customary to write the indices without horizontal separation (g^j_k rather than $g^j{}_k$ or $g_k{}^j$), which is permissible because of symmetry.

26.6 Geodesics in a Riemannian Manifold

The rest of this chapter, except for the last section, deals with geodesics. It is, however, not quite yet geometry, because we shall be concerned mainly with analytic tools and relations. The geometry proper starts in the next chapter with notions like parallel transport along a curve.

Let $\mathscr{C}$ be a smooth curve in a Riemannian manifold $\mathfrak{M}$ with initial and terminal points P_1 and P_2. We assume first that $\mathscr{C}$ lies in a single coordinate chart and is described in that chart by the functions $x^j(w)$, for $a \leq w \leq b$, which are assumed to be of class C^2. Transformations from one chart to another will be considered later. The quantity

$$L = \int_a^b \sqrt{g_{kl} \dot{x}^k \dot{x}^l} \, dw = \int_a^b ds \qquad (26.6\text{-}1)$$

is called the *length* of the curve $\mathscr{C}$. (If $\mathfrak{M}$ is immersed in a Euclidean space E^N of higher dimension, as described in Section 26.4, then L is precisely the length of $\mathscr{C}$ as a curve in E^N.) The radicand above is positive, because the metric tensor is positive definite in a Riemannian space—for pseudo-Riemannian spaces, see the next section. The radicand is understood to be the function of w given by

$$g_{kl}(x^1(w), \ldots, x^n(w)) \frac{dx^k(w)}{dw} \frac{dx^l(w)}{dw}$$

$$\overset{\text{def}}{=} \Phi(x^1, \ldots, x^n, \dot{x}^1, \ldots, \dot{x}^n), \qquad (26.6\text{-}2)$$

where the dot denotes d/dw and where each argument of $\Phi(\ldots)$ is understood as the corresponding function of w.

We note in passing that if the curve $\mathscr{C}$ lies in the intersection of two charts, the same length L is obtained from either coordinate system, because the expression (26.6-2) is a scalar; for any w, the value of Φ is independent of the coordinate system; hence, the value of L given by (26.6-1) is an invariant. If $\mathscr{C}$ is piecewise smooth, its total length is understood as the sum of the lengths of its pieces.

If a smooth curve $\mathscr{C}_0$ from P_1 to P_2, namely

$$\mathscr{C}_0 \colon x^k = y^k(w) \quad \text{(given)}, \qquad (26.6\text{-}3a)$$

where $a \leq w \leq b$, can be found such that the integral L is smaller for $\mathscr{C}_0$ than for any other curve from P_1 to P_2, then $\mathscr{C}_0$ is clearly the shortest such curve. To compare $\mathscr{C}_0$ with neighboring curves, consider curves $\mathscr{C}$ of the form

$$\mathscr{C} \colon x^k = y^k(w) + \varepsilon z^k(w), \qquad (26.6\text{-}3b)$$

where ε is a small parameter and the $z^k(\cdot)$ are arbitrary class C^2 functions such that $z^k(a) = z^k(b) = 0$ $(k = 1, \ldots, n)$; then

$$L = L_0 + \tfrac{1}{2}\varepsilon \int_a^b \Phi^{-1/2} \left(\frac{\partial \Phi}{\partial x^k} z^k + \frac{\partial \Phi}{\partial \dot{x}^k} \dot{z}^k \right) dw + O(\varepsilon^2), \qquad (26.6\text{-}4)$$

where L_0 is the length of $\mathscr{C}_0$. In Φ and its derivatives, the arguments x^k and $\dot{x}^k$ are understood to be given the values on $\mathscr{C}_0$, namely $y^k(w)$ and $\dot{y}^k(w)$. For L to be a minimum, the above integral must vanish for all choices of the functions $z^k(w)$, and this leads to differential equations for the functions $y^k(w)$ that describe the curve $\mathscr{C}_0$. If the integral vanishes for all z^k, i.e., if the y^k satisfy these differential equations, then the curve $\mathscr{C}_0$ is called a *geodesic* (*or geodesic curve*) in $\mathfrak{M}$; $\mathscr{C}_0$ may or may not be the *shortest* curve from P_1 to P_2 (see examples, below), but in any case the value of L is *stationary* on $\mathscr{C}_0$.

In particular, if $\mathfrak{M}$ is a Euclidean n-space, and the variables $x^1, \ldots, x^n$ are curvilinear coordinates, so that the metric tensor is given by (26.3-1), then $\mathscr{C}_0$ is a segment of a straight line, expressed in the curvilinear coordinates.

To simplify the following steps, it is convenient to choose the parameter w on the curve $\mathscr{C}_0$ in such a way that Φ is constant on $\mathscr{C}_0$ (but not necessarily on the neighboring curves $\mathscr{C}$; in fact, w cannot be chosen so that Φ is the same constant on $\mathscr{C}_0$ and on the neighboring curves, because $w = a$ and $w = b$ at P_1 and P_2, respectively, on these curves; hence, if w could be so chosen, all the curves would have the same length). This can be done by introducing a new parameter $\lambda = \lambda(w)$ on $\mathscr{C}_0$ according to the equation

$$\lambda(w_1) = \int_a^{w_1} \sqrt{g_{kl}\dot{y}^k\dot{y}^l}\, dw \tag{26.6-5}$$

for any w_1 between a and b (λ is *arclength* along $\mathscr{C}_0$); using λ rather than w as the variable of integration, the above equation is

$$\lambda_1 = \int_0^{\lambda_1} \sqrt{g_{kl}\dot{y}^k\dot{y}^l}\, d\lambda;$$

therefore, by differentiation with respect to λ_1,

$$\sqrt{g_{kl}\dot{y}^k\dot{y}^l} = \sqrt{\Phi} \equiv 1 \quad \text{on } \mathscr{C}_0.$$

The factor $\Phi^{-1/2}$ can now be dropped from the integral (26.6-4). That is, *when the parameter λ is chosen as arclength on $\mathscr{C}_0$, the variational problems*

$$\delta \int_a^b \sqrt{\Phi}\, d\lambda = 0 \quad \text{and} \quad \delta \int_a^b \Phi\, d\lambda = 0 \tag{26.6-6}$$

have the same solutions. Although $\Phi = 1$ on $\mathscr{C}_0$, the partial derivatives of Φ in (26.6-4) do not vanish, because they involve differentiations in other directions than merely along $\mathscr{C}_0$.

Integrating by parts in the second term in (26.6-4) after deleting $\Phi^{-1/2}$ (the integrated parts vanish because $z^k = 0$ at P_1 and P_2) and equating the integral to zero give

$$\int_a^b \left(\frac{\partial \Phi}{\partial x^k} - \frac{d}{d\lambda} \frac{\partial \Phi}{\partial \dot{x}^k} \right) z^k(\lambda)\, d\lambda = 0.$$

Owing to the arbitrary nature of the $z^k(w)$, therefore,

$$\frac{\partial \Phi}{\partial x^k} - \frac{d}{d\lambda} \frac{\partial \Phi}{\partial \dot{x}^k} = 0 \quad \text{on } \mathscr{C}_0 \qquad (k = 1, \ldots, n); \qquad (26.6\text{-}7)$$

"on $\mathscr{C}_0$" means that the functions $x^k(\lambda)$ that appear in Φ (see 26.6-2) are to be taken as the functions $y^k(\lambda)$ that describe the curve $\mathscr{C}_0$. Equations (26.6-7) are the *Euler variational equations* of the problem

$$\delta \int_a^b \Phi \, d\lambda = 0. \qquad (26.6\text{-}8)$$

From the expression (26.6-2) for Φ, it is seen that the Euler equations are specifically [k is replaced by m in (26.6-7)]

$$\frac{\partial g_{kl}}{\partial x^m} \dot{x}^k \dot{x}^l - \frac{d}{d\lambda} (g_{ml} \dot{x}^l + g_{km} \dot{x}^k) = 0$$

(the two terms in parentheses are of course equal) or

$$\left(\frac{\partial g_{kl}}{\partial x^m} - \frac{\partial g_{ml}}{\partial x^k} - \frac{\partial g_{km}}{\partial x^l} \right) \dot{x}^k \dot{x}^l - 2 g_{ml} \ddot{x}^l = 0; \qquad (26.6\text{-}9)$$

these equations ($m = 1, \ldots, n$) must be satisfied by the functions $x^k(w) = y^k(w)$ in order that $\mathscr{C}_0$ be a geodesic.

The equations can be conveniently written in terms of the *Christoffel 3-index symbols* of the first and second kind, which are

$$[kl, m] = \frac{1}{2} \left(-\frac{\partial g_{kl}}{\partial x^m} + \frac{\partial g_{ml}}{\partial x^k} + \frac{\partial g_{km}}{\partial x^l} \right) \qquad (26.6\text{-}10)$$

and

$$\{^r_{kl}\} = g^{rm}[kl, m] \quad \text{(summed on } m\text{)}, \qquad (26.6\text{-}11)$$

respectively. In terms of them, the equations of a geodesic become [after multiplying (26.6-9) by g^{rm} and summing on m]

$$\ddot{x}^r + \{^r_{kl}\} \dot{x}^k \dot{x}^l = 0 \qquad (r = 1, \ldots, n) \qquad (26.6\text{-}12)$$

(here, as elsewhere, the summation convention is used). The symbol $\{^r_{kl}\}$ is a function of $x^1, \ldots, x^n$ and is understood to be taken at $x^1(\lambda), \ldots, x^n(\lambda)$. If $\mathscr{C}: x^k = x^k(\lambda)$ is any curve satisfying (26.6-12), then $\mathscr{C}$ is called a *geodesic* and λ is called a *natural parameter* (or *affine* or *preferred parameter*) on $\mathscr{C}$. It is evident that the equations (26.6-12) are invariant under a transformation $\lambda \to a\lambda + b$, where a and b are constants and $a \neq 0$; hence the natural parameter is only defined up to such a linear transformation. In a Riemannian manifold, λ can be taken as arclength.

Whether L is actually a minimum on $\mathscr{C}_0$ (rather than a maximum or merely a stationary value) and whether the minimum is unique cannot be decided

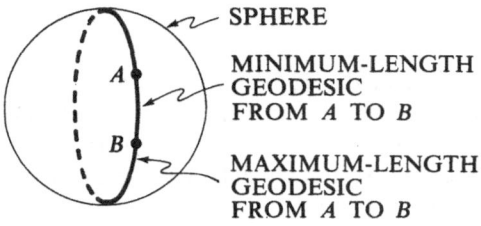

SPHERE

MINIMUM-LENGTH
GEODESIC
FROM A TO B

MAXIMUM-LENGTH
GEODESIC
FROM A TO B

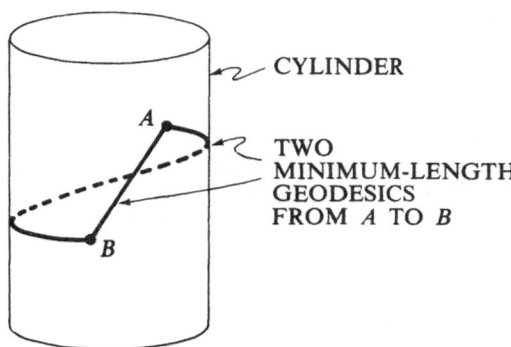

CYLINDER

TWO
MINIMUM-LENGTH
GEODESICS
FROM A TO B

Figure 26.2   Geodesics.

without further consideration (see examples in Figure 26.2, and note also that there are infinitely many geodesics on a sphere from a given point to its antipode, all having the same length). It will be seen below, however, that given any point A of the manifold, there is a neighborhood $\mathfrak{N}_A$ of A such that if B is in $\mathfrak{N}_A$, then there is only one solution $\mathscr{C}_0$ of (26.6-12) from A to B lying in $\mathfrak{N}_A$, and L is a minimum on this curve $\mathscr{C}_0$.

Notes. The quantities $[kl, m]$ are not the components of a 3rd rank tensor; neither are the quantities $\{^r_{km}\}$, for they don't satisfy the appropriate transformation laws; for example, in a Euclidean space, these quantities are all identically zero in Cartesian coordinates, but not in curvilinear coordinates. The quantities $\dot{x}^k$ are the components of a contravariant vector, but the quantities $\ddot{x}^k$ are not. Nevertheless, equation (26.6-12) has a certain invariant character; namely, if it is satisfied, for a given curve $P(\lambda)$, in one coordinate system, then it is satisfied in any other, because it was derived from the invariant equation (26.6-8). "Geodesic" and "natural parameter" are invariant concepts. The n quantities in the left members of (26.6-12), whether evaluated for a geodesic or not, are, at any point of a curve $\mathscr{C}$, the components of a contravariant vector (obtained by the so-called absolute differentiation of the vector $\dot{x}^k$—see Section 27.6), although the individual terms of the left members of (26.6-12) are not, by themselves, the components of a vector. If a curve $\mathscr{C}$ runs through several coordinate charts and satisfies (26.6-12) in each of them, then $\mathscr{C}$ is also called a *geodesic* in the manifold.

26.7 Geodesics in a Pseudo-Riemannian Manifold $\mathfrak{M}$

In this case Φ can be negative; hence L, defined as $\int_a^b \Phi^{1/2} \, dw$, is meaningless as a length. Even if L is redefined as $\int_a^b |\Phi|^{1/2} \, dw$, it is still meaningless as a length in the ordinary sense, because, given any two points P and Q, a (piecewise smooth) curve can always be found from P to Q for which $L = 0$. Nevertheless, a curve $\mathscr{C}: x^k = x^k(\lambda)$ that satisfies (26.6-12) is still called a *geodesic*, and λ is called a *natural parameter*. The quantity $\Phi = g_{jk} \dot{x}^j \dot{x}^k$ is constant on $\mathscr{C}$, and three cases arise:

$$
\begin{aligned}
&\text{If } \Phi > 0, \quad \mathscr{C} \text{ is called a } \textit{spacelike} \text{ geodesic};\\
&\text{if } \Phi = 0, \quad \mathscr{C} \text{ is called a } \textit{null} \text{ geodesic}; \qquad\qquad (26.7\text{-}1)\\
&\text{if } \Phi < 0, \quad \mathscr{C} \text{ is called a } \textit{timelike} \text{ geodesic}.
\end{aligned}
$$

Since Φ is quadratic in the $\dot{x}^k$, this classification is independent of the choice of the natural parameter. The parameter λ can be so chosen that $\Phi = 1$ in the first case and $\Phi = -1$ in the third; then, λ is called *distance* and *proper time*, respectively, along $\mathscr{C}$. Geodesics play a role in general relativity.

26.8 Geodesics; the Initial-Value Problem; the Lipschitz Condition

Let $a^1, \ldots, a^n$ be the coordinates, in a given chart, of an arbitrary point P_0 in a Riemannian or pseudo-Riemannian manifold $\mathfrak{M}$. The initial-value problem of a geodesic through P_0 with tangent vector $\xi^1, \ldots, \xi^n$ at P_0, and with natural parameter λ, is the following:

$$
\text{diff. eq.} \begin{cases} \dfrac{dx^j}{d\lambda} = p^j \\[2mm] \dfrac{dp^j}{d\lambda} = -\{^{\;i}_{kl}\} p^k p^l \end{cases} \qquad (j = 1, \ldots, n), \qquad (26.8\text{-}1)
$$

$$
\text{initial cond.} \begin{cases} x^j(0) = a^j \\ p^j(0) = \xi^j \end{cases} \qquad (j = 1, \ldots, n). \qquad (26.8\text{-}2)
$$

It will be proved in the next section that this initial-value problem always has a unique solution for λ in some interval $[-\lambda_0, \lambda_0]$.

It is convenient to call

$$
\begin{aligned}
y^k &= x^k - a^k \qquad (k = 1, \ldots, n),\\
y^{n+k} &= p^k \qquad\quad\;\; (k = 1, \ldots, n),
\end{aligned}
$$

and to rewrite the differential equations as

$$
\frac{dy^k}{d\lambda} = f^k(y^1, \ldots, y^{2n}) \qquad (k = 1, \ldots, 2n), \qquad (26.8\text{-}3)
$$

where f^k denotes the function on the right side of the kth differential equation of the set (26.8-1), for $k = 1, \ldots, 2n$, the Christoffel symbols now being regarded as functions of $y^1, \ldots, y^{2n}$. The functions f^k are defined for all $y^{n+1}, \ldots, y^{2n}$, and for all $y^1, \ldots, y^n$ such that the corresponding point $x^1, \ldots, x^n$ lies in the given chart. It is assumed that the $\{_{kl}^j\}$ and their first partial derivatives are continuous throughout the chart, and it is asserted that the functions f^k are Lipschitz continuous in any compact region of the $y^1, \ldots, y^{2n}$ space in which they are defined. That is, suppose that δ is a constant such that the f^k are defined for all $y^1, \ldots, y^{2n}$ in the cube W determined by $|y^i| \leq \delta$ $(i = 1, \ldots, 2n)$. Then there is a constant $L = L(\delta)$ such that if $\{y^i\}$ and $\{\bar{y}^i\}$ are two points in W, then

$$|f^k(y^1, \ldots, y^{2n}) - f^k(\bar{y}^1, \ldots, \bar{y}^{2n})| \leq L \max_{(j=1, \ldots, 2n)} |y^j - \bar{y}^j|,$$

$$k = 1, \ldots, 2n. \tag{26.8-4}$$

The proof is left as an exercise. It depends on the explicit form of the f^k given by (26.8-1) and on the assumption that the Christoffel symbols are functions of class C^1 of the variables $y^1, \ldots, y^n$. In this new notation, the initial value problem is

$$\frac{dy^k}{d\lambda} = f^k(y^1, \ldots, y^N)$$

$$\tag{26.8-5}$$

$$y^k(0) = y_0^k \quad \text{(given)} \qquad (k = 1, \ldots, N = 2n).$$

It will be proved in the next section that this problem has a unique solution near $\lambda = 0$, hence we have the following:

Theorem 1. *The initial-value problem* (26.8-1, 2) *for a geodesic starting at a point P_0 with initial tangent vector $\{\xi^k\}$ has a unique solution $x^k(\lambda)$ in some interval* $-\lambda_0 \leq \lambda \leq \lambda_0$ $(\lambda_0 > 0)$.

Corollary 1. *If $\mathscr{C}$: $P(\lambda)$ is a geodesic in $\mathfrak{M}$, with λ a natural parameter, satisfying an initial condition* (26.8-2), *then $P(\lambda)$ is unique on its entire length.*

PROOF. Suppose $P_1(\lambda)$ and $P_2(\lambda)$ are two such curves, and call λ_1 the point at which they diverge; that is,

$$\lambda_1 = \sup\{\lambda: P_1(\lambda') = P_2(\lambda') \text{ for } 0 \leq \lambda' \leq \lambda\}.$$

Let $P_0 = P_1(\lambda_1) = P_2(\lambda_1)$, and use Theorem 1 (with the origin displaced to λ_1) to show that if $P_1(\lambda)$ and $P_2(\lambda)$ are defined beyond $\lambda = \lambda_1$, then they coincide in some interval $(\lambda_1 - \lambda_0, \lambda_1 + \lambda_0)$, hence that they do not diverge at λ_1, after all.

If the initial tangent vector $\{\xi^k\}$ is replaced by a vector of different length in the same direction, then the geodesic obtained is the same as before, as a point set in $\mathfrak{M}$, but it has a different choice of the natural parameter λ on it; that is, it follows from the structure of the equations (26.8-1) that if $\{x^k(\lambda), p^k(\lambda)\}$ is a solution, for λ in $[0, \lambda_0]$, and if a is > 0, then $\{x^k(a\lambda), ap^k(a\lambda)\}$ is a solution for λ in $[0, \lambda_0 a]$. The initial tangent vector is changed from $\{\xi^k\}$ to $\{a\xi^k\}$. If a is taken $= \lambda_0$, it follows that, given any direction, there is always a solution, valid for all λ in $[0, 1]$, which starts in the given direction, provided that the components of the initial tangent vector are all less than some positive constant. It can also be proved that the constant depends continuously on the direction of ξ^k, and hence has a positive lower bound or minimum. Therefore,

Theorem 2. *Let* $a^1, \ldots, a^n$ *be the coordinates, in some chart, of the point* P_0, *as above. Then there is a positive constant* $K = K(P_0)$ *such that the geodesic initial value problem* (26.8-1, 2) *has a solution in the interval* $0 \le \lambda \le 1$ *for all vectors* $\{\xi^k\}$ *such that* $|\xi^k| < K$ $(k = 1, \ldots, n)$.

This theorem will be used in the next chapter to prove the existence of the so-called Riemannian coordinates about a point.

26.9 The Integral Equation; Picard Iterations

This section, which is somewhat peripheral to the study of manifolds, is devoted to a fundamental fact about ordinary differential equations, namely the existence of a solution of the initial-value problem (26.8-5) when the Lipschitz condition (26.8-4) is satisfied. That fact and the method used to prove it are useful in many parts of physics and mathematics.

We first rewrite the initial value problem (26.8-5) as an integral equation of the Volterra type (that is, one in which the independent variable λ appears as the upper limit of the integral):

$$y^k(\lambda) = y^k(0) + \int_0^\lambda f^k(\mathbf{y}(\lambda'))d\lambda', \qquad k = 1, \ldots, N, \qquad (26.9\text{-}1)$$

where $\mathbf{y}$ denotes the vector with components $y^1, \ldots, y^N$. It is convenient to introduce the norm

$$\|\mathbf{x}\| = \max_{(j)} |x^j|;$$

then the Lipschitz condition takes the form

$$|f^k(\mathbf{y}) - f^k(\bar{\mathbf{y}})| \le L\|\mathbf{y} - \bar{\mathbf{y}}\|, \qquad k = 1, \ldots, N. \qquad (26.9\text{-}2)$$

The solution of the integral equation (26.9-1) is obtained by the *Picard method of iterations*, in which $y^k(\lambda, q)$ denote the qth iteration, $y^k(\lambda, 0)$ is taken as $= y^k(0)$ (independent of λ), and the successive improvements are given, for $q = 1, 2, \ldots$, by

$$y^k(\lambda, q + 1) = y^k(0) + \int_0^\lambda f^k(\mathbf{y}(\lambda', q))d\lambda'. \qquad (26.9\text{-}3)$$

To investigate the rate of convergence, we call

$$\Delta y^k(\lambda, q) = y^k(\lambda, q) - y^k(\lambda, q - 1),$$
$$\Delta f^k(\lambda, q) = f^k(\mathbf{y}(\lambda, q)) - f^k(\mathbf{y}(\lambda, q - 1)),$$

for $q = 1, 2, \ldots$. By subtracting successive equations of the set (26.9-3), we have

$$\Delta y^k(\lambda, q + 1) = \int_0^\lambda \Delta f^k(\lambda', q) d\lambda'. \qquad (26.9\text{-}4)$$

This equation holds for $q = 0$ (as well as for $q = 1, 2, \ldots$), if we define $y^k(\lambda, -1) = 0$ and note that $f^k(0) = 0$.

The rate of convergence is given by the following:

Lemma. *Suppose that*

$$\|\mathbf{y}(0)\| < \frac{\delta}{2} \qquad (26.9\text{-}5)$$

[i.e., suppose that $\mathbf{y}(0)$ lies not merely in the cube W but in the central part of W] and suppose that $|\lambda| \le L/\log 2$, where L is the Lipschitz constant in (26.9-2). Then, for $q = 1, 2, \ldots$

$$\|\Delta \mathbf{y}(\lambda, q)\| \le \frac{\delta}{2} \frac{1}{q!} (L|\lambda|)^q \qquad (26.9\text{-}6)$$

and

$$\|\mathbf{y}(\lambda, q)\| < \delta. \qquad (26.9\text{-}7)$$

PROOF. The case $q = 1$ follows from (26.9-4) and (26.9-5). The other cases follow by induction on q, because $\int_0^\lambda \lambda^q \, d\lambda = \lambda^{q+1}/(q + 1)$ and

$$\|\mathbf{y}(\lambda, q)\| = \left\| \sum_{r=0}^q \Delta \mathbf{y}(\lambda, r) \right\|$$

$$\le \frac{\delta}{2} \sum \frac{1}{r!} (L|\lambda|)^r < \frac{\delta}{2} e^{\log 2} = \delta. \qquad (26.9\text{-}8)$$

This last permits the use of the Lipschitz conditions for each q.

It is clearly the denominator $q!$ in (26.9-6) that gives the method its great power. Since the sum in (26.9-8) is majorized by the power series for $e^{L|\lambda|}$, the sum converges absolutely and uniformly with respect to λ in any finite interval. Hence the sum can be integrated term by term, and we see that the function

$$\mathbf{y}(\lambda) = \lim_{q \to \infty} \mathbf{y}(\lambda, q)$$

satisfies the integral equation (26.9-1); hence it satisfies the initial value problem (26.8-5). Theorem 1 of the preceding section is thereby proved.

26.10 Geodesics: the Two-Point Problem

In Euclidean geometry, any two points P and Q can be connected by a unique straight line segment $\overline{PQ}$. The same is true locally in a Riemannian or pseudo-Riemannian manifold $\mathfrak{M}$, with "straight line" replaced by "geodesic."

Consider the problem of finding functions $x^j(\lambda)$ $(j = 1, \ldots, n)$ such that

$$\ddot{x}^j = -\{{}^{\,j}_{kl}\}\dot{x}^k\dot{x}^l \qquad (0 \leq \lambda \leq 1, j = 1, \ldots, n),$$
$$x^j(0) \quad \text{and} \quad x^j(1) \quad \text{given} \quad (j = 1, \ldots, n). \tag{26.10-1}$$

The following theorem was proved by J. H. C. Whitehead in 1932:

Theorem. *Any point P_0 in $\mathfrak{M}$ has a neighborhood $\mathfrak{N}$ such that if $x^j(0)$ and $x^j(1)$ $(j = 1, \ldots, n)$ are the coordinates of any two points Q_0 and Q_1 in $\mathfrak{N}$, then there is a unique geodesic segment joining Q_0 to Q_1 and lying in $\mathfrak{N}$.*

The requirement that the geodesic lie entirely in $\mathfrak{N}$ is often necessary to obtain uniqueness—see Figure 26.2, in which a geodesic on a cylinder goes the long way around, instead of following the shortest path from its initial point to its final point.

A somewhat weaker form of the theorem is easily proved; namely, there exist neighborhoods $\mathfrak{N}_1$ and $\mathfrak{N}_2$, with $P_0 \in \mathfrak{N}_1 \subset \mathfrak{N}_2$ such that if Q_0 and Q_1 are in $\mathfrak{N}_1$, then there is a unique geodesic from Q_0 to Q_1 lying in $\mathfrak{N}_2$. Whitehead's proof that $\mathfrak{N}_1$ and $\mathfrak{N}_2$ can be taken as the same involves a topological argument.

Two further results, which follow from the proof of Whitehead's theorem, are now stated, also without proof.

Corollary 1. *If the curve $x^j(\lambda)$ is continuous for $a \leq \lambda \leq b$ and satisfies the geodesic equation for $a < \lambda < b$, then it satisfies this equation also for $\lambda = a$ and $\lambda = b$.*

Corollary 2. *If, to indicate the dependence on the endpoints of the segment, the solution in Whitehead's theorem is written as $x^j(\lambda, Q_0, Q_1)$, then for any λ in $[0, 1]$, $x^j(\lambda, Q_0, Q_1)$ depends continuously on the coordinates $x^j(0)$ and $x^j(1)$ of the initial and terminal points Q_0 and Q_1.*

The existence proof for the two-point problem is similar to the one for the initial-value problem, in that the system (26.10-1) is converted into an integral equation, which is then solved by the Picard iterative scheme. The procedure is somewhat more complicated, because the integral equation is of the Fredholm type (the upper limit of the integrals is 1, not λ), so that one has neither the factor $1/q!$ nor the explicit dependence on λ that appeared in (26.9-6). For details, see Whitehead 1932.

26.11 Continuation of Geodesics

It can be proved that, in a Riemannian manifold, if the curve $x^j(\lambda)$ satisfies the geodesic equation for $a < \lambda < b$ and lies in a compact region of the manifold, then the limits of $x^j(\lambda)$ exist for $\lambda \to a$ and $\lambda \to b$; hence, in particular, Corollary 1 above applies. In a pseudo-Riemannian manifold, that does not hold, as the following example shows: Let $\mathfrak{M}$ be the surface of a cylinder, let $x^1 = z$ and $x^2 = \theta$ be cylindrical coordinates, and let the metric be given by

$$(g_{jk}) = \begin{pmatrix} 1 & 1 \\ 1 & 2z^2 \end{pmatrix} \qquad \left(-\frac{1}{\sqrt{2}} < z < \frac{1}{\sqrt{2}}\right).$$

With z so restricted, $\mathfrak{M}$ is a pseudo-Riemannian manifold. The reader can easily verify that the curve

$$z = \lambda, \qquad \theta = \frac{1}{\lambda} \qquad (0 < \lambda < \tfrac{1}{2})$$

is a geodesic. As $\lambda \to 0$, this geodesic winds infinitely many times around the cylinder and approaches the circle $z = 0$.

However, if a curve $x^j = x^j(\lambda)$ $(j = 1, \ldots, n)$ satisfies the geodesic equation for $a < \lambda < b$ and is continuous for $a \leq \lambda \leq b$, then Corollary 1 of the preceding section applies, and the quantities $x^j(b)$, $\dot{x}^j(b)$ can be used as initial data for the existence theorem of Section 26.8 to extend the geodesic to some values of $\lambda > b$; it can be similarly extended to some values of $\lambda < a$.

If a *geodesic* (as opposed to a geodesic segment) is understood as a solution of the geodesic equations that has been extended as far as possible (perhaps through many coordinate charts), then the above result may be paraphrased by saying that in a Riemannian or pseudo-Riemannian manifold a *geodesic cannot have a beginning or an end in the manifold*.

Note. This does not imply that $\lambda \to \pm\infty$ on the geodesic; also it does not imply that the geodesic may not have a beginning or end in some space in which the manifold is immersed.

26.12 Affinely Connected Manifolds

Since the transformation laws for tensors are known, and g_{ik} is a tensor, it is easy to find the transformation law for the three-index symbol $\{{}^i_{jk}\}$ from equations (26.6-10, 11). If $\{{}^i_{jk}\}'$ refers to coordinates $x'^1, \ldots, x'^n$, and $\{{}^i_{jk}\}$ to the unprimed coordinates, then the transformation law is

$$\{{}^i_{jk}\}' = \left[\{{}^r_{st}\}\frac{\partial x^s}{\partial x'^j}\frac{\partial x^t}{\partial x'^k} + \frac{\partial^2 x^r}{\partial x'^j\,\partial x'^k}\right]\frac{\partial x'^i}{\partial x^r}. \tag{26.12-1}$$

Since only the $\{{}^i_{jk}\}$ appear in the geodesic equation (26.6-12), and not the g_{jk} directly, a more general kind of geometry, called *affine geometry*, is obtained if one does not assume the existence of a metric tensor g_{jk} at all but only the existence of a set of quantities which transform like the $\{{}^i_{jk}\}$ and which appear in place of the $\{{}^i_{jk}\}$ in the geodesic equation. Then there may or may not exist a tensor g_{jk} from which these quantities can be obtained via equations (26.6-10, 11).

An *affine connection* of a manifold $\mathfrak{M}$ is defined in analogy with a tensor as a collection of sets of n^3 functions each, $\Gamma^i_{jk} = \Gamma^i_{jk}(x^1, \ldots, x^n)$, one set associated with each chart on $\mathfrak{M}$, such that in the overlap of two charts the two sets of functions are related by the transformation law

$$\Gamma'^{i}_{jk} = \left[\Gamma^{r}_{st}\frac{\partial x^{s}}{\partial x'^{j}}\frac{\partial x^{t}}{\partial x'^{k}} + \frac{\partial^{2}x^{r}}{\partial x'^{j}\,\partial x'^{k}}\right]\frac{\partial x'^{i}}{\partial x^{r}}, \qquad (26.12\text{-}2)$$

which is the same as (26.12-1). Since the Γ^{i}_{jk} are not assumed to be derived from a metric tensor as the $\{^{i}_{jk}\}$ were, it is now necessary to verify directly that this transformation law is transitive (see Section 26.1), in order to be sure that the definition of a connection is self-consistent. Verification is left as an exercise.

In an affinely connected manifold $\mathfrak{M}$ (i.e., a manifold with an affine connection defined on it), a smooth curve $\mathscr{C}: x^{i} = x^{i}(\lambda)$ $(i = 1, \ldots, n)$ is called a *geodesic*, and λ is called a *natural parameter* on $\mathscr{C}$, if the equations

$$\ddot{x}^{r} + \Gamma^{r}_{kl}\dot{x}^{k}\dot{x}^{l} = 0 \qquad (r = 1, \ldots, n) \qquad (26.12\text{-}3)$$

are satisfied on $\mathscr{C}$; the dot denotes differentiation with respect to λ. Compare with (26.6-12).

As in Riemannian geometry, the geodesic equations (26.12-3) are invariant under coordinate changes in the sense that when a curve $\mathscr{C}$ lies in the overlap of two charts, it satisfies the equations in one of the coordinate systems if and only if it satisfies them in the other. The equations are also invariant under a transformation $\lambda \to a\lambda + b$ $(a \neq 0)$ of the natural parameter on a given geodesic.

The theorems of Section 26.8 on the initial-value problem of geodesics continue to hold, provided that the components Γ^{i}_{jk} of the connection are functions of class C^{1}, which, according to (26.12-2), requires that the manifold be of class C^{3}.

Whitehead's theorem also continues to hold (in fact, Whitehead stated and proved it originally for an affinely connected space): each point P has a neighborhood $\mathfrak{N}$ such that any two points in $\mathfrak{N}$ can be connected by a unique geodesic in $\mathfrak{N}$.

The question whether, given an affine connection Γ^{i}_{jk} on a manifold, a metric tensor g_{jk} can be found which is consistent with the metric, i.e., which is such that $\Gamma^{i}_{jk} = \{^{i}_{jk}\}$, is discussed briefly at the end of Section 27.10.

The geometric structure based on the geodesics in an affinely connected manifold is called the *geometry of paths*; it is discussed in some detail in the next chapter. It is clear from (26.12-3) that for that purpose the connection may be assumed to be symmetric in the subscripts:

$$\Gamma^{i}_{jk} = \Gamma^{i}_{kj}. \qquad (26.12\text{-}4)$$

More generally, a further geometric structure is sometimes introduced, called *torsion*, based on the antisymmetric part $\frac{1}{2}(\Gamma^{i}_{jk} - \Gamma^{i}_{kj})$ of the connection—see Flanders 1963. Since torsion has no effect on the geodesics, it has to be regarded as something outside of and superposed on the geometry of the manifold as determined by its geodesics.

26.13 Riemannian and Pseudo-Riemannian Covering Manifolds

Let $\mathfrak{M}$ be a covering manifold of $\mathfrak{N}$, and ψ a projection of $\mathfrak{M}$ onto $\mathfrak{N}$. If $f(P)$ is any function (say, of class C^k) in $\mathfrak{N}$, then the function $\tilde{f}(Q) \stackrel{\text{def}}{=} f(\psi(Q))$, defined in $\mathfrak{M}$, may be said to be *lifted from* $\mathfrak{N}$ to $\mathfrak{M}$, in analogy with the lifting of curves and surfaces, discussed in Section 24.2.

Now suppose that $\mathfrak{N}$ is a Riemannian manifold. Let $\varphi^i(P)$ be the coordinates in a good neighborhood $\mathfrak{B}$ in $\mathfrak{N}$, and let g_{jk} be the components of the metric tensor in these coordinates. The φ^i and the g_{ik} are all functions in $\mathfrak{N}$, which can be lifted to $\mathfrak{M}$ as functions $\tilde{\varphi}^i$ and $\tilde{g}_{jk}$. Each connected component $\mathfrak{U}_i$ of $\psi^{-1}(\mathfrak{B})$ thus becomes a coordinate chart with a metric tensor in it, and it is easily established that $\mathfrak{M}$ is thereby made into a Riemannian manifold, called a *Riemannian covering manifold* of $\mathfrak{N}$. If $\mathfrak{M}$ is the universal covering manifold of $\mathfrak{N}$, then $\mathfrak{M}$ is called its *universal Riemannian covering manifold*.

Pseudo-Riemannian covering manifolds are similarly defined.

Consider now the problem of constructing manifolds covered by a given Riemannian manifold $\mathfrak{N}$. For a general manifold $\mathfrak{M}$, such an $\mathfrak{N}$ was constructed in Section 24.6 by means of a homeomorphism σ of $\mathfrak{M}$ onto itself such that if P is any point of $\mathfrak{M}$, then the set of points $P, \sigma(P), \sigma(\sigma(P)), \ldots, \sigma^{-1}(P), \ldots$ is discrete (has no limit point in $\mathfrak{M}$); it was constructed by identifying all such points, for each P, i.e., by regarding the set of these points as a single point of the manifold $\mathfrak{N}$ being constructed. In order for this process to give a *Riemannian* manifold $\mathfrak{N}$, it is necessary only to require that σ be an *isometric* homeomorphism, that is, one that preserves the metric. That is, let $\{\mathfrak{U}, \varphi, N\}$ be any chart in $\mathfrak{M}$, and let $\mathfrak{U}'$ be the set of all $Q = \sigma(P)$, for P in $\mathfrak{U}$. Define $\varphi'(Q)$ as $\varphi(\sigma^{-1}(Q))$. Clearly, $\{\mathfrak{U}', \varphi', N\}$ is a possible chart in $\mathfrak{M}$ (i.e., is compatible with the charts already there), by definition of a homeomorphism. Then, the g'_{jk} are required to be the same functions of the coordinates $x'^i = \varphi'^i$ in the second chart as the g_{jk} are of the coordinates $x^i = \varphi^i$ in the first chart. In this case, all the charts obtained from $\{\mathfrak{U}, \varphi, N\}$ by means of σ and its iterates have identical metric tensors in them and can be identified as being a single chart of $\mathfrak{N}$.

These ideas are used in Chapter 28 in the study of the global properties of Einstein manifolds.

CHAPTER 27

Riemannian, Pseudo-Riemannian, and Affinely Connected Manifolds

Metric, pseudometric, connection, and topology; geodesic or Riemannian coordinates; geometry in the sense of Klein; approximate congruence; congruence of stars; covariant derivative; absolute derivative; parallel transport; orientability, Lorentz orientability; Laplacian; d'Alembertian; Riemann tensor; Ricci tensor; Riemann curvature scalar; determination of the second derivatives of the metric tensor from the Riemann tensor; conditions for an affine connection to be compatible with a metric; intrinsic curvature of a manifold; flatness; Stäckel, Robertson, and Eisenhart conditions for separability of the wave equation.

Prerequisites: Chapters 23, 24, 26.

The subject of this chapter is the geometry of a manifold that has a metric, a pseudometric, or an affine connection defined on it. The dividing line between the preceding chapter and this may seem rather fine and arbitrary, since the last topic in that chapter was geodesics, and the first in this is geodesic coordinates. However, there is a fundamental difference between the two. The preceding chapter was mainly analytic, the only geometric notion being that of the distance between two points, whereas this one is mainly geometric, and even in the sense of Euclid, except that the concepts are somewhat extended and the formulation is analytic. The fundamental concepts, such as parallelism, length, curvature, and angle, are really geometric, and must be so regarded. The use of analytic methods does not detract from the geometric nature of those concepts any more than did the introduction of numerical coordinates into Euclidean geometry by Descartes. From that point of view, one of the main results of the preceding chapter, Whitehead's theorem, serves the same purpose as Euclid's *postulate* that through any two distinct points there can be drawn one and only one straight line, even though, in Whitehead's theorem, the two points must not be too far apart.

In modern mathematics, where all branches have merged to some extent, the question what is geometry and what is analysis is somewhat abstruse, but surely those notions that can be traced directly back to Euclid ought to be called geometric. In Euclid's time, geometry was regarded as the science that dealt with the physical space around us. If general relativity is correct, then Riemannian, pseudo-Riemannian, and affine geometry do also; even if relativity has to be modified, its primary ideas, which have stood for over 60 years, will surely continue to be basic for physical space-time.

The best understanding of the subject is obtained if one keeps geometric ideas (like parallel transport of a vector along a curve) in the foreground of one's thinking, and detailed analytic formulas a little bit in the background.

27.1 Topology and Metric

According to Chapter 23, the topology of a manifold is determined by its coordinate charts.

The main difference between Riemannian and pseudo-Riemannian (or affinely connected) manifolds is that in the latter the relation between the topology and the metric is less intimate than in the former. In a connected Riemannian manifold, the distance $d(P, Q)$ between any two points P and Q is the infimum of the length of a curve between them, the infimum being taken with respect to all curves from P to Q. The manifold is then a metric space, which means that a distance function is defined in it and satisfies the relations:

$$d(P, Q) \geq 0 \quad \text{for all } P, Q;$$

$$d(P, Q) = 0 \quad \text{if and only if} \quad P = Q;$$

$$d(P, Q) \leq d(P, R) + d(R, Q).$$

A topology in the space is determined by the metric, because the sets

$$S(\varepsilon, P) = \{Q: d(P, Q) < \varepsilon\}$$

(the open balls) are open sets, and the other open sets are unions of open balls. In a Riemannian manifold, this topology is identical with the topology previously defined by the coordinate charts.

In a pseudo-Riemannian manifold, with (g_{jk}) indefinite, the distance $d(P, Q)$, as just defined, is always $= 0$, and hence is not a metric. Nevertheless, the manifold has a well-defined topology determined by the coordinate charts, as explained in Section 23.3, and the same is true of an affinely connected manifold, where there is not even a concept of distance along a curve.

27.2 Geodesic or Riemannian Coordinates

The key to the investigation of the local geometric properties of a Riemannian manifold in a neighborhood of a point P_0, when the metric tensor g_{jk} is given in a coordinate system $x^1, \ldots, x^n$ containing P_0, is to choose new coordinates $x'^1, \ldots, x'^n$, which are as nearly Cartesian as possible near P_0. We should like the transformed metric tensor g'_{jk} to be equal to δ_{jk} at P_0, which can always be achieved, and to remain close to δ_{jk} nearby. It will be seen that all the first partial derivatives of the g'_{jk} can be made to vanish at P_0, and certain linear combinations of those derivatives can be made to vanish in a neighborhood of P_0. In that sense, the new coordinates will be as nearly Cartesian as possible. The remaining departure of the g'_{jk} from δ_{jk} will then reflect the non-Euclidean character of the geometry.

The pseudo-Riemannian case is similar, except that there we wish g'_{jk} to be $\pm\delta_{jk}$ at P_0, with the $+$ and $-$ signs determined according to the signature of the tensor g_{jk}.

In an affinely connected manifold, we wish all components Γ'^i_{jk} of the connection to vanish at P_0 and remain as small as possible nearby.

We consider in this section a manifold with an affine connection, with $\Gamma^i_{jk} = \Gamma^i_{kj}$, and we specialize the results to the other cases later. Let $\{\mathfrak{U}, \boldsymbol{\varphi}, N\}$ be a chart containing a point P_0, and let $x^1, \ldots, x^n$ denote the coordinates in that chart. We consider the family of all geodesics through P_0, with λ a natural parameter so chosen on each that $\lambda = 0$ at P_0. The coordinates of such a geodesic are given by the solutions $x^k(\lambda)$ of the initial-value problem

$$\frac{d^2x^k}{d\lambda^2} + \Gamma^k_{lm}\frac{dx^l}{d\lambda}\frac{dx^m}{d\lambda} = 0, \qquad (27.2\text{-}1)$$

$$x^k(0) = \varphi^k(P_0), \qquad (27.2\text{-}2)$$

$$\frac{dx^k}{d\lambda}(0) = y^k, \qquad (27.2\text{-}3)$$

where $y^1, \ldots, y^n$ are real numbers, not all zero, which specify the direction in which the geodesic starts out from P_0. According to Theorem 2 of Section 26.8, these equations have a solution $x^k(\lambda; y^1, \ldots, y^n)$, $k = 1, \ldots, n$, valid for $0 \le \lambda \le 1$, for all vectors $y^1, \ldots, y^n$ in some neighborhood of the origin. If this solution is expanded in powers of λ and the coefficients of the first three terms are obtained from the above equations, we find

$$x^k(\lambda; y^1, \ldots, y^n) = x^k(0) + \lambda y^k - \tfrac{1}{2}\lambda^2\Gamma^k_{lm}|_0 y^l y^m + O(\lambda^3|y|^3),$$

where $|y|$ denotes $\max_{(k)}|y^k|$. The subscript 0 indicates that the connection Γ is to be evaluated at P_0. The above may equally well be regarded as a power series expansion in the small quantities $\lambda y^1, \ldots, \lambda y^n$; hence, without loss of generality, we may simply set $\lambda = 1$, if the y^k are themselves sufficiently small. That is,

$$x^k(1; y^1, \ldots, y^n) = x^k(0) + y^k - \tfrac{1}{2}\Gamma^k_{lm}|_0 y^l y^m + O(|y|^3). \quad (27.2\text{-}4)$$

From this equation it is seen that the Jacobian matrix

$$\frac{\partial x^k(1; y^1, \ldots, y^n)}{\partial y^l}$$

is equal to δ_l^k at P_0, i.e., when all the y^k are zero, and it follows that the equation

$$x^k = x^k(1; y^1, \ldots, y^n)$$

can be inverted, in some neighborhood of P_0, to give the y's as functions of the x's. The first few terms of the corresponding expansion are

$$y^k = x^k - x^k(0) + \tfrac{1}{2}\Gamma^k_{lm}|_0(x^l - x^l(0))(x^m - x^m(0)) + \cdots. \quad (27.2\text{-}5)$$

The $y^1, \ldots, y^n$ are called *geodesic* (or *Riemannian*) coordinates about P_0.

The connection components Γ^k_{lm} in these equations refer to the original coordinates. The corresponding components in the geodesic coordinate system will be denoted by $\mathring{\Gamma}^k_{lm}$; we can find their values as follows. A geodesic is given in the geodesic coordinates by $y^k = y^k(\lambda) = \lambda \xi^k$, where $\xi^1, \ldots, \xi^n$ are constants; hence, $d^2 y^k/d\lambda^2 = 0$, and by comparison with the general equation (27.2-1) with x^k replaced by y^k and Γ^k_{lm} by $\mathring{\Gamma}^k_{lm}$, we see that $\mathring{\Gamma}^k_{lm}\xi^l\xi^m$ is $= 0$ on the geodesic, for all λ and all $\xi^1, \ldots, \xi^n$. Therefore

$$\mathring{\Gamma}^k_{lm} = 0 \quad \text{for all } k, l, m, \text{ at } P_0, \quad (27.2\text{-}6)$$

and generally

$$\mathring{\Gamma}^k_{lm}(y^1, \ldots, y^n)y^l y^m = 0 \quad \text{for all } k, \quad (27.2\text{-}7)$$

throughout the neighborhood of P_0 in which the geodesic coordinates are defined.

Note. Except in flat space, geodesic coordinates about P_0 are not in general geodesic coordinates about a neighboring point Q_0. Stated differently, a straight line $y^k = \lambda \xi^k + a^k$ in the coordinate space $\mathbb{R}^n$ of $y^1, \ldots, y^n$ does not generally represent a geodesic unless the constants $a^1, \ldots, a^n$ are all zero, i.e., unless that straight line passes through the origin of $\mathbb{R}^n$.

The new coordinates about P_0 are still not unique, and the remaining degree of nonuniqueness is described by the following theorem:

Theorem. *An arbitrary transformation of the x^i induces a homogeneous linear transformation of the associated geodesic coordinates y^i about a given point P_0.*

PROOF. Let $x'^1, \ldots, x'^n$ be coordinates in any other chart containing P_0, and let $y'^1, \ldots, y'^n$ be the corresponding geodesic coordinates about P_0. Any geodesic $P(\lambda)$ through P_0, with $P(0) = P_0$, when expressed in the geodesic coordinates, takes the forms

$$P(\lambda): y^j = \lambda \xi^j \quad \text{and} \quad y'^j = \lambda \xi'^j,$$

where the ξ^j and the ξ'^j are constants. To lowest order in small quantities ($\lambda \approx 0$), the x^j agree with the y^j and the x'^j with the y'^j. Therefore

$$\xi'^j = \left. \frac{\partial x'^j}{\partial x^k} \right|_{P_0} \xi^k;$$

hence for any given point $P(\lambda)$ on a geodesic through P_0, we have

$$y'^j = \frac{\partial x'^j}{\partial x^k}\bigg|_{P_0} y^k, \tag{27.2-8}$$

which is the required linear transformation.

EXERCISE

Show that the next term in the series (27.2-4) can be written as

$$-\frac{1}{3!}\, \Gamma^k_{lmr}|_{P_0}\, y^l y^m y^r, \tag{27.2-9}$$

where

$$\Gamma^k_{lmr} = \frac{1}{3}\sum \left(\frac{\partial}{\partial x^l}\Gamma^k_{mr} - 2\Gamma^k_{sl}\Gamma^s_{mr}\right), \tag{27.2-10}$$

the summation being over the cyclic permutations of the subscripts l, m, and r. Still higher-order terms, with coefficients $\Gamma^k_{lm\ldots v}$ are discussed in Eisenhart 1926.

In the Riemannian and pseudo-Riemannian cases, where the connection is obtained from a metric as $\Gamma^k_{lm} = \{^k_{lm}\}$, where $\{^k_{lm}\}$ is given by (26.6-10, 11), it is seen from (27.2-6) that the first partial derivatives of the metric tensor vanish at the origin of geodesic coordinates, because

$$\frac{\partial g_{km}}{\partial x^l} = [kl, m] + [ml, k].$$

27.3 Normal Coordinates in Riemannian and Pseudo-Riemannian Manifolds

When there is a metric tensor, the geodesic coordinates can be further specialized by means of a linear transformation of the y^i so chosen that, at the point P_0,

$$(g_{jk}) = \begin{pmatrix} +1 & & & & & \\ & +1 & & & (0) & \\ & & \ddots & & & \\ & & & +1 & & \\ & & & & -1 & \\ & (0) & & & & \ddots \\ & & & & & & -1 \end{pmatrix} \tag{27.3-1}$$

(in the Riemannian case, all the signs are $+$). That can be done as follows: First, the y^i can be transformed by an orthogonal matrix such that the matrix (g_{jk}) becomes diagonal at P_0. Let the diagonal elements be called $d_1, \ldots, d_n$, where the positive ones come first (they are all $\neq 0$). Then, by a further transformation $y'^i = \sqrt{|d_i|}\, y^i$ (here the summation convention is suspended), the matrix (g_{jk}) is reduced to the above standard form. Then, the y^i are unique, except for an orthogonal transformation (a rotation with possibly an inversion) in the Riemannian case, and except for a Lorentz transformation in the pseudo-Riemannian case. The y^i are then called *normal* or *normal geodesic* or *normal Riemannian coordinates*.

If a Riemannian manifold is immersed as an n-dimensional surface S in a Euclidean space E^N of higher dimension, as described in Section 26.4, then the normal coordinates are Cartesian coordinates in the n-dimensional hyperplane tangent to S at P_0, the projection from the surface to the hyperplane being such that geodesics through P_0 go into tangent lines, and distances along the geodesics go into distances along the tangent lines.

(In some of the older literature, any coordinates in which the first partial derivatives of the g_{jk} vanish at P_0 are called "geodesic coordinates" about P_0.)

27.4 Geometric Concepts; Principle of Equivalence

Geometry deals with points and lines and objects constructed from them. It deals specifically with spaces of points in which certain sets of points called "straight lines" or "geodesics" are singled out. In the geometries considered here, it is also assumed that the space is an n-dimensional continuum, that is, that it has the topological properties that make it an n-dimensional manifold.

A basic geometric notion is that of congruence. There is generally a group of transformations in the space, called the *congruence group*, under which geometric relations are preserved. Examples are the rigid motion group in Euclidean space and the Poincaré group (nonhomogeneous Lorentz group) in Minkowski space. Then, two figures (point sets) are said to be *congruent* if one can be transformed into the other by a transformation of the group. (In Felix Klein's "Erlangen Program" of 1872, the point of view was reversed: When a group of transformations is given in a space, it is taken to determine a geometry consisting of those relations that are preserved under the transformations of the group.)

By analogy with Euclidean and Minkowskian geometries, one might suppose that, in a space with a metric tensor g_{jk}, the group ought to consist of those transformations under which the tensor g_{jk} is mapped into itself. However, there are generally no such transformations (except the identity), unless the space is flat or has constant curvature. Hence, generally, the ordinary concept of congruence is lost.

An approximate concept of congruence can be defined for very small figures. We consider first a Riemannian manifold. Let $y^1, \ldots, y^n$ be normal coordinates about a point P_0 and $w^1, \ldots, w^n$ normal coordinates about another point Q_0. Let ψ denote the mapping of a neighborhood of P_0 onto a neighborhood of Q_0 obtained by equating the normal coordinates. That is, a point P with coordinates $y^1, \ldots, y^n$ is mapped onto a point Q with coordinates $w^1, \ldots, w^n$ if $y^i = w^i$, $i = 1, \ldots, n$. A small figure near P_0 is said to be approximately congruent to a small figure near Q_0 if ψ carries the first figure into the second. Since a normal coordinate system is unique only modulo orthogonal transformations, the result is an approximate Euclidean geometry. Geodesics through P_0 are carried by ψ into geodesics through Q_0, and it is easily seen that the angle between two geodesics (see formula below) is preserved under the mapping. (Geodesics not through P_0 generally do not go into geodesics at all, unless the space is flat.) If three or more smooth curves intersect at P_0, the star formed by their tangent vectors at P_0 is mapped into a similar star at Q_0 with preservation of all angles. If $P_0 AB$ is a small triangle with one vertex at P_0, i.e., if $P_0 A$, $P_0 B$, and AB are short geodesics, and its image under ψ is a figure $Q_0 CD$, then $Q_0 C$ and $Q_0 D$ are geodesics having the same length as $P_0 A$ and $P_0 B$; CD is nearly a geodesic, if the triangle is small, and has nearly the same length as AB. If P_0 and Q_0 are the same point and the y^k and w^k are two normal coordinate systems about that point, then the congruences are rotations and reflections that keep P_0 fixed.

Angles are defined by the formula that gives angles in curvilinear coordinates in Euclidean space. If two smooth curves, given by $x^i = x^i(\lambda)$ and $x^i = \bar{x}^i(\mu)$, intersect at a point with coordinates $x^i(\lambda_0) = \bar{x}^i(\mu_0)$, their respective directions at that point are given by the tangent vectors

$$\xi^i = \dot{x}^i(\lambda_0), \qquad \bar{\xi}^i = \dot{\bar{x}}^i(\mu_0),$$

and the angle θ between those directions is given by

$$\cos\theta = \frac{g_{jk}\xi^j\bar{\xi}^k}{\|\xi\|\,\|\bar{\xi}\|}, \tag{27.4-1}$$

where

$$\|\xi\| = |g_{jk}\xi^j\xi^k|^{1/2}, \qquad \|\bar{\xi}\| = |g_{jk}\bar{\xi}^j\bar{\xi}^k|^{1/2}. \tag{27.4-2}$$

The absolute value signs in (27.4-2) are unnecessary here but are included for later use in the pseudo-Riemannian case.

Similar considerations apply to a pseudo-Riemannian manifold. As before, let $y^1, \ldots, y^n$ and $w^1, \ldots, w^n$ be normal coordinates about points P_0 and Q_0, respectively, i.e., geodesic coordinates such that the metric tensor takes the standard form (27.3-1). These coordinates are unique only modulo Lorentz transformations; hence the geometry is approximately Minkowskian rather than Euclidean. Under the mapping ψ given as before by $w^i = y^i$, geodesics through P_0 are mapped into geodesics through Q_0 with preservation of type (spacelike, null, or timelike), and the angle between two geodesics, given by (27.4-1), is preserved. Now, however, $\cos\theta$ can be >1 or <-1; hence θ can be imaginary. If either ξ or $\bar{\xi}$ is a null vector, $\cos\theta$ is infinite or undefined, according as $g_{jk}\xi^j\bar{\xi}^k$ is $\neq 0$ or $=0$.

It should be noted that, even if ξ and $\bar{\xi}$ are both spacelike or both timelike, the value of $\cos\theta$ given by (27.4-1) may lie outside the interval $[-1, +1]$.

EXERCISE

Show that a necessary and sufficient condition for $|\cos \theta|$ to be <1 is that all linear combinations of ξ and $\bar{\xi}$ be spacelike, or all timelike. If $\cos \theta = \pm 1$, there is a null vector of the form $a\xi + b\bar{\xi}$ with a and b not both zero. (In the Riemannian case, that null vector is the zero vector; hence $\bar{\xi}$ is a scalar multiple of ξ.)

For figures in an affinely connected manifold, angles and lengths are not defined; nevertheless, certain geometric relations are invariant under the mapping ψ given by $y^i = w^i$ as above. Here, y^i and w^i are arbitrary geodesic coordinates; since there is no notion of normal coordinates, they are defined only modulo nonsingular linear or affine transformations. For example, if $\xi, \bar{\xi}$, and $\bar{\bar{\xi}}$ are the tangent vectors at P_0 to three curves, and if $\bar{\bar{\xi}}$ lies in the plane determined by ξ and $\bar{\xi}$, i.e., if the three vectors are linearly dependent, then the same is true after the mapping. In fact, a relation of the form $\bar{\bar{\xi}} = a\xi + b\bar{\xi}$ is preserved under ψ, for given a and b. More generally, any set of such vectors is linearly dependent if and only if the set of their images under ψ is linearly dependent.

The further geometric concept of *parallel displacement* or *parallel transport* along a curve, which will be described in Section 27.7, was introduced by Levi-Civita in 1917. The idea is this: In flat geometries, Euclidean, Minkowskian, or affine, where the congruence group, when referred to Cartesian coordinates, consists of transformations of the form

$$\mathbf{x} \to M\mathbf{x} + \mathbf{a},$$

M being an orthogonal, Lorentz, or general nonsingular matrix, the subgroup of the pure translations

$$\mathbf{x} \to \mathbf{x} + \mathbf{a}$$

plays a special role. (Note that in the Minkowski case, no relative *motion* is involved, only a displacement.) Namely, if one figure can be mapped into a second by a pure translation, the figures are said to be not only congruent but also to have the same *orientation* in space, or to be obtainable from each other by parallel displacement. The corresponding concept in a curved space is the parallel transport of a small figure along a curve, with generally different results, however, depending on what curve is used to connect the initial and final points.

Our approach to these questions will be in analogy with the "equivalence principle" of general relativity, according to which certain laws can be formulated by simply asserting that the corresponding laws of special relativity hold in an inertial or "freely falling" reference frame. Geodesic coordinate systems about a point play the role of inertial frames. For example, the parallel transport of a vector along a curve will be so defined that at the instant when the vector is passing a point P of the curve, if $y^1, \ldots, y^n$ are geodesic coordinates about P, then the vectors components relative to the y^j system are undergoing no change at that instant, i.e., have zero derivatives with respect to a parameter on the curve.

27.5 Covariant Differentiation

In this and the following two sections, three closely related concepts are discussed (covariant differentiation, absolute differentiation, and parallel transport), any one of which might be taken as fundamental and the others derived from it. We consider a general affinely connected manifold.

The partial derivatives of the components of a tensor transform like the components of a tensor of one-higher rank under *linear* transformations, but not under more general ones. For example, in the Euclidean plane, a vector field with constant Cartesian components has variable polar-coordinate components; hence the nonvanishing of the partials with respect to r and θ of the latter is a peculiarity of the r, θ coordinate system. To eliminate effects of this kind, we define a tensor of higher rank by giving its components at any point P in a *geodesic* coordinate system about P as the appropriate partial derivatives; then, to find its components in other coordinate systems, we must use the transformation laws.

Let $P{:}x^i = a^i$ be any point, and let $y^1, \ldots, y^n$ be the corresponding geodesic coordinates. As in Section 27.2,

$$x^i = a^i + y^i - \tfrac{1}{2}\Gamma^i_{jk}y^jy^k + \cdots, \tag{27.5-1}$$

$$y^i = x^i - a^i + \tfrac{1}{2}\Gamma^i_{jk}(x^j - a^j)(x^k - a^k) + \cdots. \tag{27.5-2}$$

In these equations, the connection coefficients refer to the x^i coordinate system and are to be taken at the point P. If $v_i(x^1, \ldots, x^n)$ is any covariant vector field, we denote by $\hat{v}_i(y^1, \ldots, y^n)$ its components in the geodesic system. A second rank covariant tensor $v_{i;\,j}$, the *covariant derivative* of v_i, is defined at P by giving its components in the y^j system as

$$\hat{v}_{i;\,j}|_P = \left(\frac{\partial}{\partial y^j}\,\hat{v}_i\right)_P \qquad (i, j = 1, \ldots, n). \tag{27.5-3}$$

We wish to transform this equation back to the x^i system. At the point P, we have $\hat{v}_i = v_i$ and $\hat{v}_{i;\,j} = v_{i;\,j}$, because $\partial x^i/\partial y^j$ is $= \delta_{ij}$ at P. However, in the right member of (27.5-3), $\hat{v}_i$ has to be known near P as well as at P, for purposes of differentiation. Near P,

$$\hat{v}_i = v_k\,\frac{\partial x^k}{\partial y^i}\,;$$

hence

$$v_{i;\,j}|_P = \hat{v}_{i;\,j}|_P = \left(\frac{\partial}{\partial y^j}\left(v_k\,\frac{\partial x^k}{\partial y^i}\right)\right)_P = \left(\frac{\partial v_k}{\partial x^l}\,\frac{\partial x^l}{\partial y^j}\,\frac{\partial x^k}{\partial y^i}\right)_P + v_k\,\frac{\partial^2 x^k}{\partial y^j\,\partial y^i}\bigg|_P.$$

At P, $\partial x^l/\partial y^j = \delta_{lj}$, etc., and the second derivative is $= -\Gamma^k_{ij}$ by (27.5-1). Therefore,

$$v_{i;\,j} = \frac{\partial}{\partial x^j}\,v_i - \Gamma^k_{ij}v_k \tag{27.5-4}$$

(summed on k); this is the law of covariant differentiation in an arbitrary coordinate system; $v_{i;\,j}(x^1, \ldots, x^n)$ is a twice covariant tensor field.

This law shows that if $v_{i;\,j}$ is defined to be given at P by equation (27.5-3) in a particular geodesic coordinate system about P, then it is also correctly given by the same equation in any other geodesic system about P. Namely, if x^i is also a geodesic system, then the Γ^k_{ij} are $=0$ at P, and (27.5-4) agrees with (27.5-3).

An immediate application of covariant differentiation is that if v_i is the gradient of a scalar f in a Euclidean space, then the Laplacian of f is given in any curvilinear coordinate system by

$$g^{ij}v_{i;\,j} = g^{ij}\left(\frac{\partial^2 f}{\partial x^j\,\partial x^i} - \Gamma^k_{ij}\frac{\partial f}{\partial x^k}\right),$$

because this quantity is a scalar (an invariant) and is obviously equal to the Laplacian of f in Cartesian coordinates.

EXERCISE

1. (a) Show that if $v^i = v^i(x^1, \ldots, x^n)$ is a contravariant vector field, then the formula

$$v^i_{;\,j} = \frac{\partial}{\partial x^j} v^i + \Gamma^i_{jk} v^k \tag{27.5-5}$$

defines a second-rank mixed tensor field. (b) Show that if T_{ij} is a second-rank covariant tensor field, then the formula

$$T_{ij;\,k} = \frac{\partial}{\partial x^k} T_{ij} - \Gamma^l_{ik} T_{lj} - \Gamma^l_{jk} T_{il} \tag{27.5-6}$$

defines a third-rank covariant field.

The covariant derivative of a general tensor has one additional term (i.e., in addition to the partial derivative) as in (27.5-4) for each covariant index and one additional term as in (27.5-5) for each contravariant index. The covariant derivative of a scalar field f is simply its gradient: $f_{;\,k} = \partial f/\partial x_k$.

The operation of covariant differentiation provides the basis for the invariant formulation of field theories and generally of physical theories where partial differential equations occur.

EXERCISE

2. Show that the operation satisfies the product rule of differentiation. First, if T_{ij} in (27.5-6) is a product $v_i w_j$, then

$$(v_i w_j)_{;\,k} = v_{i;\,k} w_j + v_i w_{j;\,k}. \tag{27.5-7}$$

More generally, if $T^{\cdots}_{\cdots}$ and $S^{\cdots}_{\cdots}$ are arbitrary tensors, then

$$(T^{\cdots}_{\cdots} S^{\cdots}_{\cdots})_{;\,k} = T^{\cdots}_{\cdots;\,k} S^{\cdots}_{\cdots} + T^{\cdots}_{\cdots} S^{\cdots}_{\cdots;\,k}, \tag{27.5-8}$$

where the indicated product may be an outer or arbitrary inner product, i.e., contracted any number of times.

Note. In the older literature, covariant differentiation is often indicated by a comma in place of the semicolon. Now, the comma often indicates ordinary partial differentiation, so that $v_{i;k} = v_{i,k} - \Gamma^l_{ik} v_l$.

EXERCISE

3.   Show that in a Riemannian or pseudo-Riemannian manifold the metric tensor behaves like a constant with respect to covariant differentiation, that is,

$$g_{ij;k} = 0, \qquad g^{ij}_{;k} = 0, \qquad \delta^i_{j;k} = 0. \tag{27.5-9}$$

(Recall that in such a manifold $\Gamma^i_{jk} = \{^i_{jk}\}$.)

A corollary to this exercise is that covariant differentiation commutes with the raising and lowering of indices: if $v_i = g_{ij} v^j$, then $v_{i;k} = g_{ij} v^j_{;k}$.

It should be noted that the result of successive covariant differentiations is not generally symmetric in the corresponding indices; $v_{j;k;l}$ is not $= v_{j;l;k}$, except in flat space, where they both reduce, in cartesian coordinates, to $\partial^2 v_j / \partial x^k \, \partial x^l$. The asymmetry is a manifestation of the intrinsic curvature of the space and appears, in Section 27.9 below, in the definition of the Riemann curvature tensor.

27.6 Absolute Differentiation Along a Curve

Suppose we are given a smooth curve $\mathscr{C}: P = P(\lambda)$, λ any real parameter, in a Riemannian, pseudo-Riemannian, or affinely connected manifold and we are given a vector or tensor field, defined at least on $\mathscr{C}$. We wish to define another vector or tensor on $\mathscr{C}$ which represents, at each point of $\mathscr{C}$, the rate of change of the first vector or tensor with respect to λ, in some absolute or invariant sense.

Specifically suppose that $\mathscr{C}$ lies in a coordinate chart $\{\mathfrak{U}, \boldsymbol{\varphi}, N\}$, hence is given by coordinates $x^i = x^i(\lambda) = \varphi^i(P(\lambda))$, and suppose that $v_i(\lambda)$ are the components, in that coordinate system, of a covariant vector defined on $\mathscr{C}$. We wish to define another covariant vector $w_i(\lambda)$ which represents the rate of change of $v_i(\lambda)$. Clearly we cannot simply write $w_i(\lambda) = dv_i(\lambda)/d\lambda$, because those quantities don't transform properly. Hence we appeal to the principle of equivalence. Let $P_0 = P(\lambda_0)$ be a point on $\mathscr{C}$, and let $y^1, \ldots, y^n$ be the geodesic coordinates about P_0 that agree with the x^i to first order. Let $\hat{v}_i(\lambda)$ be the components of v_i in the geodesic coordinates, so that $\hat{v}_i(\lambda) = v_i(\lambda)$ at $\lambda = \lambda_0$ but not generally for $\lambda \neq \lambda_0$. We define

$$w_i(\lambda_0) = \frac{d\hat{v}_i(\lambda)}{d\lambda}\bigg|_{\lambda=\lambda_0}.$$

We now transform this equation back to the original coordinates.

$$w_i(\lambda) = \frac{d}{d\lambda}\left(v_k(\lambda)\frac{\partial x^k}{\partial y^i}\right) \qquad \text{at } \lambda = \lambda_0$$

$$= \frac{dv_k(\lambda)}{d\lambda}\frac{\partial x^k}{\partial y^i} + v_k(\lambda)\frac{\partial^2 x^k}{\partial y^i \, \partial y^j}\frac{\partial y^j(\lambda)}{d\lambda} \qquad \text{at } \lambda = \lambda_0.$$

At $\lambda = \lambda_0$, the x's and the y's agree to first order, and the second derivative, according to (27.5-1), is $= -\Gamma_{ij}^k$. That is,

$$w_i(\lambda) = \frac{dv_i(\lambda)}{d\lambda} - v_k(\lambda)\Gamma_{ij}^k \frac{dx^j(\lambda)}{d\lambda}$$

Since the geodesic coordinates no longer appear, this equation is valid at any point of $\mathscr{C}$, and in any coordinate system. It is customary to denote w_i by $\delta v_i/\delta\lambda$, called the *absolute derivative* of v_i along $\mathscr{C}$; hence, along $\mathscr{C}$

$$\frac{\delta v_i}{\delta\lambda} = \frac{dv_i}{d\lambda} - \Gamma_{ij}^k v_k \frac{dx^j}{d\lambda}. \tag{27.6-1}$$

If v_i is given not only on $\mathscr{C}$ but as a vector field defined in a region containing $\mathscr{C}$, then

$$\frac{\delta v_i}{\delta\lambda} = v_{i;j} \frac{dx^j}{d\lambda}. \tag{27.6-2}$$

The absolute derivatives along $\mathscr{C}$ of other tensors are similarly defined, in complete analogy with covariant derivatives. For example, if a contravariant vector v^i is given on $\mathscr{C}$, then

$$\frac{\delta v^i}{\delta\lambda} = \frac{dv^i}{d\lambda} + \Gamma_{jk}^i v^j \frac{dx^k}{d\lambda}. \tag{27.6-3}$$

For absolute, as for covariant differentiation, the product rule holds, the metric tensor behaves like a constant, and the absolute derivative of a scalar f is its ordinary derivative $df/d\lambda$ (another scalar).

In particular, if v^i is the *tangent vector* to $\mathscr{C}$, given by

$$v^i(\lambda) = \frac{dx^i}{d\lambda},$$

then a comparison of (27.6-3) with the equation (26.12-13) of a geodesic shows that $\delta v^i/\delta\lambda = 0$ on $\mathscr{C}$ if and only if $\mathscr{C}$ is a geodesic and λ a natural parameter. In that sense, a geodesic may be characterized as a curve whose tangent vector is a constant along its length.

27.7 Parallel Transport

The problem of parallel transport was mentioned in Section 27.4. A vector is given at a point P_0 of a curve $\mathscr{C}$, and we wish to define a vector at any other point of the curve in such a way that it may be regarded as obtained from the given vector by parallel displacement along $\mathscr{C}$. According to the principle of equivalence, if P_1 is any other point of the curve and if we set up geodesic coordinates about P_1, the vector's components relative to that coordinate system should be unchanging as we move past the point P_1. That is, the absolute derivative of the vector along $\mathscr{C}$ should be zero. Therefore, if $\xi_1, \ldots, \xi_n$

are the components of the given vector at P_0, the transported vector is given at any point $P(\lambda)$ of the curve as the solution $v_i(\lambda)$ of the initial value problem

$$\frac{\delta v_i}{\delta \lambda} = 0 \quad \text{on } \mathscr{C}, \qquad v_i(\lambda_0) = \xi_i \qquad (i = 1, \ldots, n). \qquad (27.7\text{-}1)$$

We note that if the vector $v_i(\lambda)$ is transformed to another coordinate system $x'^1, \ldots, x'^n$, it satisfies the corresponding initial value problem

$$\frac{\delta v'_i}{\delta \lambda} = 0 \quad \text{on } \mathscr{C}, \qquad v'_i(\lambda_0) = \xi'_i \qquad (i = 1, \ldots, n),$$

where the ξ'_i are obtained from the ξ_i by the transformation law for a vector at P_0. That is because $\delta v_i/\delta \lambda$ is a vector, so that if all its components vanish in one coordinate system they also vanish in any other.

Parallel transport of a contravariant vector or a general tensor along a curve is similarly defined.

Now consider, in particular, a Riemannian or pseudo-Riemannian manifold. If $v^i(\lambda)$ and $w^i(\lambda)$ are any smooth vector-valued functions on $\mathscr{C}$, then, since $\delta g_{jk}/\delta \lambda = 0$,

$$\frac{d}{d\lambda}(g_{jk}v^j w^k) = g_{jk}\left(\frac{\delta v^j}{\delta \lambda} w^k + v^j \frac{\delta w^k}{\delta \lambda}\right).$$

Therefore, if $v^j(\lambda)$ and $w^k(\lambda)$ are the result of parallel transport of two given vectors along $\mathscr{C}$, it follows that $g_{jk}v^j w^k$ is constant on $\mathscr{C}$. That is, *in a Riemannian or pseudo-Riemannian manifold, magnitudes of vectors and angles between vectors are preserved under parallel transport.*

27.8 Orientability

If $\mathscr{C}_1$ and $\mathscr{C}_2$ are curves, each going from a point P to a point Q, parallel transport along $\mathscr{C}_1$ generally has a different effect from parallel transport along $\mathscr{C}_2$. Stated differently, parallel transport around a closed curve from P back to P generally transforms each vector at P into another vector at P. In a Riemannian manifold, that transformation is an orthogonal transformation at P, because it preserves all magnitudes and angles. If it is a proper orthogonal transformation (its matrix has determinant $= +1$) for all closed curves, the manifold is said to be *orientable*. That is the case in particular if the manifold is simply connected, because if a closed curve is contracted continuously to a point, the orthogonal transformation in question is changed continuously into the identity transformation, whose determinant is $= +1$.

EXERCISE

A Möbius strip made of flat paper without stretching can be regarded as a 2-dimensional Riemannian manifold, which can be covered by two or more coordinate charts, in each of which g_{jk} is $= \delta_{jk}$ throughout. Show that this manifold is not orientable.

Comment. Orientability is a topological, not a metric concept, and can be defined for general manifolds, where no metric tensor is assumed. The foregoing definition has been adopted because it can be generalized to Einsteinian manifolds, as will now be shown, in a sense that goes beyond topological orientability.

If $\mathfrak{M}$ is a pseudo-Riemannian manifold of dimension $n = 4$ and signature $s = 2$, as in relativity, then the transformation at a point P that results from parallel transport of vectors around a closed curve beginning and ending at P is a Lorentz transformation. It was shown in Section 19.4 that the full Lorentz group has four components; hence, there are four possibilities for the transformation in question. It may be a proper Lorentz transformation, or it may involve time reversal or spatial inversion or both. If it is a proper Lorentz transformation for all closed curves, then $\mathfrak{M}$ is called *Lorentz-orientable*. In that case it is possible to establish a positive direction of time (distinction between past and future) and a chirality (distinction between right-handed and left-handed coordinate systems or helices) throughout $\mathfrak{M}$. If the manifold is simply connected, it is Lorentz-orientable; hence, any Einsteinian manifold $\mathfrak{M}$ always has some covering manifold $\mathfrak{M}'$ (in particular, its universal covering manifold) that is Lorentz-orientable. Since $\mathfrak{M}$ and $\mathfrak{M}'$ are locally indistinguishable, it is never necessary, in general relativity, to consider space-time models that are not Lorentz-orientable. (That observation was made to the author by Lipman Bers, in a conversation over a cup of tea at the Courant Institute.)

An affinely connected manifold is orientable if the affine transformation of the geodesic coordinates at a point induced by parallel transport around a closed curve has a positive determinant for all such curves.

27.9 The Riemann Tensor, General; Laplacian and d'Alembertian

Let $\mathfrak{M}$ be affinely connected. It was noted in Section 27.5 that the result of differentiating a vector field twice covariantly is not generally symmetric in the corresponding indices. A direct calculation, using the formulas in that section, shows that

$$v_{j;k;l} - v_{j;l;k} = R^i_{jkl} v_i, \tag{27.9-1}$$

where R^i_{jkl} denotes the quantity

$$R^i_{jkl} = \frac{\partial}{\partial x^k} \Gamma^i_{jl} - \frac{\partial}{\partial x^l} \Gamma^i_{jk} + \Gamma^i_{mk}\Gamma^m_{jl} - \Gamma^i_{ml}\Gamma^m_{jk}. \tag{27.9-2}$$

Since v_i is arbitrary, the quotient law, applied to (27.9-1), shows that the n^4 quantities R^i_{jkl} are the components of a tensor of rank 4, which is called the *Riemann tensor* or the *Riemann curvature tensor*.

In Euclidean space, the Riemann tensor vanishes identically in any coordinate system, because it clearly vanishes in Cartesian coordinates, and if all components of a tensor vanish in one coordinate system, then they all vanish in any other. It will be seen in Section 27.12 that the vanishing of the Riemann tensor is also *sufficient* for a Riemannian manifold to be Euclidean, for a pseudo-Riemannian one to be Minkowskian, and for an affinely connected one to be flat. In each case, this statement refers just to the metric, but if the manifold is simply connected, it can be extended to a complete Euclidean, Minkowskian, or flat space.

In the special case of a Riemannian or pseudo-Riemannian manifold, where there is a metric and where indices can be raised and lowered, the Riemann tensor can be expressed in other forms—see next section.

If R^i_{jkl} is contracted with respect to its first and fourth indices, we obtain the *Ricci tensor*

$$R_{jk} = R^l_{jkl}, \tag{27.9-3}$$

which plays a role in relativity.

The remainder of this section is devoted to the Laplacian and d'Alembertian operators. In n-dimensional Euclidean space, in Cartesian coordinates, the Laplace operator is given by

$$\nabla^2 = \partial_1^2 + \partial_2^2 + \cdots + \partial_n^2,$$

where

$$\partial_k = \frac{\partial}{\partial x^k}.$$

In the Minkowski space of special relativity, the d'Alembertian operator is given by

$$\Box^2 = \partial_1^2 + \partial_2^2 + \partial_3^2 - \partial_4^2.$$

Both operators can be written as $g^{jk} \partial_j \partial_k$, where (g^{jk}) is the diagonal metric tensor. In physical theories, these operators are applied to scalar and vector field. When applied to a scalar field f, their generalization to a Riemannian or pseudo-Riemannian manifold is simply $g^{jk} f_{;j;k}$, but when they are applied to a vector field v_i, it is necessary to introduce a tensor denoted by $v_{i;jk}$, called the *second symmetric extension* of v_i (see T. Y. Thomas 1961), defined as the tensor whose components are given at the origin P of geodesic coordinates $y^1, \ldots, y^n$ by

$$v_{j;kl}|_P = \frac{\partial^2 \hat{v}_j}{\partial y^k \partial y^l}\bigg|_P, \tag{27.9-4}$$

where $\hat{v}_j(y^1, \ldots, y^n)$ are the components of the vector field relative to the geodesic coordinates. According to Exercise 4 below, this is not in general equal to the result of symmetrizing the second covariant derivative $v_{j;k;l}$ with respect to the indices k and l. According to the principle of equivalence, we therefore define the Laplacian or d'Alembertian operator, as applied to the vector field, as

$$g^{kl} v_{j;kl}, \tag{27.9-5}$$

rather than in terms of the second covariant derivative—but see Exercise 3 in the next section.

Higher symmetric extensions are similarly defined, for instance,

$$v_{j;klm}\big|_P = \frac{\partial^3 \bar{v}_j}{\partial x^k \, \partial x^l \, \partial x^m}\bigg|_P.$$

EXERCISES

1. Show that if φ is a scalar field, then

$$\varphi_{;k;l} = \varphi_{;l;k} = \varphi_{;kl},$$

whereas, if v_j is a covariant vector field, then, generally

$$v_{j;k;l} \neq v_{j;l;k}.$$

2. Show that, in an arbitrary coordinate system, the second symmetric extension is given by

$$v_{j;kl} = \frac{\partial^2 v_j}{\partial x^k \, \partial x^l} - \Gamma^m_{kl}\frac{\partial v_j}{\partial x^m} - \Gamma^m_{lj}\frac{\partial v_m}{\partial x^k} - \Gamma^m_{jk}\frac{\partial v_m}{\partial x^l} - v_m\Gamma^m_{jkl}, \qquad (27.9\text{-}6)$$

where Γ^m_{jkl} is the coefficient given by (27.2-10).

3. Show that, at the origin of geodesic coordinates,

$$\frac{\partial}{\partial y^j}\mathring{\Gamma}^m_{kl} + \frac{\partial}{\partial y^k}\mathring{\Gamma}^m_{lj} + \frac{\partial}{\partial y^l}\mathring{\Gamma}^m_{jk} = 0, \qquad (27.9\text{-}7)$$

where the superscript $^\circ$ indicates that the connection components are those that refer to the geodesic coordinates $y^1, \ldots, y^n$.

4. Show that if v_j is a smooth vector field, then

$$v_{j;kl} - \tfrac{1}{2}(v_{j;k;l} + v_{j;l;k})$$
$$= \tfrac{1}{6}(R^m{}_{lkj} + R^m{}_{klj})v_m. \qquad (27.9\text{-}8)$$

5. Show that at the origin of geodesic coordinates

$$\frac{\partial}{\partial y^j}\mathring{\Gamma}^m_{kl} = \tfrac{1}{3}(\mathring{R}^m{}_{ljk} + \mathring{R}^m{}_{kjl}). \qquad (27.9\text{-}9)$$

The last shows that the Riemann tensor gives all the intrinsic information about a space in the immediate neighborhood of a point P that is given by the connection components and their first derivatives at P. Namely, by a suitable choice of coordinates (geodesic coordinates), the Γ^m_{kl} can all be made $= 0$ at P; hence they give no intrinsic or coordinate-free information, and then their first derivatives are all determined by the Riemannian tensor as in (27.9-9). In the next section it will be seen in a similar fashion that when there is a metric tensor g_{jk}, the Riemann tensor gives all the intrinsic information that is conveyed by the g_{jk} and their first and second partial derivatives at P.

27.10 The Riemann Tensor in a Riemannian or Pseudo-Riemannian Manifold

We now suppose that a metric tensor g_{kl} is defined in the manifold. Then, the connection components Γ^i_{jl}, etc., that appear in the definition (27.9-2) of the Riemann tensor are to be identified with the Christoffel symbols of the second kind, $\{^i_{jl}\}$, etc., defined by (26.6-10, 11). A straightforward calculation then shows that the Riemann tensor R_{ijkl} (the first index has been lowered) can be expressed in terms of g_{kl} and its derivatives as follows:

$$R_{ijkl} = \frac{1}{2}\left(\frac{\partial^2 g_{il}}{\partial x^j\,\partial x^k} + \frac{\partial^2 g_{jk}}{\partial x^i\,\partial x^l} - \frac{\partial^2 g_{ik}}{\partial x^j\,\partial x^l} - \frac{\partial^2 g_{jl}}{\partial x^i\,\partial x^k}\right)$$

$$+ g^{pq}([jk, p][il, q] - [ik, p][jl, q]), \qquad (27.10\text{-}1)$$

where the square brackets denote the Christoffel symbols of the first kind, given by (26.6-10).

The number of independent components of this tensor is less than n^4 because of the following symmetry relations, which follow from the above equation:

$$R_{ijkl} = -R_{jikl} = -R_{ijlk} = R_{klij}, \qquad (27.10\text{-}2)$$

$$R_{ijkl} + R_{iklj} + R_{iljk} = 0. \qquad (27.10\text{-}3)$$

We shall now show that, in consequence of these relations, the number of independent components of the Riemann tensor is

$$\frac{n^2(n^2 - 1)}{12}, \qquad (27.10\text{-}4)$$

which is $= 1, 6$, and 20, respectively, in 2, 3 and 4 dimensions. First, it follows from the three relations (27.10-2) that any nonvanishing component can be written (by change of sign, if necessary) as R_{ijkl}, where $i < j$ and $k < l$, and where, if (ij) and (kl) are regarded as two-digit base-n integers, $(ij) \le (kl)$. The number of possible values of (ij) or (kl) is then $\frac{1}{2}n(n - 1) = m$ and the number of possible pairs (ij), (kl) is

$$\tfrac{1}{2}m(m + 1) = \frac{n(n - 1)(n^2 - n + 2)}{8}. \qquad (27.10\text{-}5)$$

Next, unless i, j, k, and l are all different, (27.10-3) reduces to a combination of the preceding identities (27.10-2), and if they are different, it can be assumed, without loss of generality, that $i < j < k < l$, for if i', j', k', l' is any permutation of i, j, k, l, the identity (27.10-3) for i, j, k, l can be obtained from the same one for i', j', k', l' by use of the preceding identities (27.10-2). Therefore the number of independent identities of type (27.10-3) is $n(n - 1)(n - 2)(n - 3)/4!$, and subtraction of this form (27.10-5) gives (27.10-4). It can be proved that the R_{ijkl} satisfy no further algebraic identities independent of (27.10-2, 3).

As in the preceding section, if R^i_{jkl} is contracted once, the Ricci tensor

$$R_{jk} = R^l_{jkl} \qquad (27.10\text{-}6)$$

results. In Riemannian and pseudo-Riemannian spaces, this tensor is symmetric: $R_{jk} = R_{kj}$. A second contraction yields the *Riemann curvature scalar*

$$R = g^{jk}R_{jk}. \qquad (27.10\text{-}7)$$

The covariant derivative of the Riemann tensor satisfies the *Bianchi identity*

$$R_{ijkl;m} + R_{ijlm;k} + R_{ijmk;l} = 0, \qquad (27.10\text{-}8)$$

as is easily verified from (27.10-1) by using geodesic coordinates about a point, so that the covariant derivatives become ordinary derivatives at that point.

Inner multiplication of the last equation with $g^{ik}g^{jl}$ gives

$$-R_{;m} + R^k_{m;k} + R^l_{m;l} = 0.$$

Here the second and third terms are equal, and the first term is just $-\partial R/\partial x_m$. It follows from this equation that the symmetric tensor

$$L_{jk} = R_{jk} - Rg_{jk}, \qquad (27.10\text{-}9)$$

which appears in the Einstein field equation, is divergence-free:

$$L^j_{k;j} = 0, \qquad L_j{}^k_{;k} = 0. \qquad (27.10\text{-}10)$$

We now consider briefly the question, under what conditions an affinely connected manifold can be made Riemannian by introduction of a metric compatible with the connection, that is, the question whether, when the $\Gamma^r_{kl}(x^1, \ldots, x^n)$ are given, with $\Gamma^r_{kl} = \Gamma^r_{lk}$, there exist functions $g_{kl}(x^1, \ldots, x^n)$, with $g_{kl} = g_{lk}$, such that $\Gamma^r_{kl} = \{^r_{kl}\}$, i.e., such that

$$\Gamma^r_{kl} = \frac{1}{2} g^{rm}\left(-\frac{\partial g_{kl}}{\partial x^m} + \frac{\partial g_{ml}}{\partial x^k} + \frac{\partial g_{km}}{\partial x^l}\right), \qquad (27.10\text{-}11)$$

where, as usual, the matrix (g^{rm}) is the inverse of the matrix (g_{kl}). This equation can be solved to give

$$\frac{\partial g_{ks}}{\partial x^l} = g_{rs}\Gamma^r_{kl} + g_{rk}\Gamma^r_{sl}. \qquad (27.10\text{-}12)$$

This equation is just $g_{ks;l} = 0$, by definition of the covariant derivative, and is known to be satisfied in a Riemannian or pseudo-Riemannian manifold. Therefore, the necessary and sufficient condition for the existence of a metric is that this system (27.10-12) of equations have a solution $g_{kl}(x^1, \ldots, x^n)$.

EXERCISES

1. Show from the compatibility condition of the system (27.10-12) that the solution g_{kl}, if it exists, must satisfy the equation

$$R^r_{kjl}g_{rs} + R^r_{sjl}g_{rk} = 0, \qquad (27.10\text{-}13)$$

which is just the first of the relations (27.10-2). Derive further conditions from this equation by covariant differentiation.

2.   Consider a manifold in which the affine connection is given by

$$\Gamma^i_{j2}(x^1,\ldots,x^n) = \Gamma^i_{2j}(x^1,\ldots,x^n) = \delta^i_j x^1;$$

otherwise $\Gamma^i_{jk} = 0.$

Show that, in this manifold, (27.10-13) has no symmetric solution at the origin. Show that the Ricci tensor is not symmetric: $R_{12} \neq R_{21}$.

3.   Show that in a Riemannian or pseudo-Riemannian manifold, if the Ricci tensor R_{ij} vanishes, as in empty space-time according to general relativity, the two natural definitions of the Laplacian or d'Alembertian operator, as applied to a vector field, agree:

$$g^{kl}v_{j;kl} = g^{kl}v_{j;k;l}.$$

Otherwise, as noted in the preceding section, it is the expression on the left that gives the correct form at the origin of geodesic coordinates.

4.   Show that in a Riemannian or pseudo-Riemannian manifold, at the origin of geodesic coordinates,

$$\frac{\partial^2}{\partial y^l \, \partial y^m} \mathring{g}_{jk} = -\tfrac{1}{3}(\mathring{R}_{jklm} + \mathring{R}_{kljm})$$

$$= -\tfrac{1}{3}(\mathring{R}_{jklm} + \mathring{R}_{jmkl}). \tag{27.10-14}$$

It follows that the Riemann tensor gives all the information about the geometry in the immediate neighborhood of a point that is given by the metric tensor and its first and second partials at the point, because there is a coordinate system (geodesic coordinates) such that the standard equations $g_{jk} = \delta_{jk}$ and $\partial g_{jk}/\partial x^l = 0$ hold at the point; hence these quantities give no information, and then the second derivatives are given by the above equation.

27.11   The Riemann Tensor and the Intrinsic Curvature of a Manifold

Suppose that an n-dimensional manifold $\mathfrak{M}$ is immersed in (i.e., is regarded as an n-dimensional surface in) a Euclidean space E^N, with its metric inherited from E^N. The geometry of $\mathfrak{M}$ then has both intrinsic and extrinsic aspects, the former being those aspects that are determined by the metric tensor in $\mathfrak{M}$ and are independent of the particular way $\mathfrak{M}$ is immersed in E^N. (For example, a plane sheet of unstretchable material can be rolled up as shown in Figure 27.1, and its inherited metric is then still identical with that of the Euclidean plane.) A geodesic in $\mathfrak{M}$ is generally a curve in E^N, and its curvature is an extrinsic property; the center of curvature lies outside $\mathfrak{M}$.

On the other hand, the kind of curvature possessed by the surface of a sphere can be determined from the metric. The curvature of the earth can in principle be determined by measurements made entirely on its surface. One constructs a large triangle whose sides are geodesics (shortest paths between

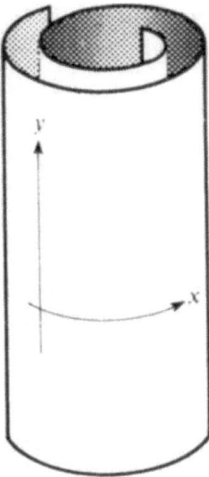

Figure 27.1

vertices); one measures the area A of the triangle and the sum Σ of its angles. Then the radius r of the earth is given by the formula

$$\frac{A}{r^2} = \Sigma - \pi. \qquad (27.11\text{-}1)$$

Clearly, $\Sigma - \pi$ is the angular deviation of a vector from its initial direction after being carried round the triangle by parallel transport. One may imagine the vector as an arrow lying on the deck of a ship that steams around the triangle. When the ship changes bearing at a vertex by turning through a certain angle to the right, the arrow is then turned through the same angle to the left with respect to the ship before proceeding.

Note that a surface having this kind of curvature can be immersed in various ways in E^3. A hemisphere can be bent, without stretching, by distorting the equator, as shown in Figure 27.2. According to the differential geometry of surfaces in E^3, the product of the two principal radii of curvature at a given point is invariant under such bending without stretching.

A formula that generalizes (27.11-1) will now be derived for a general n-dimensional affinely connected manifold $\mathfrak{M}$, not necessarily immersed in a flat space of higher dimension. Let $x^1, \ldots, x^n$ be coordinates in a chart in $\mathfrak{M}$. Consider a very small closed curve $\mathscr{C}$ in $\mathfrak{M}$ obtained by letting the point x^i

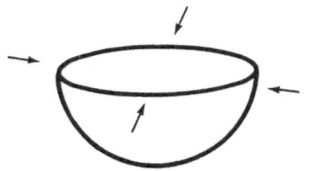 

Figure 27.2

trace out a parallelogram in the coordinate space $\mathbb{R}^n$ with vertices at

$$x^i = \begin{cases} x^i(0), \\ x^i(0) + a^i, \\ x^i(0) + a^i + b^i, \\ x^i(0) + b^i, \\ x^i(0), \end{cases}$$

where the a^i and b^i are small quantities. A simple calculation shows that if a vector v^i is carried by parallel transport round $\mathscr{C}$ starting with $v^i = \xi^i$ at the first vertex, the net deviation Δv^i is given, to lowest order in small quantities, by

$$\Delta v^i = \tfrac{1}{2}\xi^j R^i{}_{jkl}(a^l b^k - a^k b^l). \tag{27.11-2}$$

This is the required generalization of (27.11-1). From it, the Riemann tensor can be found in terms of the deviations Δv^i produced in various vectors ξ^i by parallel transport round various small loops.

This formula, as written, is valid only for small loops $\mathscr{C}$, whereas (27.11-1) was valid for an arbitrarily large triangle. That is because a sphere has constant curvature. If the same procedure were used for a general ellipsoid, it would have to be restricted to very small triangles.

27.12 Flatness and the Vanishing of the Riemann Tensor

Given two Riemannian manifolds of the same number of dimensions, the question arises whether they may not be identical spaces described by different coordinate systems. This raises global topological questions and also the following more local one: Given two Riemannian metrics, each determined by a set of n^2 smooth functions,

$$g_{jk}(x^1, \ldots, x^n) \qquad (j, k = 1, \ldots, n)$$

and

$$g'_{jk}(x'^1, \ldots, x'^n) \qquad (j, k = 1, \ldots, n),$$

defined in regions N and N', respectively, of $\mathbb{R}^n$; the question is whether there is a transformation

$$x'^j = x'^j(x^1, \ldots, x^n) \qquad (j = 1, \ldots, n)$$

of N onto N' or a part of N onto a part of N', such that the g_{jk} transform into the g'_{jk} under the transformation law for second-rank covariant tensors. If so, the two coordinate systems can be interpreted as determining two charts in a single manifold. This is a difficult question in general, but can be answered completely if the functions of one set, say the $g'_{jk}(\cdots)$, are constants, in which case the space is Euclidean and the Riemann tensor vanishes.

It will be shown that the vanishing of the Riemann tensor is sufficient as well as necessary for a space to be flat. Consequently, it is now assumed that $R^i{}_{jkl}$ vanishes everywhere in an affinely connected manifold, and it will be shown that coordinates $y^1, \ldots, y^n$ can be so chosen in any simply connected region that the corresponding connection coefficients Γ^i_{jk} vanish identically. If there is a metric tensor, the y's can be so chosen that the g_{jk} assume the standard constant values $\pm\delta_{jk}$ throughout.

Let $x^1, \ldots, x^n$ be coordinates in a simply connected chart and consider the following initial-value problem for a covariant vector field $v_i(x^1, \ldots, x^n)$:

$$\text{DE: } v_{i;\,j} = 0, \quad \text{that is, } \quad \frac{\partial v_i}{\partial x^j} - \Gamma^k_{ij} v_k = 0, \qquad (27.12\text{-}1)$$

$$\text{IC: } v_i(0, \ldots, 0) \quad \text{given.} \qquad (27.12\text{-}2)$$

As is known, this problem has a solution for arbitrary initial data (27.12-2) provided that the compatibility condition

$$\frac{\partial}{\partial x^l}\,(\Gamma^k_{ij} v_k) - \frac{\partial}{\partial x^j}\,(\Gamma^k_{il} v_k) = 0 \qquad (27.12\text{-}3)$$

is satisfied. Evaluation of the derivatives in this equation by the product rule and use of the partial differential equation (27.12-1) once more gives the equation

$$\left(\frac{\partial}{\partial x^l}\,\Gamma^k_{ij} - \frac{\partial}{\partial x^j}\,\Gamma^k_{il}\right) + \Gamma^k_{ij}\Gamma^m_{lk} v_m - \Gamma^k_{il}\Gamma^m_{jk} v_m = 0.$$

By renaming the dummy indices, this can be written as

$$R^k{}_{jli} v_k = 0,$$

where $R^k{}_{jli}$ is given by (27.9-2). Therefore, the compatibility condition is satisfied, because the Riemann tensor vanishes. Let $v_i(x^1, \ldots, x^n; p)$, $p = 1, \ldots, n$, be n solutions of the initial value problem obtained by taking n linearly independent initial vectors $v_i(0, \ldots, 0; p)$, $p = 1, \ldots, n$; each of these solutions satisfies $v_{i;\,j} = 0$. Now consider, for each $p = 1, \ldots, n$, the further initial value problem

$$\frac{\partial}{\partial x_i}\,y(x^1, \ldots, x^n) = v_i(x^1, \ldots, x^n; p),$$

$$y(0, \ldots, 0) = 0.$$

The compatibility condition of this problem is

$$\frac{\partial}{\partial x^j}\,v_i - \frac{\partial}{\partial x^i}\,v_j = 0,$$

that is

$$v_{i;\,j} - v_{j;\,i} = 0$$

(the other terms in the covariant derivatives cancel), and hence is satisfied, because the covariant derivatives are zero. We denote the solution of the new initial-value problem by $y^p(x^1, \ldots, x^n)$, for each p, and we choose new coordinates by setting $y^p = y^p(x^1, \ldots, x^n)$, $p = 1, \ldots, n$, which is possible because the Jacobian of the y's with respect to the x's at the origin is $\neq 0$ [because it is the determinant whose columns are the vectors $v_i(0, \ldots, 0; p)$, which were chosen to be linearly independent]; according to the implicit function theorem, we can solve for the x's in terms of the y's in some neighborhood N_0 of $\mathbb{R}^n$, so that the y's become new independent variables in N_0.

With respect to the new coordinates, the vector fields v_j have the components

$$\mathring{v}_j(y^1, \ldots, y^n; p) = v_i(x^1, \ldots, x^n; p)\frac{\partial x^i}{\partial y^j}$$

$$= \frac{\partial y^p}{\partial x^i}\frac{\partial x^i}{\partial y^j} = \delta^p_j.$$

Since these vector fields were so constructed as to have vanishing covariant derivatives, we have

$$0 = \mathring{v}_{j;k} = 0 - \mathring{\Gamma}^i_{jk}\mathring{v}_i = -\mathring{\Gamma}^i_{jk}\delta^p_i$$

$$= -\mathring{\Gamma}^p_{jk}.$$

Therefore the connection coefficients are all zero in the new coordinates, as was claimed.

Now suppose the manifold is Riemannian or pseudo-Riemannian, and hence has a metric tensor. The initial vectors $v_i(0, \ldots, 0)$ in (27.12-2) can be chosen orthonormal, so that

$$g^{jk}(x^1, \ldots, x^n)v_j(x^1, \ldots, x^n; p)v_k(x^1, \ldots, x^n; q) = \pm\delta_{pq}$$

at $\mathbf{x} = 0$; but g^{jk} and the v_j all have vanishing covariant derivatives; hence this equation holds for all $\mathbf{x}$. The metric tensor (contravariant form) in the y^j coordinate system has components

$$\mathring{g}^{pq} = g^{jk}\frac{\partial y^p}{\partial x^j}\frac{\partial y^q}{\partial x^k}$$

$$= g^{jk}v_j(\ldots; p)v_q(\ldots; q);$$

hence $\mathring{g}^{pq} = \pm\delta_{pq}$ throughout, and hence also $\mathring{g}_{pq} = \pm\delta_{pq}$ throughout, as was claimed, because the matrix $(\mathring{g}_{pq})$ is the inverse of the matrix $(\mathring{g}^{pq})$.

The chart determined by the new coordinates $y^1, \ldots, y^n$ in the region N_0 of $\mathbb{R}^n$ is in general only a part of the original manifold $\mathfrak{M}$, but it can be extended to a complete flat manifold $\mathfrak{M}'$ by extending N_0 to all of $\mathbb{R}^n$ and requiring the Γ^i_{jk} and the g_{kl} to be constant in all $\mathbb{R}^n$. Then $\mathfrak{M}'$ is the same as $\mathfrak{M}$ if $\mathfrak{M}$ was complete and simply connected.

An example of a nonsimply connected flat manifold $\mathfrak{M}$ is the 2-torus with angles θ and φ as the coordinates and with (g_{jk}) equal to the 2×2 unit matrix everywhere. (That is, of course, *not* the metric that is inherited when the torus is embedded in the usual way in E^3.) In this case, the manifold $\mathfrak{M}'$ is the universal covering manifold of $\mathfrak{M}$.

27.13 Eisenhart's Analysis of the Stäckel Systems

The main application of Riemannian (rather, pseudo-Riemannian) geometry to physics is in general relativity. An example of an application outside relativity theory, which uses the result of the preceding section, is found in the paper of Eisenhart 1934 on separable coordinate systems for the wave equation.

If the function $\Psi = \Psi(x^1, \ldots, x^n, t)$ satisfies the wave equation

$$\left(\nabla^2 - \frac{1}{c^2}\frac{\partial^2}{\partial t^2}\right)\Psi + V\Psi = 0, \qquad (27.13\text{-}1)$$

where ∇^2 is the n-dimensional Laplacian and $V = V(x^1, \ldots, x^n)$ is a given scalar function, then, since V does not depend on t, the first step of the separation process can always be taken: One looks for solutions in the form

$$\Psi = \psi(x^1, \ldots, x^n)e^{i\omega t}, \qquad (27.13\text{-}2)$$

and then ψ satisfies the *reduced wave equation*

$$\nabla^2\psi + (\lambda + V)\psi = 0, \qquad (27.13\text{-}2)$$

where $\lambda = \omega^2/c^2$. The same equation is also obtained from the Schrödinger equation. If $x^1, \ldots, x^n$ are general curvilinear coordinates, the Laplacian can be written, when the metric tensor is known, as $g^{ij}\psi_{;i;j}$, and then the reduced wave equation takes the form

$$g^{ij}\left[\frac{\partial^2\psi}{\partial x^i\,\partial x^j} - \{^k_{ij}\}\frac{\partial\psi}{\partial x^k}\right] + (\lambda + V)\psi = 0. \qquad (27.13\text{-}4)$$

Whether this equation for ψ can be solved by the method of separation of variables depends on the forms of the functions $g_{ij}(\cdots)$ and $V(\cdots)$.

In the separation procedure (see, for example, Morse and Feshbach 1953), one starts by looking for special solutions in the form of a product

$$\psi(x^1, \ldots, x^n) = X_1(x^1)X_2(x^2)\ldots X_n(x^n).$$

When this product is substituted into (27.13-4), it turns out that, for certain choices of the coordinate system (hence of the g_{ij}) and of the function V, one obtains a system of second-order ordinary differential equations, one for each of the functions X_i, depending on n arbitrary so-called separation constants, of which λ in the above equations is the first. In this way one obtains a large enough family of special solutions (depending on the separation constants and constants of integration that appear in the solution of the ordinary differential equations) to serve as a complete set for the expansion of an arbitrary function of $x^1, \ldots, x^n$.

In brief outline, the investigations of various authors showed that three conditions on the metric are necessary for separability, i.e., for success of the procedure just described. First, the coordinates must be orthogonal; that is, the matrix (g_{ij}) must be diagonal. The other two conditions on the g_{ij} are the so-called Stäckel and Robertson conditions. It was shown by H. P. Robertson 1927 that those conditions are also sufficient for separability, if the function V has a suitable form.

At that point, the theory was not yet in a satisfactory state, because a tensor $g_{ij}(x^1, \ldots, x^n)$ satisfying the above conditions might not be the metric tensor in a Euclidean space, or in any other space of interest, for that matter. Eisenhart 1934 therefore added to the above three conditions the further condition that the Riemann tensor R_{ijkl} vanish identically. Any tensor g_{jk} satisfying all these conditions is then the metric tensor in Euclidean space for *some* curvilinear coordinate system. For $n = 3$, Eisenhart gave a full classification of all separable systems in E^3, and for each of the systems he determined explicit formulas for the separation procedure.

The separable systems in E^3 are the general ellipsoidal coordinates and special cases obtained therefrom by such limiting processes as making two semiaxes equal, letting one focus go to infinity, and so on, including prolate and oblate spheroidal, parabolic, paraboloidal, elliptic cylinder, parabolic cylinder, circular cylinder, spherical, and Cartesian coordinates. For detailed descriptions of these systems, see Morse and Feshbach.

CHAPTER 28

The Extension of Einstein Manifolds

Special-relativistic and general-relativistic field equations; stress-energy
tensor; the cosmological constant; Einstein manifold; the Schwarzschild and
Finkelstein charts; Birkhoff's theorem; the meaning of spherical symmetry;
the Kruskal manifold; maximal and geodesically complete manifolds; other
extensions of the Schwarzschild charts; time reversal; the Kerr manifold; the
Cauchy problem of the Einstein field equations.

Prerequisites: Chapters 23, 24, 26, and 27, and some knowledge of general
relativity.

Of the many mathematical problems connected with general relativity, the
extension problem has been chosen for discussion in this chapter, because it is
concerned with the global geometrical and topological properties of Einstein
manifolds, and those properties seem to me to constitute the most basically
mathematical aspect of the theory. Although no use is made of formulas or
results outside the preceding chapters of this book, the present chapter will
probably be intelligible only to readers with some knowledge of relativity.
In particular, the first two sections do not pretend to be in any sense a derivation of the principles of relativity, but merely a discussion of them.

28.1 Special Relativity

It is recalled that the electromagnetic theory of J. C. Maxwell not only
perfected and completed classical theoretical physics, but also led to three
major crises: (1) Unlike all previous physical theories, it failed to be invariant
under Galilean or Newtonian transformations from one inertial frame to
another; hence it seemed to imply the existence of an absolute frame of
reference in the universe. (2) When the principles of statistical physics were
applied to it, it led to the so-called ultraviolet catastrophe. (3) When applied to
electrons, regarded as point charges, it led to infinite self-force and self-
energy. The first crisis was resolved by the special theory of relativity, the
second by the quantum hypothesis, both around 1900–1905, and the third

(not yet perfectly) by the renormalization program of quantum electro-dynamics in the early 1950's.

To resolve the first difficulty, Lorentz and Fitzgerald introduced the hypothesis that when an object is in motion relative to the absolute frame of reference, it becomes contracted by a certain factor in the direction of motion, and its actions become slowed by a similar factor, owing somehow to the electromagnetic effects of the extra terms in Maxwell's equations caused by the motion of its frame of reference. The hypothesis contradicted no known facts, because very little was then known of the structure of matter, although it was clear that electromagnetism was involved. The consequence of the hypothesis was that all purely electromagnetic phenomena were invariant under the so-called Lorentz transformations, which take the contraction and time-dilation into account (in fact Maxwell's equations are invariant under them); hence one could never detect motion through the absolute frame by electromagnetic effects.

Einstein then introduced the further hypothesis that *all* physical laws are invariant under Lorentz transformations (those laws had of course to be modified, when velocities comparable to c were involved, to achieve this). All inertial frames thereby acquired equal status; hence, the laws of physics were simplified in that it was no longer necessary to consider how electromagnetic effects of motion through the absolute frame might affect mechanics, thermo-dynamics, atomic structure, and so on. As soon as the laws were known in one frame, the principle of Lorentz invariance then determined them in every other. In the next section, similar simplifications of physical laws by general relativity are pointed out.

28.2 The Einstein Gravitational Field Equations

The classical gravitational field equation is

$$\nabla^2 \varphi = 4\pi G \rho, \tag{28.2-1}$$

where $\varphi = \varphi(\mathbf{x})$ is the gravitational potential, $\rho = \rho(\mathbf{x})$ is the matter density, and G is the universal constant of gravitation. The equation can be made Lorentz-invariant, as follows: In special relativity, ρ is $1/c^2$ times the 4,4 component of a symmetric second-rank tensor $T_{\mu\nu}$, the stress-energy tensor. (As in Section 19.4, x^1, x^2, x^3 are space coordinates, and x^4 is ct.) Hence, φ is taken as the 4,4 component of a tensor $\varphi_{\mu\nu}$, and equation (28.2-1) is replaced by

$$\left(\nabla^2 - \frac{1}{c^2}\frac{\partial^2}{\partial t^2}\right)\varphi_{\mu\nu} = \frac{4\pi G}{c^2} T_{\mu\nu},$$

or

$$h^{\rho\sigma}\frac{\partial}{\partial x^\rho}\frac{\partial}{\partial x^\sigma}\varphi_{\mu\nu} = \frac{4\pi G}{c^2} T_{\mu\nu}, \tag{28.2-2}$$

where

$$(h^{\rho\,\sigma}) = (h_{\rho\,\sigma}) = \begin{bmatrix} 1 & & & \\ & 1 & & (0) \\ & (0) & 1 & \\ & & & -1 \end{bmatrix} \tag{28.2-3}$$

is the metric tensor of flat Minkowsky space.

[As in Section 19.4, Greek indices go from 1 to 4, and Latin indices from 1 to 3. Physicists often let the Greek indices go from 0 to 3, and they then write the tensors $h^{\rho\sigma}$ and $h_{\rho\,\sigma}$ with the first diagonal element $=1$ and the others $= -1$, so that $ds^2 = (dx^0)^2 - (dx^1)^2 - (dx^2)^2 - (dx^3)^2$. The conventions used in this chapter are thought to be a little closer to those of the mathematical literature.]

The law of motion for a point mass in a gravitational field, which says classically that the acceleration is equal to $-\nabla\varphi$, takes the Lorentz-invariant form

$$\ddot{x}^\mu + h^{\mu\nu} \frac{\partial \varphi_{\rho\sigma}}{\partial x^\nu} \dot{x}^\rho \dot{x}^\sigma = 0,$$

where the dot stands for differentiation with respect to proper time τ along the trajectory and the $x^\mu(\tau)$ are the coordinates of the particle. The equation resembles that of a geodesic, but of course the second term comes from the gravitational field and has nothing to do with geometry.

Such a theory may be called the *special relativity theory* of gravitation. It was never considered seriously, to any extent, because Einstein discovered the general theory before observational tests of relativistic effects became feasible. In the special theory, when there is no gravitational field, so that $\varphi_{\mu\nu} \equiv 0$, free bodies move along straight lines, i.e., geodesics, in the 4-dimensional Minkowski space, whereas, when $\varphi_{\mu\nu} \neq 0$, the trajectories depart from straight lines, and relative accelerations are observed.

In Einstein's general theory, the trajectories are assumed to be always geodesics (in the absence of nongravitational forces), and the relative accelerations are attributed to the curvature of space time. That assumption simplified physics in that it then became unnecessary to consider how a gravitational field might modify the laws of electromagnetism, quantum theory, and so on, for those laws were assumed to take their field-free form in a local inertial or "freely falling" frame of reference, in which the gravitational field has been "transformed away." A study of the equations of the geodesics shows that, when space-time is almost flat (almost Minkowskian), a coordinate system can be found in which the tensor $g_{\mu\nu}$ differs only slightly from the $h_{\mu\nu}$ and differs in such a way that the motion is the same as though space time were exactly flat and there were a gravitational potential given by $\varphi_{\mu\nu} = (c^2/2)(g_{\mu\nu} - h_{\mu\nu})$. In this limit, the Einstein field equations must reduce to

$$g^{\rho\,\sigma} \frac{\partial}{\partial x^\rho} \frac{\partial}{\partial x^\sigma} g_{\mu\nu} = \frac{8\pi G}{c^4} T_{\mu\nu}. \tag{28.2-4}$$

Einstein therefore chose an equation of the form

$$W_{\mu\nu} = \frac{8\pi G}{c^4} T_{\mu\nu}, \tag{28.2-5}$$

where $W_{\mu\nu}$ is a tensor containing the components of the metric tensor and their first and second derivatives with respect to the x^μ and where $W_{\mu\nu}$ was required to have the following properties: (1) It must be symmetric and divergence-free, i.e.,

$$W_{\mu\nu} = W_{\nu\mu}, \qquad g^{\nu\sigma} W_{\mu\nu;\sigma} = 0;$$

this is necessary because the stress-energy tensor $T_{\mu\nu}$ of the matter has these properties. (2) When the fields are weak (i.e., space-time is nearly flat), it must reduce to the left member of (28.2-4) in the coordinates referred to. Einstein found that the only choice of the $W_{\mu\nu}$ that has all these properties is the left member of the following equation, which is called the *Einstein field equation*:

$$R_{\mu\nu} - \tfrac{1}{2}R g_{\mu\nu} - \Lambda g_{\mu\nu} = \frac{8\pi G}{c^4} T_{\mu\nu}; \tag{28.2-6}$$

here, $R_{\mu\nu}$ is the Ricci tensor defined in Section 27.9, and Λ is a constant. The second term in (28.2-6) is necessary to make the left member divergence-free, according to equations (27.10-9, 10).

In applications to small systems like the solar system or a single galaxy, one would like to suppose that space-time is asymptotically flat at large distances; this requires that the so-called cosmological constant Λ be zero. Einstein assumed, however, that Λ is not exactly zero but very small. He believed that that was necessary in order to obtain closed (i.e., finite) models of the universe and thus avoid the various forms of Olbers's paradox that appear in an infinite and asymptotically uniformly populated universe. A. A. Friedman showed in 1922, however, that closed models could also be obtained for $\Lambda = 0$. Since then, Λ has usually been taken as zero.

In the remainder of this chapter, Λ will be taken as zero, and the discussion will be further restricted to empty regions of space time, where $T_{\mu\nu} = 0$. In this case, since $g^{\mu\nu}g_{\mu\nu} = 4$, contraction of (28.2-6) shows that the scalar curvature R is zero; hence, the gravitational equation reduces to $R_{\mu\nu} = 0$. An *Einstein manifold* is therefore defined as a 4-dimensional manifold of signature 2 in which the Ricci tensor $R_{\mu\nu}$ is zero throughout. More general definitions are clearly possible and are sometimes found, for example in Petrov 1969.

The problem considered in this chapter is how to extend a given Einstein manifold (usually given by a single chart) so as to obtain a larger Einstein manifold, in fact, to obtain a maximal extension, in some sense, of the given manifold.

An extension of that kind played an important role in the development of relativity. The famous Schwarzschild solution for the field around a spherical mass, which will be discussed in the next section, appeared to indicate a singularity of space-time at a certain distance (the so-called "Schwarzschild radius") from the center. The nature of this "singularity" was much discussed in the early days of relativity. It is now known that it is only a singularity of the coordinate system used and that the manifold can be extended, by adopting additional coordinate charts, to a manifold with no singularities, except at the central point. It is clear that one must understand a "solution" of the Einstein equations to mean a manifold, not merely a single formula for a line element.

28.3 The Schwarzschild Charts

In 1916, K. Schwarzschild found a stationary spherically symmetric solution of the Einstein equation $R_{\mu\nu} = 0$ for empty space, which is asymptotic to the flat Minkowski metric at large distances. The result is a manifold consisting of a single chart, which will now be described; it presumably represents space-time around a central spherical mass.

To say that the solution is stationary means that coordinates can be chosen in the chart so that the $g_{\mu\nu}$ are independent of one coordinate, say x^4, which can be regarded as the time. Clearly, then g_{44} must be <0, while the metric form $g_{jk}\, dx^j\, dx^k$ obtained by setting $dx^4 = 0$ must be positive definite.

To say that the solution is spherically symmetric means that there is a continuous group G of transformations in the manifold, not involving x^4, which is isomorphic to the rotation group SO(3), and under which the metric is invariant. At large distances, the transformations of G are assumed to take the usual form of rotations in x^1, x^2, x^3; otherwise, we can't say much about them in advance, since we have no theory of nonlinear representations of groups. For example, we cannot say that there is a point in the manifold (center of rotation) that is invariant under all transformations of G; there is in fact no such point in the manifold.

It is assumed that the manifold $\mathfrak{M}_3$ obtained by taking a fixed value of x^4 is asymptotically Euclidean at large distances, in the sense that $\mathfrak{M}_3$ contains a chart whose coordinate domain N_3 consists of all points in $\mathbb{R}^3$ such that

$$(x^1)^2 + (x^2)^2 + (x^3)^2 > a^2, \qquad (28.3\text{-}1)$$

where a is a constant, and then the metric tensor g_{jk} is asymptotic to δ_{jk} at large distances in N_3. It is assumed that, at least at large distances, there is a family of concentric spheres, each invariant under all transformations of the group G. Consider a geodesic starting at a point P of such a sphere S_0, starting inward along a normal to S_0, and followed inward as far as possible in the chart. It is invariant under those rotations that keep the point P on S_0 fixed, because the metric is invariant; hence geodesics are mapped into geodesics, but the geodesic in question is uniquely determined by its initial direction at P. If such geodesics from all points P of S_0 are followed in for a fixed distance, their terminal points determine another invariant surface or sphere S_1, and the geodesics can be used to map S_0 onto S_1, so that, in particular, spherical polar coordinates θ, φ on S_0 determine coordinate θ, φ on S_1, which transform under the transformations of the group G just as they transform on S_0. Lastly, let r be a variable, increasing smoothly outward, used to distinguish one of the invariant spheres from another. Then we take r, θ, φ as coordinates in the chart, and then the metric has spherical symmetry with respect to those coordinates in the usual sense.

We now have a 4-dimensional chart in $\mathfrak{M}$, whose coordinate domain N in $\mathbb{R}^4$ is given by

$$
\begin{aligned}
a &< r < \infty, \\
0 &< \theta < \pi, \\
-\pi &< \varphi < \pi, \\
-\infty &< x^4 < \infty,
\end{aligned}
\tag{28.3-2}
$$

where a is a constant not yet known.

As usual, this chart fails to cover the north and south poles $\theta = 0, \pi$ and the "international date line" $\varphi = \pm\pi$ on each surface $r = \text{const.}$, $x^4 = \text{const.}$, but it can be supplemented by another exactly similar chart oriented in such a way that the two completely cover all such surfaces. Henceforth, this process will be automatically assumed, whenever polar angles θ and φ are used; a reference to "a chart" under such circumstances really refers to a manifold containing these two charts.

It is now shown that when the metric is restricted to a surface $r = \text{const.}$, $x^4 = \text{const.}$, it necessarily takes the form

$$
ds^2 = A(d\theta^2 + \sin^2\theta\, d\varphi^2),
\tag{28.3-3}
$$

where A is a positive constant, which may depend on r. To show this, let r, x^4, and φ all be held constant, and consider the metric form at $\theta = 0, \varphi = 0$. It will be just $ds^2 = A\, d\theta^2$, where $A = g_{22}(r, 0, 0, x^4)$; to find ds^2 for other values of θ and φ, it is only necessary to apply a rotation which transforms θ, φ into $0, 0$. Since the metric is invariant, ds^2 must transform into $A\, d\theta^2$, but the line element (28.3-3) is known to be invariant under rotations of the coordinate system θ, φ; hence, (28.3-3) is valid on the whole surface.

Because of the symmetry, i.e., the invariance of the metric under the group G, it is easily shown that the $g_{\mu\nu}$ can depend only on r, and that there can be no terms in $g_{\mu\nu}\, dx^\mu\, dx^\nu$ that connect either of θ or φ with either of r or x^4. Hence the metric is of the form

$$
ds^2 = g_{11}\, dr^2 + A(r)(d\theta^2 + \sin^2\theta\, d\varphi^2) + 2g_{14}\, dr\, dx^4 + g_{44}(dx^4)^2.
\tag{28.3-4}
$$

At large distances, $g_{11} \approx 1$, $g_{44} \approx -1$, $g_{14} \approx 0$. Hence, in particular, by continuity, g_{44} is $\neq 0$ down to some "radius" r_0 (a very small radius, it will turn out). For $r > r_0$, at least, then, the term in g_{14} can be removed by a transformation to a new variable

$$
x'^4 = x^4 + \int \frac{g_{14}}{g_{44}}\, dr,
$$

θ, φ, and x^4 remaining unchanged. Solving for dx^4 gives $dx^4 = dx'^4 - (g_{14}/g_{44})dr$, and if this is put into (28.3-4), the terms in $dr\,dx'^4$ cancel, at the expense of two new terms in dr^2, which, however, can be incorporated into the first term $g_{11}\,dr^2$ by changing g_{11}. Then, only diagonal elements appear in ds^2; since the first three terms are positive and the fourth is negative, we can write

$$ds^2 = e^\alpha\,dr^2 + e^\gamma(d\theta + \sin^2\theta\,d\varphi^2) - e^\beta(dx^4)^2, \qquad (28.3\text{-}5)$$

where α, β, and γ are functions of r and where the prime has been omitted from x^4.

A final transformation is now made by replacing r by $r' = e^{\gamma/2}$. Then, e^β is replaced by $e^{\beta'}$, and $e^\alpha(dr)^2$ is replaced by $e^{\alpha'}(dr')^2$, where α' and β' are functions of r'. Primes are again omitted, and the result is

$$ds^2 = e^\alpha\,dr^2 + r^2(d\theta^2 + \sin^2\theta\,d\varphi^2) - e^\beta(dx^4)^2. \qquad (28.3\text{-}6)$$

This is the form with which Schwarzschild started.

The determination of the functions $\alpha(r)$ and $\beta(r)$ now proceeds as follows: One computes the Christoffel 3-index symbols $[\mu\nu, \sigma]$ and $\{{}^{\sigma}_{\mu\nu}\}$, then the components of the Riemann tensor $R_{\mu\nu\rho\sigma}$ and, by contraction, those of the Ricci tensor $R_{\mu\nu}$. These are all expressions in α, β, and their first derivatives, so that setting $R_{\mu\nu} = 0$ gives differential equations for $\alpha(r)$ and $\beta(r)$. For the details of this some somewhat lengthy calculation, the reader is referred to any book on general relativity (for example, Tolman 1934 or Weber 1961). It is found that

$$e^\beta = e^{-\alpha} = 1 - \frac{r_0}{r}, \qquad (28.3\text{-}7)$$

where r_0 is an arbitrary constant, called the *Schwarzschild radius*, which will be assumed to be >0. (The corresponding chart with $r_0 < 0$ will appear presently.)

It was pointed out in Section 28.2 that when space-time is nearly flat, the metric can be interpreted in terms of an equivalent (classical) gravitational field. For $r \gg r_0$, the above metric is equivalent to a Coulomb field with potential

$$\varphi \approx -\frac{GM}{r},$$

M being another constant. The relation between the constants M and r_0 is

$$r_0 = \frac{2GM}{c^2}. \qquad (28.3\text{-}8)$$

The application of the Schwarzschild metric to problems of the solar system, and the observational tests of relativity based on this metric, form an important chapter in the theory of relativity. If M in (28.3-8) is taken equal to the solar mass, $2 \cdot 10^{33}$ g, r_0 is found to be ≈ 3 km. For $r \gg r_0$, space-time is very nearly flat. For points inside the sun, i.e., for $r < R_\odot \approx 6.5 \cdot 10^5$ km, a different metric, the *Schwarzschild interior metric*, must be used, which takes into account the nonvanishing of the stress-energy tensor $T_{\mu\nu}$. Since $r_0 \ll R_\odot$, space-time is very nearly flat (Minkowskian) at all points outside the sun, i.e., for $r > R_\odot$ (also at all points inside, it turns out), and the gravitational field is very nearly a Coulomb field. In fact, as the reader is doubtless aware, the observational tests of general relativity require the measurement of exceedingly small effects.

This chapter is concerned with a primarily mathematical problem that comes out of the theory, namely that of finding the maximal extensions of the empty space solutions, such as the Schwarzschild exterior solution or the Kerr solution for the field around a rotating mass, into further regions of space-time devoid of matter. The astronomical interpretation of these solutions in terms of "black holes in space" or cosmological models, is outside the scope of the present discussion.

Henceforth, units of length and time will be used such that $r_0 = c = 1$, and t will be written for x^4. Then, the Schwarzschild line element is

$$ds^2 = \left(1 - \frac{1}{r}\right)^{-1} dr^2 + r^2(d\theta^2 + \sin^2\theta\, d\varphi^2) - \left(1 - \frac{1}{r}\right)dt^2. \quad (28.3\text{-}9)$$

Three coordinate charts can be constructed, using this metric. For the present, each of them is to be thought of as a separate Einstein manifold. To define a coordinate chart, it is only necessary to specify the region N in the coordinate space $\mathbb{R}^4$ in which r, θ, φ, and t vary. Clearly, singularities of the $g_{\mu\nu}$ must be avoided; hence there are three possibilities:

$$
\begin{aligned}
N_\mathrm{I}: \qquad & 1 < r < \infty, \\
& 0 < \theta < \pi, \\
& 0 < \varphi < 2\pi, \\
& -\infty < t < \infty; \\
N_\mathrm{II}: \qquad & 0 < r < 1 \\
& \text{(same as above, for } \theta,\, \varphi,\, t); \\
N_\mathrm{III}: \qquad & -\infty < r < 0 \\
& \text{(same as above, for } \theta,\, \varphi,\, t).
\end{aligned}
$$

The resulting charts will be called the (*Schwarzschild*) *Charts I, II, and III*, respectively.

If r is replaced by $-r$, it is seen that Chart III is simply the solution around a negative point mass. This is an Einstein manifold, according to the definition adopted here, and it is even geodesically complete, in the sense of Section 28.6, below. By itself, it is uninteresting, since negative masses presumably do not exist. However, it will be seen in Section 28.8, that a metric very similar to that of Chart III appears in part of the Kerr manifold, which represents the field around a rotating mass.

The following points concerning the time dependence are noted in passing: It was shown in 1923 by G. D. Birkhoff that the metric of Chart I is obtained even if one drops the assumption of stationarity. That is, the metric of Chart I is the only spherically symmetric one that is asymptotically flat at infinite distances. This comes about as follows: If the functions α and β that appear above are allowed to depend on t as well as on r, then a more complicated general solution appears. However, this solution can always be transformed into the stationary one (28.3-9) by a pure transformation of coordinates. It follows, for example, that the gravitational field around a radially pulsating star is static. In electromagnetic terminology, there is no monopole radiation of gravitational waves. There is no dipole radiation, either, because there are no negative masses; however, quadrupole radiation can occur, and this is believed to provide one mechanism for loss of energy by pulsars. Finally, it is noted that the metric of Chart II is not stationary in the sense of the definition at the beginning of this section; the timelike variable is r, not t.

The definition of spherical symmetry adopted above needs one comment. If P is any point in the manifold $\mathfrak{M}_3$ (x^4 being constant), and if $S(P)$ denotes the set of all points into which P is carried by the transformations of the group G, then $S(P)$ is a surface $r = $ const., which is 2-dimensional and is a sphere (a 2-sphere) by any ordinary criterion. That need not be true in a general spherically symmetric manifold. Let $\mathfrak{M}$ be the manifold of the group SO(3), and let G be the group of left translations $\varphi(h)$: $g \rightarrow hg$ ($\forall g$) in $\mathfrak{M}$. Then, the mapping $h \rightarrow \varphi(h)$ is an isomorphism of SO(3) onto G. However, if P is any point of $\mathfrak{M}$, then the set $S(P)$ of all points into which P is carried by transformations of G is not 2-dimensional—it is all of $\mathfrak{M}$, hence 3-dimensional. In the Schwarzschild Chart I, the asymptotic flatness at infinite distances evidently puts an additional restriction on the effect of the group G on the manifold, so that we have only 2-dimensional invariant sets. The manifold of SO(3) is compact, so no such restriction can be applied to it.

28.4 The Finkelstein Extensions of the Schwarzschild Charts

New coordinates were introduced in the Schwarzschild Chart I by D. Finkelstein (1958), by writing

$$t' = t + \log(r - 1),$$
$$r, \theta, \varphi \quad \text{unchanged,} \tag{28.4-1}$$

whereupon the metric form (28.3-9) becomes

$$ds^2 = \left(1 + \frac{1}{r}\right)dr^2 + r^2(d\theta^2 + \sin^2\theta\, d\varphi^2) - \left(1 - \frac{1}{r}\right)dt'^2 + \frac{2}{r}\, dt'\, dr.$$

$$\tag{28.4-2}$$

For $r > 1$, t' goes from $-\infty$ to $+\infty$ when t does; hence, the range of variation of r, θ, φ, t' is the same as that of r, θ, φ, t. The resulting chart covers the same region of space time as the Schwarzschild Chart I, but in different coordinates.

FINKELSTEIN   I

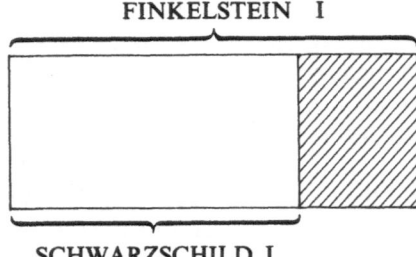

SCHWARZSCHILD I   **Figure 28.1**

However, (28.4-2) contains no singularity at $r = 1$; hence this chart can by extended, by letting r go down to zero, to give the *Finkelstein Chart I*, defined by (28.4-2) together with

$$N_{\mathrm{F,I}}: \qquad 0 < r < \infty,$$
$$0 < \theta < \pi,$$
$$-\pi < \varphi < \pi,$$
$$-\infty < t' < \infty;$$

it covers a larger portion of space time than the Schwarzschild Chart I, as indicated schematically in Figure 28.1.

By the further transformation

$$t'' = t' - \log(1 - r) \qquad (28.4\text{-}3)$$

applied in the shaded part of the Finkelstein Chart (the part where $0 < r < 1$), the Schwarzschild Chart II is obtained (in the variables r, θ, φ, t''); hence the Finkelstein Chart contains one copy each of the Schwarzschild Charts I and II. However, it is not simply a matter of eliminating the barrier at $r = 1$ in the metric (28.3-9); to see that, consider the t, r and t', r coordinate spaces (the coordinates θ and φ being fixed), as in Figure 28.2. When the straight line $\mathscr{C}'$ in the Finkelstein Chart is followed down to $r = 1$, the corresponding curve $\mathscr{C}$ in the Schwarzschild Chart ($\mathscr{C}$ and $\mathscr{C}'$ represent the same curve in space time) escapes off to $t = +\infty$, hence cannot be continued further, while the original curve $\mathscr{C}'$ can be continued down to $r = 0$.

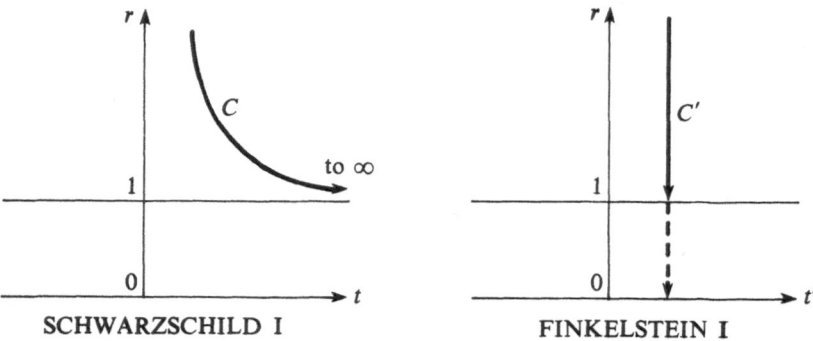

SCHWARZSCHILD I   FINKELSTEIN I

Figure 28.2

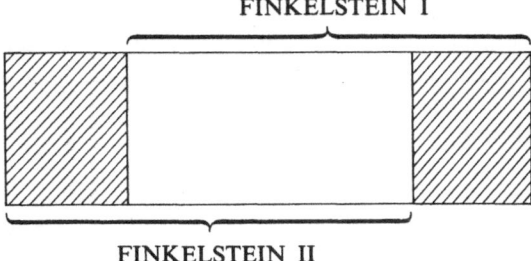

FINKELSTEIN I

FINKELSTEIN II   **Figure 28.3**

Furthermore, in place of (28.4-1) there is a further transformation

$$t''' = t - \log(r - 1),$$

which, when applied to the Schwarzschild metric, gives the same as (28.4-2), but with $+(2/r)dt'\,dr$ replaced by $-(2/r)dt'''\,dr$. The resulting chart (with $0 < r < \infty$) is called the *Finkelstein Chart II*. In this case, the curve corresponding to $\mathscr{C}$ of Figure 28.4(2) escapes to $t = -\infty$ instead of $t = +\infty$; hence, this extension of the Schwarzschild Chart I leads to a still different part of space-time, as indicated schematically in Figure 28.3. Indefinite further extensions can be obtained by alternate successive use of transformations of the type $t \to t \pm \log(r - 1)$ and $t \to t \pm \log(1 - r)$ in the intervals $(1, \infty)$ and $(0, 1)$ of r. Throughout the manifolds described so far, the Einstein field equations are satisfied in the form $R_{\mu\nu} = 0$, and the signature is 2.

28.5 The Kruskal Extension

In the Schwarzschild Chart I, with metric (28.3-9), M. Kruskal introduced (1960) new coordinates u, θ, φ, v by the equations

$$u = \sqrt{r - 1}\, e^{r/2} \cosh \frac{t}{2},$$

$$\theta, \varphi \text{ unchanged,} \qquad\qquad\qquad (28.5\text{-}1)$$

$$v = \sqrt{r - 1}\, e^{r/2} \sinh \frac{t}{2},$$

whereupon the metric form becomes

$$ds^2 = f(u, v)^2(du^2 - dv^2) + r(u, v)^2(d\theta^2 + \sin^2\theta\, d\varphi^2), \quad (28.5\text{-}2)$$

where f and r are certain functions, of which r is defined as the positive solution of the equation

$$[r(u, v) - 1]e^{r(u, v)} = u^2 - v^2, \qquad\qquad (28.5\text{-}3)$$

and then f is given by

$$f(u, v)^2 = \frac{4}{r(u, v)}\, e^{-r(u, v)}. \qquad\qquad (28.5\text{-}4)$$

The Schwarzschild Chart I corresponds to the ranges $u > 0$, $-u < v < u$ of the variables (θ and φ have their usual ranges); this corresponds to the shaded quadrant labelled "I" in Figure 28.4. However, the metric form (28.5-2) is nonsingular and of signature 2, and is a solution of the equation $R_{\mu\nu} = 0$, throughout the more extended range

$$N_{\mathrm{K}} \begin{cases} -\infty < u < \infty, \\ \theta, \varphi, \text{ as usual,} \\ -\sqrt{1 + u^2} < v < \sqrt{1 + u^2}, \end{cases} \tag{28.5-5}$$

which corresponds to the entire area between the solid curves in Figure 28.4, consisting of the regions labeled I, I', II, and II'. Each of the regions I and I' is a copy of the Schwarzschild Chart I and each of the regions II and II' is a copy of the Schwarzchild Chart II. Any two adjacent regions of these four constitute a copy of one of the Finkelstein Charts, and the origin ($u = v = 0$) is a point of space-time not covered by any of the preceding charts.

Kruskal showed that the manifold $\mathfrak{R}$ determined by the metric (28.5-2) and the region of the u, v plane given by (28.5-5), which will be called the *Kruskal manifold*, is a maximal extension, in a sense that will be defined, of the Schwarzschild Charts.

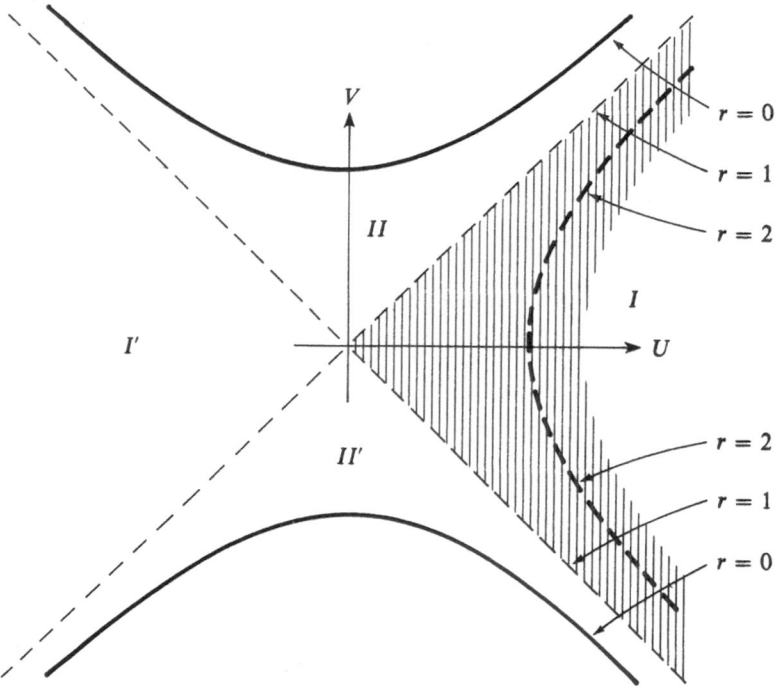

Figure 28.4 Diagram of the Kruskal manifold.

28.6 Maximal Extensions; Geodesic Completeness

The distinguishing feature of the Kruskal manifold $\Re$ is that if a geodesic $\mathscr{C}$ is started at an arbitrary point $x^\mu(0)$ and in a direction given by an arbitrary initial tangent vector $\dot{x}^\mu(0)$, then either it can be continued to infinite values of the natural parameter λ in the manifold or it encounters, for some finite λ, a genuine singularity. A manifold having this property will be called *geodesically complete* [some authors use this term a little differently—see Ellis (1972)].

By a *genuine singularity* is meant a point or set of points where some curvature invariant (that is, some scalar constructed from the functions $g_{\mu\nu}$ and their derivatives of various orders) becomes infinite. Such a singularity cannot be removed by a change of coordinates, because this scalar, being an invariant, tends to infinity in any coordinate system, as that point or set of points is approached.

The proof of the geodesic completeness of the Kruskal manifold is elementary but tedious. It is shown first that the curves $r(u, v) = 0$ in the u, v plane consists of genuinely singular points. It is easy to find curvature scalars that $\to \infty$ as $r(u, v) \to 0$. The Riemann curvature scalar R is not one of these, because $R \equiv 0$ in $\Re$. However, the scalar $R_{\alpha\beta\gamma\delta}R^{\alpha\beta\gamma\delta}$, where $R_{\alpha\beta\gamma\delta}$ is the Riemann tensor, has the value $6[r(u, v)]^{-4}$, hence $\to \infty$ as $r(u, v) \to 0$.

The value of the *function* $r(u, v)$ in the Kruskal manifold is equal to the value of the *coordinate* r in the Schwarzschild manifolds. When r is a polar coordinate, one tends to think of $r = 0$ as representing a "single point." In the Kruskal manifold, however, the equation $r(u, v) = 0$ appears to determine a whole curve, namely the hyperbola drawn in Figure 28.4. In any case, however, this "point" or "curve" is not in the manifold $\Re$, and is not in the region of space-time being described. Whether the singularity is a "point" or is "extended" is probably a physically meaningless question.

Kruskal then showed that, if a geodesic $\mathscr{C}$ has arbitrary $x^\mu(0)$ in $\Re$ and arbitrary $\dot{x}^\mu(0)$, then either $r(u, v) \to 0$ as $\lambda \to$ some finite λ_0 on $\mathscr{C}$, or else $\lambda \to \infty$ along $\mathscr{C}$. Hence $\Re$ is geodesically complete. In either of the Schwarzschild Charts I or II, there are geodesics for which $r \to 1$ and $t \to \pm\infty$ for finite values of the natural parameter λ. Since λ is an invariant having physical significance, while the coordinates r, θ, φ, t are purely arbitrary, it seems clear that it ought to be possible to continue these geodesics in some other charts, and that is just what the Kruskal extension achieves.

28.7 Other Extensions of the Schwarzschild Manifolds

Instead of the transformation (28.5-1) introduced by Kruskal, consider the transformation from r, θ, φ, t to $\xi, \theta, \varphi, \eta$ given by

$$\xi = (r - 1)e^r,$$

$$\theta, \varphi \text{ unchanged}, \tag{28.7-1}$$

$$\eta = (r - 1)e^r \sinh t.$$

The Schwarzschild metric (28.3-9) then becomes

$$ds^2 = \frac{e^{-r(\xi)}}{r(\xi)(\xi^2 + \eta^2)} [\xi(d\xi^2 - d\eta^2) + 2\eta \, d\xi \, d\eta] + r(\xi)^2[d\theta^2 + \sin^2 \theta \, d\varphi^2],$$

(28.7-2)

where the function $r(\xi)$ is given, for any $\xi > -1$, as the positive solution of the equation

$$\xi = [r(\xi) - 1]e^{r(\xi)}.$$ (28.7-3)

A manifold based on (28.7-2) which will be called $\mathfrak{N}'$, is given by the chart in which the ranges of the variables are:

$$N': \quad \left.\begin{array}{l} -1 < \xi < \infty \\ -\infty < \eta < \infty \end{array}\right\} \text{ except } \xi = \eta = 0,$$

θ, φ as usual.

This manifold contains one copy each of the Schwarzschild manifolds I and II; they are in the regions $0 < \xi < \infty$ and $-1 < \xi < 0$, respectively. It will be seen below that this manifold $\mathfrak{N}'$ is maximal—i.e., it cannot be extended to any larger one; it might therefore possibly seem superior to the Kruskal manifold, since the formulas are simpler. However, the singularity at $\xi = \eta = 0$ is in a sense a singularity of the coordinate system and not a genuine singularity; it will be seen that all curvature invariants have finite limits as the point $\xi = \eta = 0$ is approached.

The manifold $\mathfrak{N}'$ has the further property of time reversal. That is, if $\tan \alpha$ denotes the slope $d\xi/d\eta$ of a null geodesic in the ξ, η plane (a geodesic on which $ds^2 = 0$ while θ and φ are constant), and if $\tan \beta$ denotes ξ/η, it is seen from (28.7-2) that

$$\tan \beta(1 - \tan^2 \alpha) = 2 \tan \alpha,$$

which is equivalent to the equation $\tan \beta = \tan 2\alpha$; hence, if a point in the ξ, η plane encircles the origin once clockwise, β increases by 2π while α increases by π; i.e., the directions of the null geodesics have been reversed.

Clearly, the type of singularity exhibited by the manifold $\mathfrak{N}'$ at $\xi = \eta = 0$ is physically unacceptable; generally, a maximal extension of a given Einstein manifold is not physically reasonable unless it is geodesically complete.

The nature of the singularity at $\xi = \eta = 0$ is further explored in the following exercises:

EXERCISES

1. Consider the one-chart 2-dimensional manifold $\mathfrak{M}$ given by the metric form

$$ds^2 = \frac{\xi(d\xi^2 - d\eta^2) + 2\eta \, d\xi \, d\eta}{\xi^2 + \eta^2}$$ (28.7-4)

and the region

$$N = N_{\mathfrak{M}}: \qquad \left.\begin{array}{l} -\infty < \xi < \infty \\ -\infty < \eta < \infty \end{array}\right\}, \quad \text{except } \xi = \eta = 0$$

of the coordinate space. Show that $\mathfrak{M}$ is flat (i.e., that $R_{\alpha\beta\gamma\delta} \equiv 0$), and is of signature zero, hence is locally Minkowskian. Show that $\mathfrak{M}$ exhibits time reversal.

   2.   Find the geodesics on the manifold $\mathfrak{M}$ of the preceding exercise. Show that, when a geodesic $\mathscr{C}$ goes to ∞ in the ξ, η plane, the natural parameter λ on $\mathscr{C}$ goes to $\pm \infty$, whereas, when a point approaches the origin on a geodesic, λ tends to a finite value. The geodesics on which the latter happens are the half-lines $\xi = r \cos \alpha$, $\eta = r \sin \alpha$, α a constant and $r > 0$.

   Hints: These exercises become trivial if one transforms to new variables x, t such that

$$\xi = x^2 - t^2, \qquad \eta = 2xt, \tag{28.7-5}$$

which can also be written as $\xi + i\eta = (x + it)^2$, whereupon ds^2 becomes $dx^2 - dt^2$. The transformation $\xi, \eta \to x, t$ is double-valued, but can be made single-valued in any simply connected part of the region $N_{\mathfrak{M}}$.

   3.   The transformation (28.7-5) $x, t \to \xi, \eta$ makes the punctured x, t plane into a two-sheeted covering manifold of $\mathfrak{M}$. Discuss the other covering manifolds of $\mathfrak{M}$, and show that the two-sheeted one is the only one that has a geodesically complete extension (obtained by putting the origin of the x, t plane back in).

   4.   By similar use of the transformation from u, v to ξ, η given by $\xi + i\eta = (u + iv)^2$, show that the Kruskal manifold $\mathfrak{K}$, with the point $u = v = 0$ removed, is a two-sheeted covering of the manifold $\mathfrak{K}'$ defined at the beginning of this section. Show that the most general covering of $\mathfrak{K}'$ contains n copies of the Schwarzschild I and n copies of Schwarzschild II, where n is a positive integer or ∞. Show that the two-sheeted covering is the only one that has a geodesically complete extension (obtained by putting the point $u = v = 0$ back in). Since the projection from $\mathfrak{K}$ to $\mathfrak{K}'$ preserves the metric, it is evident that all curvature invariants in $\mathfrak{K}'$ have finite limits as the singularity at $\xi = \eta = 0$ is approached.

28.8   The Kerr Manifolds

An axisymmetric solution of the empty space field equation $R_{\mu\nu} = 0$ was obtained by Kerr (1963); it can be interpreted—at least, its outer part can— as the solution around a rotating star. The metric is asymptotically Minkowskian at large distance r, as in the Schwarzschild solution. At large but finite distances, the metric contains a term that represents the Newtonian potential $-GM/r$, as in the Schwarzschild solution, where M is the star's mass, and also a term α/r^2 times an angular factor, where α is the star's angular momentum. [*Note*: This is a purely relativistic effect; it arises from the presence of the momentum density ρv of the matter in certain components of $T_{\mu\nu}$ on the right side of (28.2-6); the classical Poisson equation does not give such an effect. Classically, rotation has an *indirect* effect on the gravitational field through the oblateness of a rotating star, but that effect varies as $1/r^3$.] The solution will now be discussed, without derivation or proof.

One description of the Kerr manifold is in terms of coordinates called t, x, y, z. (As in similar cases, the reader should be warned that these are arbitrary symbols; x, y, and z are not necessarily to be interpreted as Cartesian coordinates in any sense, nor t as time.) In units such that $2M$, G, and c are unity (the unit of length is then the Schwarzschild radius $2MG/c^2$), the metric form is

$$ds^2 = dx^2 + dy^2 + dz^2 + \frac{\rho^3}{\rho^4 + a^2 z^2} (k_\mu \, dx^\mu)^2 - dt^2, \qquad (28.8\text{-}1)$$

where

$$k_\mu \, dx^\mu = \frac{z}{\rho} \, dz + \frac{\rho}{\rho^2 + a^2} (x \, dx + y \, dy) + \frac{a}{\rho^2 + a^2} (x \, dy - y \, dx) - dt,$$

$$(28.8\text{-}2)$$

and where $\rho = \rho(x, y, z)$ is a function determined by the equation

$$\frac{x^2 + y^2}{\rho^2 + a^2} + \frac{z^2}{\rho^2} = 1. \qquad (28.8\text{-}3)$$

The axis of rotation is the z axis; a is a constant equal to twice the ratio of the angular momentum to the mass in the units used. For rapidly rotating stars, $a \gg 1$, while a can be of the order of 1 or smaller for slowly rotating ones.

Various charts can be based on this solution. There is a singularity on the circle $x^2 + y^2 = a^2$, $z = 0$, where $R_{\alpha\beta\gamma\delta} R^{\alpha\beta\gamma\delta} \to \infty$. This corresponds to the singularity at $r = 0$ in the Kruskal solution, and in fact the circle contracts to the origin as $a \to 0$.

Equation (28.8-3) has two roots $\rho(x, y, z)$ of opposite sign at every point x, y, z except on the disk $x^2 + y^2 < a^2$, $z = 0$, where $\rho = 0$. As the point x, y, z passes through this disk, it is necessary to switch from the one solution to the other, in order to make the derivatives of the $g_{\mu\nu}$ continuous. This is achieved by introducing charts $\mathfrak{M}_1, \ldots, \mathfrak{M}_4$ as follows. In all of them, ds^2 is given by (28.8-1), but with the sign of $\rho(x, y, z)$ specified in each case.

$$\text{For } \mathfrak{M}_1 \quad \begin{cases} N = \{\text{all } x, y, z, t\} - \{x^2 + y^2 \le a^2, z = 0\}, \\ \rho(x, y, z) > 0. \end{cases} \qquad (28.8\text{-}4)$$

The equation for N means that the closed central disk is excluded from N. In analogy with function theory, this disk is called a *branch cut*.

$$\text{For } \mathfrak{M}_2 \quad \begin{cases} N = \text{same as for } \mathfrak{M}_1, \\ \rho(x, y, z) < 0. \end{cases} \qquad (28.8\text{-}5)$$

The next two charts serve to connect $\mathfrak{M}_1$ and $\mathfrak{M}_2$ across the disk. For them, N could be any simply connected region containing the open disk (but not the circle, of course). It could be some sort of thin wafer, but for simplicity we take it to be all space with the region of the x, y plane exterior to the disk excluded as a branch cut.

For $\mathfrak{M}_3$
$$\begin{cases} N = \{\text{all } x, y, z, t\} - \{x^2 + y^2 \geq a^2, z = 0\}, \\ \text{sgn } \rho(x, y, z) = \text{sgn } z. \end{cases} \tag{28.8-6}$$

For $\mathfrak{M}_4$
$$\begin{cases} N = \text{same as for } \mathfrak{M}_3. \\ \text{sgn } \rho(x, y, z) = -\text{sgn } z. \end{cases} \tag{28.8-7}$$

The chart $\mathfrak{M}_3$ agrees with $\mathfrak{M}_1$ for $z > 0$ and with $\mathfrak{M}_2$ for $z < 0$, while $\mathfrak{M}_4$ agrees with $\mathfrak{M}_1$ for $z < 0$ and with $\mathfrak{M}_2$ for $z > 0$. A doubly connected manifold $\mathfrak{M}_0$ can now be constructed by the mappings.

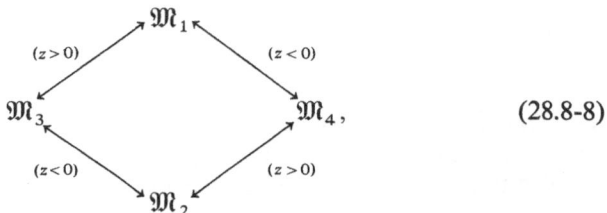

$$\tag{28.8-8}$$

where, in each case, the mapping is the identity mapping

$$t \to t, \qquad x \to x,$$
$$y \to y, \qquad z \to z.$$

in the half-space ($z > 0$ or $z < 0$) referred to in the diagram. In this manifold, a point can return to its initial position after threading twice through the circles $x^2 + y^2 = a^2$, $z = 0$.

This manifold contains two asymptotically flat regions extending to infinite distances: Both have $|\rho| \gg \max\{1, a\}$; one has $\rho > 0$ and the other $\rho < 0$. The first of these exhibits, at large distances, the gravitational field of a positive mass ($= \frac{1}{2}$ in the present units), and the second that of a negative mass ($= -\frac{1}{2}$). The two regions are connected through the central ring.

An analysis of the geodesics shows that this manifold $\mathfrak{M}_0$ is geodesically complete, for $|a| > \frac{1}{2}$, but not for $|a| < \frac{1}{2}$. The extension to a larger and geodesically complete manifold, for $|a| < \frac{1}{2}$ is carried out in Boyer and Lindquist 1967. The resulting manifold is rather complicated, and the reader is referred to Boyer and Lindquist for its description; the present discussion will be restricted to an analogy between the Finkelstein charts and the manifold $\mathfrak{M}_0$ described above.

To investigate the metric at large distances (large values of ρ), it is convenient to introduce polar coordinates r, θ, φ by writing $z = r \cos \theta$, etc., with the usual reminder that this does not necessarily make r a radial distance coordinate. For large positive ρ, (28.8-3) shows that $\rho \approx r$. Then (28.8-2) gives $k_\mu \, dx^\mu \approx dr - dt$, and (28.8-1) then gives, in the approximation $\rho = r$,

$$ds^2 = \left(1 + \frac{1}{r}\right)dr^2 + r^2(d\theta^2 + \sin^2 \theta \, d\varphi^2) - \left(1 - \frac{1}{r}\right)dt^2 - \frac{2}{r} \, dr \, dt,$$

which is the metric (28.4-2) of the Finkelstein Chart I. That is, in the limit $a = 0$, (28.8-1) is exactly the Finkelstein Chart I; hence, it is expected that, for small a, it will be necessary to supplement this chart with another copy of it and two copies of a chart that reduces to the Finkelstein Chart II for $a = 0$, joined together as the Finkelstein Charts are in the Kruskal manifold. These latter charts are obtained by changing the sign of dt in equation (28.8-2) for the quantity $k_\mu \, dx^\mu$.

There are therefore two versions of the manifold $\mathfrak{M}_0$ defined by the schema (28.8-8), according to the sign of dt in (28.8-2), say $\mathfrak{M}_0^-$ and $\mathfrak{M}_0^+$; they are both contained in the geodesically complete manifold constructed by Boyer and Lindquist.

28.9 The Cauchy Problem

The Cauchy problem of Einstein's equations (for empty space) has one feature in common with the Cauchy problem of Maxwell's equations. It is recalled that two of Maxwell's equations, namely

$$\mathbf{V} \cdot \mathbf{E} = 0, \qquad \mathbf{V} \cdot \mathbf{H} = 0 \qquad (28.9\text{-}1)$$

can be regarded as conditions solely on the initial data; if these conditions are satisfied at $t = 0$, it then follows from the other equations,

$$\frac{\partial \mathbf{E}}{\partial t} = -c\mathbf{V} \times \mathbf{H}, \qquad \frac{\partial \mathbf{H}}{\partial t} = c\mathbf{V} \times \mathbf{E} \qquad (28.9\text{-}2)$$

(these are six equations when written in component form), that the divergence conditions (28.9-1) are then automatically satisfied for all $t \geq 0$. Hence, the full system of eight equations contains enough redundancy so that only the six equations (28.9-2) in six unknowns need to be regarded as the equations of evolution.

The Einstein field equations

$$R_{\beta\gamma} = 0 \qquad (28.9\text{-}3)$$

have similar properties, but with a somewhat different consequence. First, since the tensors $g_{\beta\gamma}$ and $R_{\beta\gamma}$ are both symmetric, we may take the $g_{\beta\gamma}$ with $\beta \leq \gamma$ as the unknowns, and we only need to consider the equations (28.9-3) with $\beta \leq \gamma$. Then there are ten equations in ten unknowns. It will be seen that these equations impose four conditions or constraints on the initial data and are such that, if these conditions are satisfied at $t = x^4 = 0$, then they are automatically satisfied later. Therefore, there are only six independent equations of evolution for ten unknowns; the solution is hence under-determined and contains four arbitrary functions. That is just as it ought to be. The initial-value problem, if properly formulated, ought to determine the geometry of space-time for $x^4 > 0$, or at least for x^4 in some interval $(0, T)$, but the geometry does not uniquely determine the $g_{\mu\nu}$, owing to the possibility of coordinate changes. Any solution of the initial-value problem can be modified by subjecting the coordinates in the region $x^4 > 0$ to any arbitrary transformation that leaves the initial data on the initial surface $x^4 = 0$ unchanged.

Since the differential equations (28.9-3) are of the second order, the initial data are the values of

$$g_{\mu\nu}, \frac{\partial g_{\mu\nu}}{\partial x^4} \quad (\text{all } x^1, x^2, x^3) \quad \text{for } x^4 = 0. \tag{28.9-4}$$

The choice $x^4 = 0$ of the initial hypersurface $\mathscr{S}$ does not imply that $\mathscr{S}$ is flat, because the metric tensor g_{jk} in $\mathscr{S}$ is arbitrary (it is recalled that Latin indices take values from 1 to 3), but the initial data are assumed to be such that the 3×3 matrix (g_{jk}) is positive definite on $\mathscr{S}$. Also, $(g_{\mu\nu})$ must be nonsingular and of signature 2 on $\mathscr{S}$. It follows that $\mathscr{S}$ is spacelike, and, by use of the formula for matrix inverse, that $g^{44} < 0$ on $\mathscr{S}$. To simplify the present discussion, the functions (28.9-4) are assumed to be analytic, so that the power-series method may be used to solve the Cauchy problem. All partial derivatives of the $g_{\mu\nu}$ that do not involve differentiation with respect to x^4 more than once, are then determined on $\mathscr{S}$ by the functions (28.9-4).

The components $R_{\beta\gamma}$ of the Ricci tensor are obtained from equation (27.10-1) for $R_{\alpha\beta\gamma\delta}$ by contracting with respect to α and δ, that is, by multiplying by $g^{\alpha\delta}$ and then summing on α and δ from 1 to 4. The result is

$$R_{jk} = \frac{1}{2} g^{44} \frac{\partial^2 g_{jk}}{(\partial x^4)^2} + \cdots. \tag{28.9-5}$$

$$R_{j4} = -\frac{1}{2} g^{4k} \frac{\partial^2 g_{jk}}{(\partial x^4)^2} + \cdots, \tag{28.9-6}$$

$$R_{44} = \frac{1}{2} g^{jk} \frac{\partial^2 g_{jk}}{(\partial x^4)^2} + \cdots, \tag{28.9-7}$$

where only those terms containing second derivatives with respect to x^4 have been written; the dots stand for terms containing quantities that have been differentiated at most once with respect to x^4. [The summation convention applies in (28.9-6, 7), but the Latin indices take only the values 1, 2, 3.] Since $g^{44} \neq 0$, the differential equations $R_{jk} = 0$ and the initial data determine the second derivatives of the g_{jk} on $\mathscr{S}$. When these second derivatives are substituted into the other four equations $R_{\beta 4} = 0$, four conditions on the initial data are obtained, while the second derivatives of the $g_{\beta 4}$ with respect to x^4 are undetermined.

The equations of evolution can be separated from the auxiliary condition by a device due to Lichnerowicz (see Adler, Bazin, and Schiffer 1965, where there is an excellent discussion of the Cauchy problem). The mixed form of the Einstein tensor is

$$G^{\alpha}{}_{\beta} = R^{\alpha}{}_{\beta} - \tfrac{1}{2} R \delta^{\alpha}{}_{\beta}, \tag{28.9-8}$$

where $R^\alpha{}_\beta = g^{\alpha\gamma}R_{\gamma\beta}$ is the mixed form of the Ricci tensor and $R = g^{\mu\nu}R_{\mu\nu}$ is the curvature scalar. It is asserted that the system of differential equations

$$R_{jk} = 0 \qquad (j, k = 1, 2, 3, j \le k), \tag{28.9-9}$$

$$G^4{}_\beta = 0 \qquad (\beta = 1, \ldots, 4) \tag{28.9-10}$$

is equivalent to the original system (28.9-3).

Note. For any solution $g_{\mu\nu}$ of these equations, R has the value zero, so that G^α_β and R^α_β have the same value (namely zero), but, as expressions containing the dependent variables $g_{\mu\nu}$ and their derivatives, they are different; hence (28.9-9, 10) is a set of differential equations different from the set (28.9-3), but equivalent to it. To prove the assertion, it suffices to note that, according to (28.9-8), if the R_{jk} are set equal to zero, then,

$$\begin{aligned} G^4{}_k &= g^{4\gamma}R_{\gamma k} = g^{44}R_{4k}, \\ G^4{}_4 &= \tfrac{1}{2}g^{44}R_{44}; \end{aligned} \tag{28.9-11}$$

hence, since $g^{44} \ne 0$, the system (28.9-10) implies the vanishing of all $R_{\beta\gamma}$, and this in turn implies the vanishing of all the $G^\alpha{}_\beta$.

If the expressions (28.9-5, 6, 7) for the components of the Ricci tensor are substituted into (28.9-8), it is seen that the differential equations $G^4{}_\beta = 0$ do not contain any second derivatives with respect to x^4. Hence, these equations play the role of auxiliary conditions; in particular they are conditions that must be satisfied by the initial data.

An important property of the Einstein tensor (which has already played a role in Section 28.2) is that it is divergence-free, according to equation (27.10-10); that is,

$$G^\alpha{}_{\beta\,;\,\alpha} = 0. \tag{28.9-12}$$

This equation will be used to show that if the functions $g_{\mu\nu}$ satisfy the differential equations (28.9-9) for all x^4 in some interval $[0, T]$ and satisfy the auxiliary condition (28.9-10) for $x^4 = 0$, then they satisfy the auxiliary condition also for all x^4 in $[0, T]$. The formulas for covariant differentiation in Section 27.5 show that (28.9-12) can be written as

$$\frac{\partial}{\partial x^\alpha} G^\alpha{}_\beta + A^{\alpha\gamma}{}_{\beta\delta}G^\delta{}_\gamma = 0, \tag{28.9-13}$$

where the coefficients $A^{\alpha\gamma}{}_{\beta\delta}$ depend only on the $g_{\mu\nu}$ and their first derivatives.

Since the functions $g_{\mu\nu}$ are such that all R_{jk} are $= 0$, equations (28.9-8) and (28.9-11) show that each $G^j{}_\beta$ (as a function of the $g_{\mu\nu}$ and their first derivatives) can be expressed in terms of the $G^4{}_\gamma$ ($\gamma = 1, \ldots, 4$). Equation (28.9-13) then takes the form

$$\frac{\partial}{\partial x^4} G^4{}_\beta = B^{\gamma j}_\beta \frac{\partial}{\partial x^j} G^4{}_\gamma + C^\gamma{}_\beta G^4{}_\gamma.$$

where the coefficients $B^\gamma_{\beta}{}^j$ and C^γ_{β} depend only on the $g_{\mu\nu}$ and their first derivatives. This is a linear system of differential equations for the G^4_{β} (when the $g_{\mu\nu}$ are given), in which the time derivatives are given explicitly in terms of the spatial derivatives. According to the Cauchy-Kovalevski theorem, the solution is unique; hence if all four G^4_{β} vanish on $\mathscr{S}\,(x^3 = 0)$, then they vanish also for $x^4 > 0$, as was to be proved.

Lastly, the equations $G_{jk} = 0$ (28.9-9) can be regarded as differential equations for the functions g_{jk} ($j, k = 1, 2, 3, j \leq k$). However, they also contain the functions $g_{\alpha 4}$ [in the terms indicated by dots in (28.9-5)]. These four functions can be specified arbitrarily (but smoothly) in all space-time, provided only that they match the values $g_{\alpha 4}$ and $\partial g_{\alpha 4}/\partial x^4$ given on the surface $\mathscr{S}(x^4 = 0)$. Then, since $g^{44} \neq 0$, (28.9-5) shows that the equations $G_{jk} = 0$ (28.9-9) determine the second time derivatives of all the g_{jk}. By the Cauchy-Kovalevski theorem, again, these equations have a unique solution, in some interval $0 \leq x^4 \leq T$, for the given initial data.

28.10 Concluding Remarks

The general problem of the extension of Einstein manifolds is far from solved. Almost nothing is known about the existence or uniqueness of geodesically complete extensions. The simple manifolds in Section 28.7 have no such extensions, and the following one has many. Consider a single chart in which the $g_{\mu\nu}$ are defined to be Minkowskian

$$(g_{\mu\nu}) = \begin{pmatrix} 1 & & & \\ & 1 & & (0) \\ & & 1 & \\ (0) & & & -1 \end{pmatrix} \tag{28.9-1}$$

in a bounded region N of the coordinate space. In one geodesically complete extension, N' is all of $\mathbb{R}^4$ and the $g_{\mu\nu}$ are given by (28.9-1) everywhere. This is static, flat, empty space. In another extension, N is again all of $\mathbb{R}^4$, but some gravitational waves are present that have not yet arrived at the space-time region represented by N. Almost nothing is known about the Cauchy problem, except for existence and uniqueness in *some* time interval $0 \leq x^4 \leq T$. Even if there is a unique solution for all $x^1, \ldots, x^4$, this is still only one chart, and it may be an imcomplete manifold.

CHAPTER 29

Bifurcations in Hydrodynamic Stability Problems

Equation of evolution; Navier-Stokes equation; Poiseuille and Couette problems; Taylor vortices; wavy vortices; flows and semiflows in a Hilbert space; normal modes; completeness of the normal mode system; invariant manifolds; stable and unstable manifolds; fixed points, closed orbits, and invariant tori; bifurcation; supercritical and subcritical bifurcation; subharmonic bifurcation; Poincaré mappings.

Prerequisites: Chapters 1–8; some knowledge of fluid dynamics.

The work of Lorenz 1963 and of Ruelle and Takens 1971 initiated the introduction into hydrodynamic stability theory of concepts and principles from the currently active mathematical field of topological dynamical systems. It was immediately clear that some of the concepts, for example the concept of generic properties of systems, have many applications elsewhere in physics. Also the new concepts and principles put new light on old things, such as bifurcation phenomena. The idea of strange attractors and their connection with continuous power spectra gave a new understanding of chaotic behavior generally.

In this chapter and the next two, the new ideas are presented in the setting of the study of the onset of turbulence.

29.1 The Classical Problems of Hydrodynamic Stability

We consider the flow of an ideal incompressible viscous fluid under conditions of mild instability, so that the flow is smoother and simpler than the chaotic flow that constitutes fully developed turbulence, while the elements of instability and unpredictability are present in a simple form and can be studied analytically. In an arrangement in which the Reynolds number is slowly increased, the regime of interest to us is the early onset of turbulence.

The flow represents a balance between a steady influx of energy from an external source and the dissipation of energy by viscous friction. In the classical problems of the Poiseuille type, energy is supplied by an externally imposed pressure gradient which, for example, causes flow through a long circular pipe or in the channel between parallel plane walls; in the Couette problems the energy is supplied by laterally moving walls, such as a sliding plane wall or a rotating cylinder; in the Bénard problems it is supplied by an external heat source, which causes thermal convection. Other examples are flow past a circular cylinder (von Karman problem) or in the boundary layer over a flat plate parallel to the primary flow (Blesius problem).

A given problem can usually be characterized by a length l_0 and a velocity v_0, where l_0 may be a diameter or other dimension, and v_0 may be a mean velocity of the flow or the velocity of motion of one of the walls. The dimensionless quantity

$$R = \frac{l_0 v_0}{v}, \tag{29.1-1}$$

where v is the kinematic viscosity coefficient ($=$ viscosity coefficient divided by the fluid density), is called the *Reynolds number* of the problem. It is usually assumed that R is varied by varying v_0, keeping v and the dimensions constant. However, the fluid-dynamical equations are invariant under changes of length and speed scales (keeping all length and speed *ratios* constant) and the substitution of one fluid for another, so long as R is kept fixed.

The Taylor problem of flow between rotating concentric cylinders, sometimes called the circular Couette problem, and the Bénard problem of convection in a horizontal layer of fluid heated from below are the examples most studied so far of the early onset of turbulence.

29.2 Examples of Bifurcations in Hydrodynamics

We consider first the special case of the Taylor problem of flow between concentric rotating cylinders in which the outer cylinder is at rest. When the inner cylinder rotates slowly, the motion is laminar; the fluid velocity has only a θ component (where r, θ, and z are cylindrical coordinates), and that component depends only on r. When a certain critical velocity of rotation is exceeded, that flow, the so-called Couette flow, becomes unstable and has superposed on it a perturbation consisting of uniformly spaced ring vortices, as shown in Figure 29.1. If A denotes some measure of the strength of the vortices, say the maximum of the r or z component of velocity of the perturbation, then the dependence of A on the angular velocity Ω of the inner cylinder is as shown schematically in Figure 29.2(a). The two possible signs of A, for given Ω, correspond to the two possible directions of vortex rotation, which are equally likely and are determined by initial conditions (neighboring vortices rotate oppositely). If Ω exceeds the critical value Ω_1 only slightly, $|A|$ is approximately proportional to $\sqrt{\Omega - \Omega_1}$.

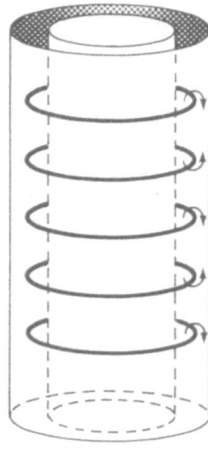

Figure 29.1   Taylor vortices.

The appearance of the new flow at $\Omega = \Omega_1$ is called a *bifurcation*. In the case just described, it is a bifurcation to another steady flow; at a fixed point in space the fluid velocity is independent of time; the ring vortices are stable and persist as long as the inner cylinder is kept rotating.

When a second critical angular velocity Ω_2 is exceeded, the ring vortices are unstable, and a second bifurcation takes place to wavy vortices, indicated schematically in Figure 29.3. The waves rotate about the common axis of the cylinders at about the mean angular velocity of the fluid in the Couette flow; hence, at a fixed point in space, the fluid velocity is now periodic in time. If A_1 denotes the amplitude of the waviness, the dependence of A_1 on Ω is as indicated in Figure 29.2(b).

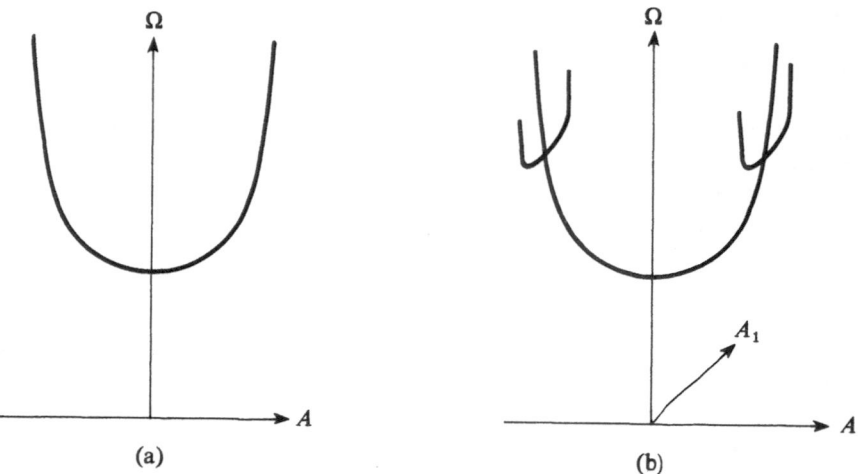

Figure 29.2   Bifurcations in the Taylor problem.

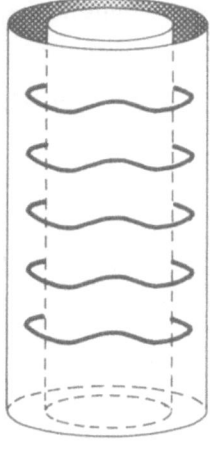

Figure 29.3   Wavy vortices in the Taylor problem.

A perhaps more familiar example of bifurcation is found in the flow past a circular cylinder. At low speed, the flow is laminar; when a critical velocity is exceeded, the laminar flow is unstable and is replaced by the so-called von Karman vortex street, in which vortices form alternately on the two sides of the cylinder, with their axes parallel to the cylinder, and then move downstream with a speed around half the ambient fluid speed. In this case the flow is periodic after the first bifurcation.

29.3 The Navier–Stokes Equations

In terms of the fluid's velocity field $\mathbf{u}(\mathbf{x}, t)$ and its pressure field $p(\mathbf{x}, t)$ the Navier–Stokes equations are

$$\frac{\partial \mathbf{u}}{\partial t} + (\mathbf{u} \cdot \nabla)\mathbf{u} + \nabla p - \nu\nabla^2\mathbf{u} = 0, \qquad (29.3\text{-}1)$$

$$\nabla \cdot \mathbf{u} = 0, \qquad (29.3\text{-}2)$$

in a region $\mathscr{R}$ of physical space, together with the boundary condition that

$$\mathbf{u} \quad \text{is given on} \quad \partial\mathscr{R}, \qquad (29.3\text{-}3)$$

and a suitable initial condition. The density has been taken $=1$ by suitable choice of units.

Let $\mathring{\mathbf{u}}(\mathbf{x}$ and $\mathring{p}(\mathbf{x})$ represent a steady solution of these equations, for example the Couette flow in the Taylor problem. For study of the effect of perturbations (either finite or infinitesimal) on that solution, it is convenient to write the total fields as

$$\mathring{\mathbf{u}}(\mathbf{x}) + \mathbf{u}(\mathbf{x}, t), \qquad \mathring{p}(\mathbf{x}) + p(\mathbf{x}, t),$$

where now **u** and p represent the departure from the steady solution. The equations are then

$$\frac{\partial \mathbf{u}}{\partial t} + (\mathbf{\mathring{u}} \cdot \nabla)\mathbf{u} + (\mathbf{u} \cdot \nabla)\mathbf{\mathring{u}} + (\mathbf{u} \cdot \nabla)\mathbf{u} + \nabla p - \nu \nabla^2 \mathbf{u} = 0, \qquad (29.3\text{-}4)$$

$$\nabla \cdot \mathbf{u} = 0, \qquad (29.3\text{-}5)$$

$$\mathbf{u} = 0 \quad \text{on} \quad \partial \mathcal{R}. \qquad (29.3\text{-}6)$$

Note that the equations have *not* been linearized. The known function $\mathbf{\mathring{u}}(\mathbf{x})$ now appears in the coefficients of the terms linear in **u**.

29.4 Hilbert Space Formulation

Corresponding to each hydrodynamic stability problem there is a problem of evolution in a Banach space $\mathfrak{H}$ (in practice a Hilbert space), consisting of an equation of evolution of the form

$$\frac{du}{dt} = Lu + B(u, u) \qquad \left\{ \begin{array}{c} u = u(t) \in \mathfrak{H} \\ t \geq 0 \end{array} \right\} \qquad (29.4\text{-}1)$$

with an initial condition

$$u(0) = u_0 \quad \text{given.} \qquad (29.4\text{-}2)$$

For each t, $u(t)$ is a point in $\mathfrak{H}$ and represents an instantaneous state of the system—a velocity field $\mathbf{u}(\mathbf{x})$ and a pressure field $p(\mathbf{x})$ in the physical space. L is a linear operator. The nonlinear terms of the Navier-Stokes equation, which (29.4-1) represents, are quadratic and are written as $B(u, u)$, where $B(\cdot, \cdot)$ is a bilinear function defined on a suitable domain in $\mathfrak{H} \times \mathfrak{H}$. We shall sometimes write $Q(u)$ for $B(u, u)$.

It is sometimes convenient to consider a more general equation

$$M \frac{du}{dt} = Lu + B(u, u), \qquad (29.4\text{-}3)$$

where M is another linear operator. If M has an inverse, then this equation can be reduced to the form (29.4-1), but often the evolution takes place not in all of $\mathfrak{H}$ but in a subspace $\mathfrak{H}_0$. Then, M has an inverse in the subspace but not necessarily in $\mathfrak{H}$, and M^{-1} may be rather complicated for calculation; hence, (29.4-3) is often convenient.

A particular Hilbert space suitable for the Taylor problem is described in the next chapter.

29.5 The Initial-Value Problem; the Semiflow in $\mathfrak{H}$

The dominant terms of the Navier–Stokes equation (29.3-1) are the first and last terms on the left, which give the equation the character of a diffusion equation. With suitable boundary conditions, the corresponding diffusion equation has a unique solution, for initial **u** in a set dense in $\mathfrak{H}$, and the solution depends continuously on the initial **u**. Because of this continuous dependence, generalized solutions can be defined for arbitrary initial **u** and they too depend continuously on the initial **u**. See Richtmyer and Morton 1967. The solution cannot in general be continued backward for negative t, although certain solutions can; in particular, normal mode solutions can.

The general behavior of the full nonlinear Navier–Stokes equation is similar (see Ladyzhenskaya 1969, 1975 and Marsden and McCracken 1976, Section 9), but the proofs are more difficult and the theory is less complete.

We shall assume that (29.4-1) or (29.4-3) has a unique solution $u(t)$ in $\mathfrak{H}$, for $t \geq 0$, for arbitrary $u(0)$ in $\mathfrak{H}$. For given initial u, we call the solution $\varphi(u, t)$, so that

$$u(t) = \varphi(u(0), t). \tag{29.5-1}$$

For fixed u, $\varphi(u, t)$ $(t \geq 0)$ is called a *motion* in $\mathfrak{H}$; for fixed $t \geq 0$, the correspondence $u \to \varphi(u, t)$ is a mapping in $\mathfrak{H}$, which is assumed continuous; for $t = 0$, it is the identity mapping, because $\varphi(u, 0) = u$. The function $\varphi(\cdot, \cdot)$ is called a *semiflow* in $\mathfrak{H}$.

Although the motions cannot generally be continued backward in time, they are unique, insofar as they can be. Stated differently, two distinct motions never coalesce, at a finite time, so as to be identical thereafter.

29.6 The Normal Modes

We look for solutions of the linearization of (29.4-1), namely of

$$\frac{du}{dt} = Lu, \tag{29.6-1}$$

in the form

$$u(t) = \psi e^{\lambda t}; \tag{29.6-2}$$

hence we look for eigenfunctions ψ and eigenvalues λ of L, given by

$$L\psi = \lambda\psi, \tag{29.6-3}$$

where, of course, $\psi \neq 0$. [We could also start from (29.4-3) and seek solutions of $L\psi = \lambda M\psi$.]

In the hydrodynamical problems, L is not self-adjoint; hence the usual spectral theory does not apply. Nevertheless, in most cases, L has a pure point spectrum, and there are denumerably many eigenvalues, so we write

$$L\psi_j = \lambda_j \psi_j \qquad (j = 1, 2, \ldots). \tag{29.6-4}$$

The completeness of the set $\{\psi_j\}$ of eigenfunctions for the expansion of an arbitrary u in $\mathfrak{H}$ has been discussed for certain hydrodynamical problems by DiPrima and Habetler 1969, using a theorem of Naimark on operators in a Hilbert space, and more generally by Sattinger 1970, using a theorem of

Carleman. Completeness is established in the sense that the finite linear combinations of the ψ_j are dense in $\mathfrak{H}$, provided that we include *generalized eigenfunctions*, if there are any, i.e., vectors ψ such that

$$(L - \lambda I)^{k+1}\psi = 0, \qquad (L - \lambda I)^k\psi \neq 0. \tag{29.6-5}$$

(If $k = 0$, ψ is an ordinary eigenfunction.)

Fluid-dynamicists have always assumed that in normal problems there are only ordinary eigenfunctions. In analogy with the finite-dimensional case, it seems likely that existence of generalized eigen-functions is a non-generic property of a fluid system. (See Appendix to Chapter 31.) An $n \times n$ matrix A can be represented by a point in a space of V of n^2 dimensions. A necessary condition for the matrix A to have a generalized eigenvector is that the characteristic equation

$$p(\lambda; A) \overset{\text{def}}{=} \det(A - \lambda I) = 0 \tag{29.6-6}$$

have a multiple root, i.e., that it have a root λ in common with the equation

$$\frac{d}{d\lambda} p(\lambda; A) = 0. \tag{29.6-7}$$

Elimination of λ from (29.6-6) and (29.6-7) gives an algebraic equation for A; hence only matrices A that lie on a surface in V, i.e., on a set of measure zero, can have generalized eigenvectors.

We shall assume that the operator L has only ordinary eigenfunctions.

Incidentally, it is easy to check for the possible existence of generalized eigenfunctions in computational problems in a Hilbert space. If one has solved the equation $L\psi = \lambda\psi$, then the adjoint equation $L^*\chi = \bar{\lambda}\chi$ has solutions; if it has a solution such that $(\chi, \psi) \neq 0$, then there are no generalized eigenvectors corresponding to the eigenvalue λ.

In the hydrodynamic problems, the eigenvalues lie mostly on or near the negative real axis; specifically they lie in a region bounded by a parabola:

$$\text{Re } \lambda \leq c_0 - c_1(\text{Im } \lambda)^2 \qquad (c_1 > 0);$$

c_0 generally increases with the Reynolds number R, but only a finite number of the eigenvalues are in the right half-plane, for any R. Because the Navier–Stokes equations are real, the eigenvalues either are real or appear in complex conjugate pairs.

29.7 Reduction to a Finite-Dimensional Dynamical System

To study the onset of turbulence, it is not necessary to know about all orbits of the system (29.4-1) in the Hilbert space $\mathfrak{H}$. It suffices to know about a special family of orbits, which, according to the physical argument given below, lie in the so-called unstable manifold that emerges from the origin in $\mathfrak{H}$.

The motions in that manifold constitute a finite-dimensional dynamical system. That idea was used by Davey, DiPrima, and Stuart 1968 and by Ruelle and Takens 1971.

Stable and unstable manifolds play an important role in the theory of differentiable dynamical systems; see Smale 1967, Abraham and Robbin 1967, and Kelley 1967. Much of the theory originated in celestial mechanics and is not directly applicable to our problem because (1) the dynamical systems of celestial mechanics are finite-dimensional to start with, while those of hydrodynamics are not, and (2) the former are Hamiltonian, while the latter are strongly dissipative and their solutions do not generally extend to negative times, so that the former determine a flow in the Hilbert space and the latter only a semiflow (but see Sell 1971, where much of the theory is so formulated as to include semiflows). We shall follow a largely intuitive approach based on physical considerations.

We assume that the Reynolds number has been maintained for a long time at a value somewhat above the first critical value, and that for that Reynolds number a finite number of the eigenvalues of the linearized problem, λ_k ($k = 1, \ldots, K$), are in the right half-plane Re $\lambda > 0$, while the rest are in the left half-plane. We assume that at a very early time (very large negative t), which we shall call the *early linear regime*, an arbitrary but very small disturbance was present, consisting of a linear combination of all the normal modes $\psi_k e^{\lambda_k t}$ ($k = 1$ to ∞). We assume that at a later time (t still negative), the *late linear regime*, all the normal modes except the first K of them have died out, and the disturbance is of the form of a linear combination

$$\sum_{k=1}^{K} a_k \psi_k e^{\lambda_k t}, \tag{29.7-1}$$

where the a_k are constants. We assume that this disturbance is still sufficiently small that these modes are growing exponentially and independently. At a still later time, the *nonlinear regime* (which we think of as including the "present" instant $t = 0$), the solution (29.7-1) has continued to grow until, owing to the nonlinearities, it is no longer of that simple form (although it still depends on the parameters $a_1, \ldots, a_K$), but may for example begin to spiral toward a closed orbit or exhibit other complicated nonlinear behavior.

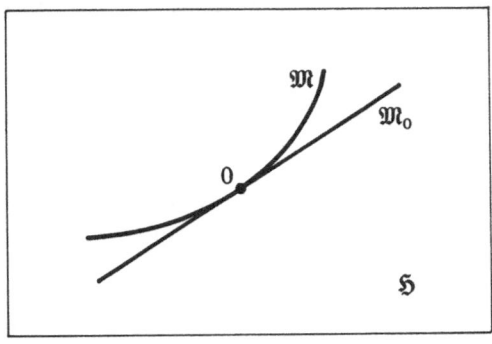

Figure 29.4

If we take a fixed t, say $t = 0$, and let the parameters $a_1, \ldots, a_K$ vary, the resulting points lie on K-dimensional surface or manifold $\mathfrak{M}$ in the Hilbert space tangent at the origin to the linear manifold $\mathfrak{M}_0$ spanned by the eigenvectors $\psi_1, \ldots, \psi_K$. (See Figure 29.4) $\mathfrak{M}$ is invariant under the semiflow in $\mathfrak{H}$ determined by (29.4-1) in the sense that any orbit that starts in $\mathfrak{M}$ remains in $\mathfrak{M}$, because, according to (29.7-1), a shift by t_0 of the origin of time is simply equivalent to changing the parameter values according to the scheme $a_k \rightarrow a_k e^{\lambda_k t_0}$.

The orbits that lie in $\mathfrak{M}$ may be regarded as describing the motions of a dynamical system with K degrees of freedom, whose properties we wish to investigate.

$\mathfrak{M}$ is called the *unstable manifold* in $\mathfrak{H}$ that issues or emerges from the origin. The unstable manifold issuing from any other fixed point in $\mathfrak{H}$ can be similarly defined after first linearizing the equation of evolution about that point.

For unstable manifolds issuing from closed or quasiperiodic orbits, see Abraham and Robbin 1967, Appendix C by Al Kelley.

The manifold $\mathfrak{M}$ is *locally attracting* in the sense that there is a neighborhood of the origin such that any orbit that remains in that neighborhood for all $t > 0$ approaches $\mathfrak{M}$ as $t \rightarrow \infty$. We shall not attempt to make that statement more precise except in one case: If $\mathfrak{M}$ results from a supercritical bifurcation at $R = R_c$, as discussed in the remaining sections of this chapter, then, for $R > R_c$, $\mathfrak{M}$ contains a new fixed point (in addition to the origin) or a new closed orbit or invariant torus, which is near the origin and which is stable, and in fact attracting, with respect to disturbances within $\mathfrak{M}$; then, for $R - R_c$ small enough, it is attracting also for arbitrary small disturbances (not necessarily in $\mathfrak{M}$). For example, Davy 1962 showed that the Taylor vortices are stable in the relevant 2-dimensional unstable manifold, and we conclude that for small $R - R_c$ they are stable with respect to *arbitrary* small disturbances, as observed in experiments.

The local stability of $\mathfrak{M}$ reflects, in the linear approximation, the exponential decrease with time of all normal modes except those appearing in the construction of $\mathfrak{M}$—see (29.7-1).

Coordinates in $\mathfrak{M}$ can be chosen in various ways. More suitable than the parameters $a_1, \ldots, a_K$ are coordinates $x_1, \ldots, x_K$ based on a projection onto the linear manifold $\mathfrak{M}_0$ to which $\mathfrak{M}$ is tangent at the origin of $\mathfrak{H}$. For any u in $\mathfrak{H}$, that projection is given by an operator P defined by

$$Pu = \sum_{k=1}^{K} (\chi_k, u)\psi_k, \tag{29.7-2}$$

where the vectors $\{\chi_k\}$ are eigenfunctions of the adjoint problem and form a biorthogonal system with the $\{\psi_k\}$. Hence, for any u in the unstable manifold $\mathfrak{M}$, we take the coordinates as

$$x_k = (\chi_k, u), \qquad k = 1, \ldots, K. \tag{29.7-3}$$

For orbits lying in $\mathfrak{M}$, the equations of motion (29.4-1) take the form

$$\dot{x}_k = F_k(x_1, \ldots, x_K), \qquad k = 1, \ldots, K. \tag{29.7-4}$$

For points near the origin, we have

$$F_k(x_1, \ldots, x_K) = \lambda_k x_k + \text{higher order terms.} \tag{29.7-5}$$

The calculations of the functions F_k is described in the next chapter. It is based on the idea that if $u(x_1, \ldots, x_K)$ is the point of $\mathfrak{M}$ (a point in $\mathfrak{H}$) corresponding to coordinate values $x_1, \ldots, x_K$, and if $x_k(t)$ ($k = 1, \ldots, K$) is any solution of (29.7-4), then the quantity

$$u(t) = u(x_1(t), \ldots, x_K(t)) \tag{29.7-6}$$

must satisfy the equation (29.4-1) of evolution in $\mathfrak{H}$. That requirement suffices to determine both the dependence of the $x_k(\cdot)$ on t and of $u(\cdots)$ on the x_k. The computational procedure assumes analyticity throughout, so that $u(x_1, \ldots, x_K)$ can be expanded in a power series in the x_k with coefficients that are elements of $\mathfrak{H}$ and the $F_k(x_1, \ldots, x_K)$ as ordinary power series. That assumption must be regarded as tentative, although the Navier–Stokes flow in $\mathfrak{H}$ is known to have at least C^∞ smoothness (see Marsden and McCracken 1976).

For the n-dimensional reversible systems of interest to celestial mechanics, one defines also the *stable manifold* that emerges from the origin (similarly from any other fixed point); it is tangent at the origin to the linear manifold spanned by the remaining eigenvectors $\varphi_{K+1}, \ldots, \varphi_n$. It can be characterized as consisting of motions $u(t)$ such that $u(t) \to 0$ as $t \to \infty$. In fact, the stable manifold is usually discussed first, and then the unstable manifold is defined as the stable manifold that would result from replacing t by $-t$. Although, in hydrodynamics, most motions cannot be reversed in time, the particular motions that lie on the unstable manifold $\mathfrak{M}$ can be, and $\mathfrak{M}$ can be characterized as consisting of these motions such that $u(t) \to 0$ as $t \to -\infty$.

For use in Section 29.10, we mention another version of the unstable manifold, which refers to mappings, rather than flows. In place of the family of mappings $u \to \varphi(u, t)$ in a Hilbert space depending on a continuous parameter t we consider a family of mappings $\mathbf{x} \to \boldsymbol{\Phi}_m(\mathbf{x})$ in an n-dimensional manifold $\mathfrak{M}$ depending on a discrete parameter m, given by iterating a mapping $\boldsymbol{\Phi}$:

$$\boldsymbol{\Phi}_m(\mathbf{x}) = \boldsymbol{\Phi}(\boldsymbol{\Phi}(\cdots \boldsymbol{\Phi}(\mathbf{x})\cdots)) \qquad (m \text{ iterations}).$$

We suppose that $\mathbf{x} = 0$ is a fixed point of $\boldsymbol{\Phi}$ and we linearize near 0:

$$\boldsymbol{\Phi}(\mathbf{x}) = M\mathbf{x} + \text{higher order terms,}$$

where M is an $n \times n$ matrix. If M has eigenvalues $\alpha_1, \ldots, \alpha_k$ ($k < n$) lying outside the unit circle $|\alpha| = 1$ and corresponding to independent eigenvectors $\mathbf{v}_1, \ldots, \mathbf{v}_k$, while the other eigenvalues all lie inside the unit circle, then there is an invariant k-dimensional manifold $\mathfrak{N}$ lying in $\mathfrak{M}$ and tangent at the origin to the linear manifold $\mathfrak{N}_0$ spanned by $\mathbf{v}_1, \ldots, \mathbf{v}_k$; see Abraham and Robin 1967 or Smale 1967.

Note on Real and Complex Hilbert Spaces. In the Hilbert space $\mathfrak{H}$ of a fluid dynamical system the functions p, u, v, w (pressure and fluid velocity components) are allowed to assume complex values, even though they are real for physical flows, so that physical solutions are restricted to a real subspace $\mathfrak{H}_0$ of $\mathfrak{H}$. Since the Navier–Stokes equations are real, the eigenvalues of the linearized problem either are real or occur in complex conjugate pairs. We can assume that an eigenfunction ψ_k is real if λ_k is real and that ψ_k and $\psi_{k'}$ are complex conjugates if λ_k and $\lambda_{k'}$ are. Then, in the representation $u = \sum c_k \psi_k$ of an element of $\mathfrak{H}$, we can require that c_k be real if λ_k is and that $c_k = \bar{c}_{k'}$ if $\lambda_k = \bar{\lambda}_{k'}$; then u lies in $\mathfrak{H}_0$. With that understanding, the manifold $\mathfrak{M}$ is a real K-dimensional surface tangent at the origin to a real linear subspace $\mathfrak{M}_0$ of $\mathfrak{H}_0$, and the coordinates x_k in $\mathfrak{M}$ given by (29.7-3) also have the property that x_k is real if λ_k is, and $x_k = \bar{x}_{k'}$ if $\lambda_k = \bar{\lambda}_{k'}$.

29.8 Bifurcation to a New Steady State

In the simpler of the two classical Hopf bifurcation theorems, it is assumed that one simple real eigenvalue $\lambda_1(R)$ crosses into the right half-plane (i.e., passes through the origin), as the Reynolds number R increases past the critical value R_c:

$$\lambda_1(R_c) = 0, \qquad \lambda_1'(R_c) = \beta > 0. \tag{29.8-1}$$

Then the unstable manifold is one-dimensional, and the equation of motion (29.7-4) takes the form

$$\dot{x} = F(x; R), \tag{29.8-2}$$

where we have suppressed the subscript 1 on x and have taken the dependence on R into account. According to (29.7-5) and (29.8-1), this equation can be

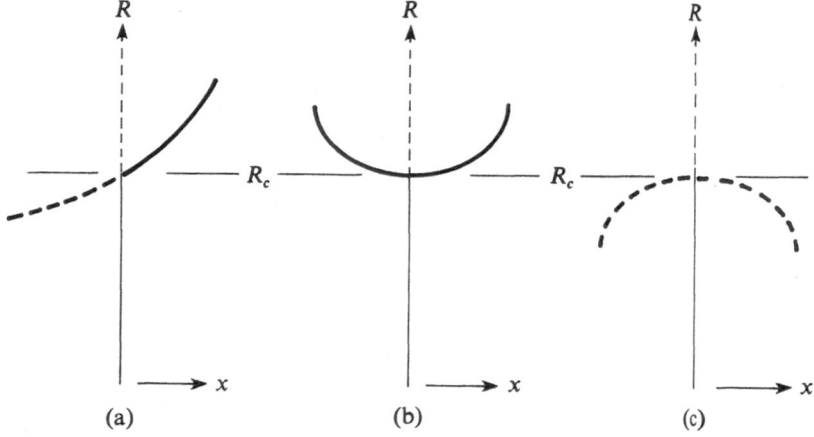

(a)   (b)   (c)

Figure 29.5

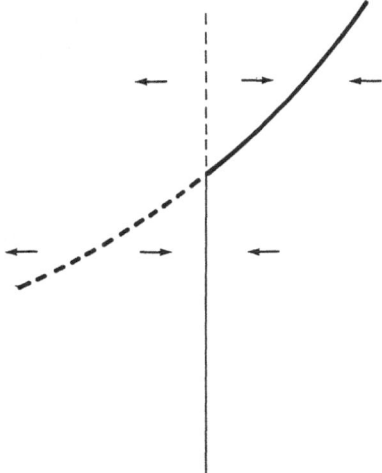

Figure 29.6

expanded as

$$\dot{x} = \beta(R - R_c)x + \text{higher order terms.} \qquad (29.8\text{-}3)$$

The stationary orbits $\dot{x} = 0$ are represented by the points on the locus $F(x; R) = 0$ in the x, R plane. The locus consists of the R axis and a curve passing through the point $x = 0$, $R = R_c$, as shown in three cases in Figures 29.5a, b, c.

If the next lowest term in (29.8-3) is ax^2, there is an unsymmetric bifurcation; if it is ax^3, there is a symmetric one, which is supercritical if $a < 0$ and subcritical if $a > 0$.

Stability is determined by the sign of $\dot{x}$ at points near the curves. For example, in the case of the unsymmetric bifurcation, the motion of points in the x, R plane is indicated by the arrows in Figure 29.6. In all cases, upturning branches are stable and downturning ones unstable, while the solution $x = 0$ is always unstable for $R > R_c$.

In the subcritical bifurcation illustrated in Figure 29.5c, there is no stable equilibrium in the neighborhood of $x = 0$ for $R > R_c$. In this case, if R is increased very slowly past R_c, a typical orbit takes the system from $x \approx 0$ to distant points of the configuration space in a relatively short time as soon as R exceeds R_c. This phenomenon is called an *explosive transition* and contrasts with the adiabatic sequence of stable states which the orbit follows in the other cases.

29.9 Bifurcation to a Periodic Orbit

In the second case of the classical Hopf bifurcation theorem, it is assumed that a single complex conjugate pair of eigenvalues crosses into the right half-plane, as R increases past R_c, while the rest remain in the left half-plane.

We write

$$\lambda_1, \lambda_2 = \sigma \pm i\omega = \sigma(R) \pm i\omega(R), \tag{29.9-1}$$

where

$$\sigma(R_c) = 0, \qquad \sigma'(R_c) > 0, \qquad \omega(R_c) \neq 0. \tag{29.9-2}$$

The manifold $\mathfrak{M}$ has two dimensions. In place of the complex conjugate coordinates x_1 and x_2 in $\mathfrak{M}$, we take real coordinates x and y such that $x_1 = x + iy, x_2 = x - iy$.

To lowest order, a motion in $\mathfrak{M}$ is given, according to (29.7-4, 5), by

$$\frac{d}{dt}(x + iy) = \lambda_1(x + iy) = (\sigma + i\omega)(x + iy).$$

Close to the origin, the orbits are approximately the spirals

$$(x + iy) \approx \text{const. } e^{\sigma t}(\cos \omega t + i \sin \omega t).$$

In polar coordinates,

$$\begin{aligned} \dot{r} &= \sigma r + O(r^2), \\ \dot{\theta} &= \omega + O(r). \end{aligned} \tag{29.9-3}$$

It follows that in some neighborhood of the origin, θ is always increasing, on any orbit, and r always positive; r may increase or decrease; close to the origin, r increases if $\sigma > 0$ and decreases if $\sigma < 0$. We now investigate what happens a little farther out.

We define the *Poincaré mapping* of the problem as a mapping $x \to \Phi(x)$ of the x axis in $\mathfrak{M}$, by saying that if an orbit has coordinates x, 0, for some t, then it has coordinates $\Phi(x)$, 0, when θ has increased by 2π. Note that x and $\Phi(x)$ can be either both positive or both negative. We write

$$\Phi(x) = x[1 + g(x)]. \tag{29.9-4}$$

The orbit structure depends on the properties of the function $g(x)$. From the spirals near the origin, we see that

$$1 + g(0) = e^{2\pi\sigma/\omega}; \tag{29.9-5}$$

in particular, $g(0) = 0$ for $R = R_c$, because then $\sigma = 0$. Hence, if we expand $g(x)$ in a Taylor series in x and $R - R_c$, we have

$$g(x) = g(x, R) = ax + b(R - R_c) + cx^2 + \cdots. \tag{29.9-6}$$

The coefficient a must be $= 0$, for otherwise $g(x, R_c)$ would have opposite signs for $x > 0$ and $x < 0$ near $x = 0$, and that would imply that an orbit crosses itself, i.e., its second turn would be farther from the origin than its first turn on one side and closer to the origin on the other. The coefficient b is positive, because, under the assumption (29.9-2) about σ, orbits near the origin spiral out for $R > R_c$ and in for $R < R_c$.

We consider the locus in an x, R plane of the equation $g(x, R) = 0$. If x_0, R_0 is a point on that locus, then, for $R = R_0, \Phi(x_0) = x_0$, that is, there is a closed orbit that cuts the x axis at $x = x_0$. The locus contains the point $x = 0, R = R_c$; near that point it is a curve that cuts the R axis horizontally; it turns upward, as in Figure 29.5b, if $c < 0$, and it turns down, as in Figure 29.5c, if $c > 0$. The bifurcation is called *supercritical* in the first of these cases and *subcritical* in the second. As in the preceding section, one must expect an explosive transition, if $c > 0$.

29.10 Bifurcation from a Periodic Orbit to an Invariant Torus

The next bifurcation, after one that results in a closed orbit, hence a periodic motion, can result in an invariant 2-dimensional torus, as shown by the example of Hopf 1948. Theorems on such a bifurcation have been given by various authors, including Naimark 1959, Sacker 1964, Ruelle and Takens 1971, and Lanford 1973. The theorems have been based largely on Floquet theory, but we shall take a more intuitive approach, based on the notion of a Poincaré mapping.

Let R_1 be the critical value of the Reynolds number R for first appearance of the periodic orbits in a supercritical bifurcation, as discussed in the preceding sections. We assume that for some $R > R_1$ the unstable manifold $\mathfrak{M}$ has dimension K. $\mathfrak{M}$ contains the 2-dimensional manifold discussed in the preceding section, and we suppose the coordinates in $\mathfrak{M}$ so chosen that the first two of them are the coordinates x, y of the preceding section; the remainder are called $x_3, \ldots, x_K$. Then, if R is not too much above R_1, the closed orbit encircles the origin in the x, y subspace and cuts the positive and negative x axes once each. Let V be the $(K - 1)$-dimensional hypersurface in $\mathfrak{M}$ given by $y = 0$; then, V is intersected twice by the closed orbit, as shown schematically in Figure 29.7, and we denote one of the intersections

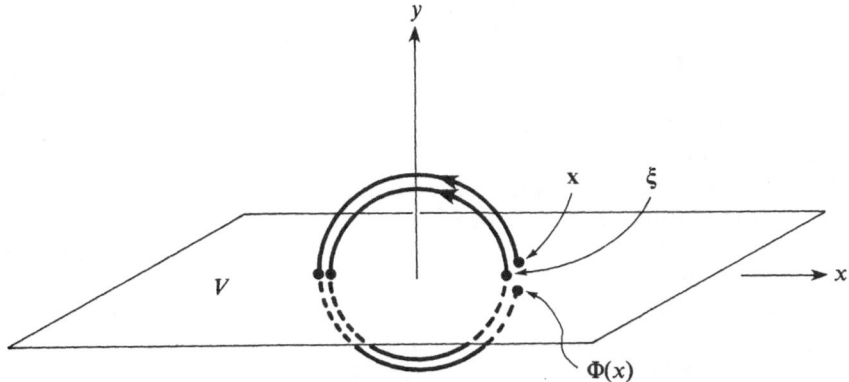

Figure 29.7

by $x = \xi = \xi(R)$, where x is a vector with $K - 1$ components $x, x_3, \ldots, x_K$. For x near ξ, let $\Phi(x)$ denote the second intersection with V of an orbit that starts at x, as shown. Then

$$\Phi: x \to \Phi(x)$$

is a Poincaré mapping defined in some neighborhood of ξ in V. Note that $\Phi(\xi) = \xi$. For x near ξ, we linearize:

$$\Phi(x) - \xi = M(x - \xi) + \text{higher order terms}, \qquad (29.10\text{-}1)$$

where M is a $(K - 1) \times (K - 1)$ matrix.

We now assume that M has one pair of eigenvalues $\alpha \ [= \alpha(R)]$ and $\bar{\alpha}$ with corresponding eigenvectors $v, \bar{v}$, such that

$$|\alpha(R_2)| = 1 \quad \text{for some } R_2 > R_1, \qquad (29.10\text{-}2)$$

$$\frac{d}{dR} |\alpha(R)|_{R = R_2} > 0, \qquad (29.10\text{-}3)$$

while the other eigenvalues of M all lie inside the unit circle. As R is increased, the closed orbit loses stability at $R > R_2$; hence, a new bifurcation occurs at $R = R_2$.

For $R > R_2$, according to Section 29.7, the mapping Φ in V has a 2-dimensional unstable manifold or surface S in V tangent at ξ to the linear manifold S_0 spanned by v and $\bar{v}$. To simplify visualization, regard V as 2-dimensional; then $S = V$. S is invariant under Φ; hence we may think of Φ as a mapping in S. We let u and v be real coordinates in S such that the projection of $x - \xi$ from S onto S_0 is $(u + iv)v + (u - iv)\bar{v}$, and we call $z = u + iv$. Then the Poincaré mapping becomes $z \to z' = \alpha z + \text{higher order}$ terms, or, more explicitly,

$$\Phi: z \to z' = \alpha z \left[1 + \sum_{j+k=1}^{\infty} q_{jk} z^j \bar{z}^k \right], \qquad (29.10\text{-}4)$$

where the q_{jk} are coefficients (generally complex) and where the summation is over all nonnegative integers j and k such that $j + k \geq 1$.

It will be shown, under certain further assumptions, that for $R > R_2$ there is a nearly circular invariant closed curve $\mathscr{C}$ in S encircling the point ξ. Then, if we let $\mathscr{C}$ be carried along by the flow in $\mathfrak{M}$, it leaves the hypersurface V and traces out an invariant tube in $\mathfrak{M}$ which is then closed, forming a torus, when $\mathscr{C}$ passes through V again near ξ.

To facilitate analysis of the Poincaré mapping, it is convenient to introduce new coordinates ξ and η in S in place of u and v, so chosen as to make the Poincaré mapping Φ take as simple a form as possible (a *normal* form), namely to eliminate some of the nonlinear terms in (29.10-4). If $|\alpha|$ were $\neq 1$, we could eliminate as many of the nonlinear terms as we wish (see Siegel, *Himmelsmechanik*, 1956, Section 21), but in order to let R vary through the value R_2, where $|\alpha(R)| = 1$, without a singularity of the transformation, we must retain the term $\alpha q_{11} z^2 \bar{z}$ in (29.10-4); otherwise, we can eliminate all terms of order ≤ 4. (In principle, we could eliminate terms of order ≤ 6 by retaining a term $\alpha q_{2\,2} z^3 \bar{z}^2$, and so on, but the elimination of terms of order ≤ 4 is just what is needed for our analysis.) We transform to $\zeta = \xi + i\eta$ by the equations

$$z = \zeta + \sum_{l+m=2}^{4} \varphi_{lm} \zeta^l \bar{\zeta}^m, \tag{29.10-5}$$

$$z' = \zeta' + \sum_{l+m=2}^{4} \varphi_{lm} \zeta'^l \bar{\zeta}'^m. \tag{29.10-6}$$

We shall choose the coefficients φ_{lm} so that when these equations are substituted into (29.10-4), the Poincaré mapping takes the *normal form*

$$\zeta \to \zeta' = \alpha\zeta(1 + \beta|\zeta|^2) + O(\zeta^5), \tag{29.10-7}$$

where $O(\zeta^5)$ contains terms of degree 5 and higher in ζ and $\bar{\zeta}$ and where β is a new constant. To determine the coefficients φ_{lm} in terms of the given coefficients q_{jk}, we substitute the right member of (29.10-5) for z into the right member of (29.10-4); then we substitute the right member of (29.10-7) for ζ' into the right member of (29.10-6) and take the result as z' for the left member of (29.10-4). Equation (29.10-4) thereby becomes an identity in ζ and $\bar{\zeta}$, and the net coefficient of $\zeta^p\bar{\zeta}^q$ can be equated to zero. The resulting equations, if taken in the right order, can be solved for the coefficients φ_{lm} provided that

$$\alpha, \alpha^2, \alpha^3, \alpha^4, \quad \text{and } \alpha^5 \quad \text{are all} \neq 1, \tag{29.10-8}$$

as explained in the Appendix to this chapter. (The equation for the case $p = 2$, $q = 1$, cannot be solved for $\varphi_{2\,1}$ when $|\alpha| = 1$, but can always be solved for β, and we simply set $\varphi_{2\,1} = 0$.)

For this method of analysis to succeed, we must assume, according to (29.10-8), that as R is increased and the point $\alpha(R)$ crosses the unit circle $|\alpha| = 1$ in the complex plane, it crosses at a point which is not a root of unity of any order less than 6. What happens if it crosses at such a point is discussed briefly in the next section.

If we let μ be a new dimensionless parameter in place of R given by $\mu = |\alpha(R)| - 1$, and if r and θ are polar coordinates, given by $\zeta = re^{i\theta}$, the Poincaré mapping takes the form

$$\mathbf{\Phi}: \begin{cases} r' = (1 + \mu)r + c_1 r^3 + f(r, \theta)r^5, & (29.10\text{-}9) \\ \theta' = \theta + c_2 + c_3 r^2 + g(r, \theta)r^4, & (29.10\text{-}10) \end{cases}$$

where f and g are smooth functions and where c_1, c_2, c_3, f, and g all depend smoothly on μ in some neighborhood of $\mu = 0$.

The constant c_1 plays the role of the Landau constant; if $c_1 < 0$, which we shall assume, we have a supercritical bifurcation.

If the higher order term containing $f(r, \theta)$ in (29.10-9) were missing, the circle $\mathscr{C}_0: r = r_0$, where

$$r_0 = \sqrt{\mu/(-c_1)},$$

would be invariant under the Poincaré mapping, for $\mu > 0$. It is centered on the closed orbit shown in Figure 29.7; hence, as it is carried along by the flow in $\mathfrak{M}$, it would generate a torus. To take into account the effect of the higher order terms, we define a sequence of curves

$$\mathscr{C}_n: \{r = r_n(\theta), 0 \le \theta \le 2\pi\}$$

inductively, starting with $\mathscr{C}_0$, by the equation $\mathscr{C}_{n+1} = \Phi(\mathscr{C}_n)$. It can be shown that for μ small enough ($R - R_2$ small enough), these curves converge, as $n \to \infty$, to a limiting curve $\mathscr{C}_\infty$, which is invariant under Φ.

The proof of convergence and the derivation of the restrictions on μ are quite long and are omitted, except for an outline and a few formulas given in the Appendix to this chapter. It is shown, for example, that if μ is small enough, the annulus

$$\tfrac{2}{3}r_0 < r < \tfrac{4}{3}r_0 \tag{29.10-11}$$

is mapped under Φ into the smaller annulus

$$(\tfrac{2}{3} + \tfrac{5}{27}\mu)r_0 < r' < (\tfrac{4}{3} - \tfrac{5}{27}\mu)r_0; \tag{29.10-12}$$

that suggests that Φ is a contracting mapping. It is also shown that if the maximum radial displacement of $\mathscr{C}_n$ under Φ is denoted by

$$\Delta_n = \max_{(\theta)} |r_{n+1}(\theta) - r_n(\theta)|,$$

then

$$\Delta_{n+1} \le K \Delta_n, \tag{29.10-13}$$

for a constant $K < 1$, if μ is small enough. Finally, it is shown that

$$\left| \frac{dr_n(\theta)}{d\theta} \right| < \mu \quad \text{for all } n, \text{ all } \theta, \tag{29.10-14}$$

if μ is small enough, so that the limiting curve $\mathscr{C}_\infty$ is at least Lipschitz continuous.

Lastly, we consider the tube in the K-dimensional space $\mathfrak{M}$ consisting of orbits that start from points of the curve $\mathscr{C}_\infty$. That tube joins onto itself, forming a closed surface, when the orbits return to the surface S, where they again pass through the curve $\mathscr{C}_\infty$. To see that the surface is homeomorphic to a torus, not a Klein bottle, we have to be sure that the orientation of $\mathscr{C}_\infty$ is preserved when the tube is closed, but that follows from the Poincaré mapping (29.10-9, 10), which shows that θ' in an increasing function of θ.

It should be noted that we have not shown that any single orbit covers the resulting torus densely, and hence yields a quasiperiodic function of time. In fact, it will be seen in Chapter 31 that that is in general unlikely, owing to Peixoto's theorem.

29.11 Subharmonic Bifurcation

If, in the notation of the preceding section, $\alpha(R)$ passes through the unit circle $|\alpha| = 1$ at a root of unity of degree less than 6, as R increases, the argument given for the existence of an invariant torus breaks down. In that case, the bifurcation may lead to one or more further periodic orbits.

We consider only the simplest case in which the higher terms of the mapping (29.10-4) are mostly already missing, and hence do not have to be eliminated. We assume $\alpha = \exp\{2\pi i p/q\}$ $(q \leq 5)$ and we take the mapping to be

$$z' = \alpha z(1 + \beta|z|^2) \qquad (\beta \text{ real}, <0)$$

or, in polar coordinates,

$$r' = (1 + \mu)r + c_1 r^3 \qquad (c_1 < 0),$$

$$\theta' = \theta + 2\pi\frac{p}{q}. \tag{29.11-1}$$

In this case there are new orbits, at a distance $r_0 = \sqrt{\mu/(-c_1)}$ (in the surface S) from the old orbit, which are closed, because each point of the circle $r = r_0$ is transformed into itself after q-fold iteration of the Poincaré mapping. For small positive μ the old orbit is unstable, if $c_1 < 0$, and the new ones stable. As μ is increased past 0, the period of the observed orbit is suddenly increased by a factor q.

In this case there *may* still be an invariant torus as result of the bifurcation, although our method of finding it breaks down. In the above example, there is such a torus, because the circle $r(\theta) = r_0$ is invariant.

Conversely, even when the invariant torus is established, there may be points on the curve $\mathscr{C}_\infty$ that are invariant under a certain number of iterations of the Poincaré mapping, and hence lead to closed orbits on the torus.

Appendix to Chapter 29—Computational Details for the Invariant Torus

We consider first the problem of determining the coefficients φ_{lm} of (29.10-5, 6) from the given coefficients q_{jk} of (29.10-4) so as to make the Poincaré mapping take the normal form (29.10-7). As noted in the text, we substitute (29.10-5, 6, 7) into (29.10-4) to give an equation in powers of ζ and $\bar\zeta$, and we equate the net coefficients of $\zeta^p\bar\zeta^q$. For $(p, q) = (0, 0), (1, 0)$, and $(0, 1)$, the resulting equations are satisfied automatically. The next twelve equations are taken in the order

$$
\begin{array}{ccccc}
(p, q) = & (2, 0) & (1, 1) & (0, 2) & \\
& (3, 0) & (2, 1) & (1, 2) & (0, 3) \\
& (4, 0) & (3, 1) & (2, 2) & (1, 3) & (0, 4)
\end{array}
$$

and are solved for the coefficients φ_{pq} in that order; then φ_{pq} is the only unknown in the equation in which it first appears, and it appears in the following way:

$$\alpha^p \bar{\alpha}^q \varphi_{pq} + \cdots = \alpha \varphi_{pq} + \cdots .$$

We can solve for φ_{pq} except when

$$\alpha^p \bar{\alpha}^q = \alpha.$$

Since $p + q > 1$, this can happen only if $|\alpha| = 1$ and then only when

$$\alpha^{|p+q-1|} = 1.$$

Unless $p + q - 1 = 0$, this last equation holds only when α is a $|p + q - 1|$th root of unity, and in this way the restrictions (29.10-8) are obtained. The case $p + q - 1 = 0$ occurs only for $(p, q) = (2, 1)$; that equation cannot be solved for φ_{21} if $|\alpha| = 1$, but it can be solved for β; hence we merely set $\varphi_{21} = 0$.

We also record, without derivation, some bounds on μ, which suffice to ensure the existence of the invariant limiting curve $\mathscr{C}_\infty$ discussed in the text. First, the annulus (29.10-11) is mapped into the annulus (29.10-12) if

$$\max|f|\mu < \tfrac{45}{1024}c_1^2 \quad \text{and} \quad \mu < \tfrac{3}{13},$$

where $\max|f|$ means the maximum of $|f(r, \theta, \mu)|$ in some region $r \leq r_1$, $|\mu| \leq \mu_1$ in which the normal form (20.10-9, 10) holds. Second, the bound (29.10-14) on $dr_n/d\theta$ holds if also

$$\max|f_r|c_4^5\mu^{3/2} + 5\max|f|c_4^4\mu + \max|f_\theta|c_4^5\sqrt{\mu} < \tfrac{1}{6}$$

and

$$2|c_3|c_4\sqrt{\mu} + \max|g_r|c_4^4\mu^2 + 4\max|g|c_4^3\mu^{3/2} + \max|g_\theta|c_4^4\mu < \tfrac{1}{6},$$

where

$$c_4 = \frac{4}{3\sqrt{-c_1}}.$$

Lastly, (29.10-13) holds with $K = 1 - \mu/6$, if also

$$2|c_3|c_4\sqrt{\mu} + \max|g_r|c_4^4\mu^2 + 4\max|g|c_4^3\mu^{3/2} + 5\max|f|c_4^4\mu$$
$$+ \max|f_r|c_4^5\mu^{3/2} < \tfrac{1}{6}.$$

CHAPTER 30

Invariant Manifolds in the Taylor Problem

Power series representation of a finite-dimensional manifold in a Hilbert space; coordinates in the manifold; determination of the expansion coefficients; dynamical system in the manifold; separation of variables; Taylor vortices, wavy vortices; helical vortices.

Prerequisites: Chapter 29.

This chapter is devoted to a method for calculating the unstable manifold that arises from the basic flow in hydrodynamical problems, devised by Davey 1962, Davey, DiPrima, and Stuart 1968, and Eagles 1971 for the study of the Taylor problem. Although the results that have been obtained with it so far are rather limited, it is the only method known at present for such problems (but see Hassard 1980 for a similar method in finite-dimensional spaces). Presumably, methods of this sort will be needed to understand the early onset of turbulence in detail.

30.1 Survey of the Taylor Problem to 1968

Let r_1 and r_2 be the radii of the inner and outer cylinders (assumed infinitely long) that confine the flow, and let Ω_1 and Ω_2 be their angular velocities. The unperturbed flow has angular velocity at radius r ($r_1 \leq r \leq r_2$) given by

$$\Omega(r) = A + B/r^2, \tag{30.1-1}$$

where A and B are determined by the no-slip conditions at the walls, namely $\Omega_1 = A + B/r_1^2$ and $\Omega_2 = A + B/r_2^2$. The problem is characterized by various dimensionless parameters, for example r_1/r_2 (which is fixed in a given apparatus) and the two Reynolds numbers

$$R_1 = \frac{\Omega_1 r_1^2}{\nu}, \qquad R_2 = \frac{\Omega_2 r_2^2}{\nu}. \tag{30.1-2}$$

The stability of the basic laminar Couette flow given by (30.1-1) against small axisymmetric disturbances was studied by G. I. Taylor 1923, both theoretically and experimentally. He found stability for values of R_1 and R_2 represented by points lying below the curve in Figure 30.1, for the case $r_1/r_2 = 0.880$. Rayleigh had shown earlier, by a simple argument, that the flow is stable for $0 < R_1 < R_2$, that is, to the right of the dashed line in the figure.

Taylor's calculations showed that, as R_1 is increased beyond its critical value (the value on the curve) for fixed R_2, the basic flow becomes unstable with respect to a normal mode whose velocity field is of the form

$$\mathbf{u} = \text{Re}[\mathbf{f}(r)e^{i\alpha z}] \tag{30.1-3}$$

in cylindrical coordinates, r, θ, z. This disturbance has the structure of a series of circular vortices, uniformly spaced in the z direction, as sketched in Figure 29.1 in the preceding chapter. Taylor found that the experimentally observed vortices (which can be made visible by means of particles suspended in the fluid) were in qualitative agreement with the calculations; in particular, their axial separation was in agreement with the value of α in (30.1-3) for which the disturbance first becomes unstable (i.e., becomes unstable for the smallest value of R_1).

For R_1 not too far above its critical value, the vortices (now called Taylor vortices) are stable; together with the basic flow, on which they are super-posed, they form a new steady flow (at each point in space, the fluid velocity vector is independent of time), which persists indefinitely, so long as the cylinders are kept rotating.

Taylor observed experimentally that when a still higher critical value of R_1 is reached, the vortices become wavy, and the waves rotate around the axis at approximately the mean angular velocity $(\Omega_1 + \Omega_2)/2$.

Taylor's analysis, being linear, said nothing about the flow except below and immediately above the first critical value of R_1 and said nothing about

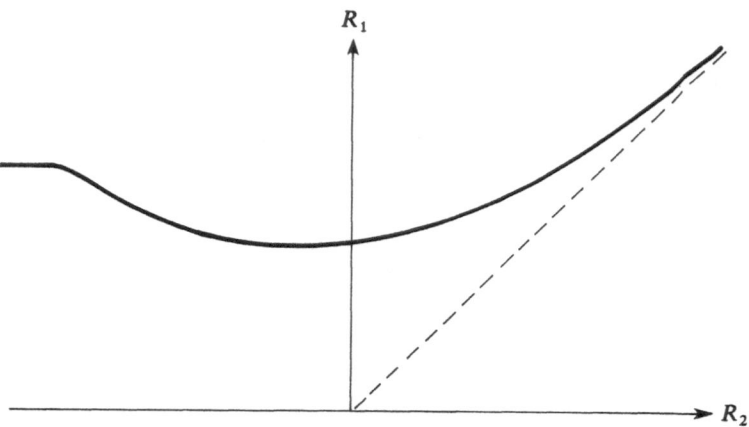

Figure 30.1   Taylor's stability diagram (schematic).

the stability or instability of the Taylor vortices, once they have formed. Evidently, for R_1 in the range where the Taylor vortices are observed, the exponential growth of the normal mode (30.1-3) levels off at a finite amplitude, owing to the effect of nonlinearities.

Subsequent improved theoretical and experimental work has generally confirmed Taylor's results. However, Taylor considered only axisymmetric disturbances, and it was shown by Krueger, Gross, and DiPrima 1966 that, for counter-rotating cylinders with Ω_2/Ω_1 sufficiently negative (beyond the range studied experimentally by Taylor), the mode that first becomes unstable is not axisymmetric, but has an angular dependence of the form $e^{im\theta}$; m increases through the values 1, 2, 3, ... as Ω_2/Ω_1 is made more and more negative. Hence the leftmost part of the curve in Figure 30.1 must be revised downward, but only by a quite small amount, because the stability of the mode depends only very slightly on m, in the case of the rather narrow gap $(r_1/r_2 = 0.880)$ studied by Taylor.

These predictions were confirmed experimentally by Snyder 1970, who showed also that the nonaxisymmetric modes, when they appear, are helical. (The linear theory cannot distinguish helical vortices from wavy ring vortices; the four normal modes containing $e^{\pm i\alpha z}e^{\pm im\theta}$ are all equally likely and can be combined to give real dependence either as $\frac{\sin}{\cos}\alpha z\,\frac{\sin}{\cos}m\theta$ or $\frac{\sin}{\cos}(\alpha z \pm m\theta)$; only a nonlinear theory can say which of these is preferred.)

In recent years, the nonlinear theory developed by Davey 1962, Davey, DiPrima, and Stuart 1968, and Eagles 1971 has led to an understanding of the structure and stability of the finite-amplitude Taylor vortices, the second bifurcation to the wavy vortices, and the structure and stability of the wavy vortices, as discussed below.

In a problem like this, involving a sequence of bifurcations, the theory consists ideally of a sequence of alternately linear and nonlinear investigations. After each bifurcation the structure and amplitude of the new flow is found by a nonlinear calculation. Its stability is then investigated by linearizing the equations about the new flow and studying the growth of infinitesimal disturbances, in order to find the next bifurcation, and so on.

30.2 Calculation of Invariant Manifolds.

In this section we describe in general terms the method for calculating the unstable manifold developed by the authors mentioned above on the basis of earlier work by Stuart and Watson.

We take the equation of evolution in the form (29.4-3), so that the eigenvalues and eigenfunctions of the linearized problem satisfy the equations

$$\lambda_j M\psi_j = L\psi_j \qquad (j = 1, 2, \ldots). \tag{30.2-1}$$

We assume, as in Section 29.6, that the ψ_j form a complete set in $\mathfrak{H}$, and that the adjoint functions, which satisfy the equation

$$\overline{\lambda}_j M^*\chi_j = L^*\chi_j \qquad (j = 1, 2, \ldots), \tag{30.2-2}$$

form a biorthogonal set with the ψ_j in the sense that

$$(\chi_j, M\psi_k) = \delta_{jk}. \tag{30.2-3}$$

We assume that the unstable manifold $\mathfrak{M}$ is K-dimensional, and we take coordinates of a point u on $\mathfrak{M}$ by projection onto $\mathfrak{M}_0$, as in Section 29.7:

$$x_k = (\chi_k, Mu) \qquad (k = 1, \ldots, K). \tag{20.2-4}$$

Therefore, for u in $\mathfrak{M}$,

$$u = \sum_{k=1}^{K} x_k \psi_k + u', \tag{30.2-5}$$

where u' is orthogonal to $\chi_1, \ldots, \chi_K$:

$$(\chi_k, Mu') = 0 \qquad (k = 1, \ldots, K). \tag{30.2-6}$$

The objective is to find a function $u(x_1, \ldots, x_K)$, or $u(\mathbf{x})$ for short, with values in $\mathfrak{H}$, which gives the point of $\mathfrak{M}$ having coordinate values $x_1, \ldots, x_K$, at least in some neighborhood of the origin. We assume that the function is analytic, and hence can be expanded in a power series

$$u(\mathbf{x}) = \sum_{(\mathbf{q} \in \mathscr{L})} \mathbf{x}^{\mathbf{q}} u_{\mathbf{q}}, \tag{30.2-7}$$

where $\mathbf{x}$ and $\mathbf{q}$ are K-component vectors, where $\mathscr{L}$ denotes the lattice

$$\mathscr{L} = \{\mathbf{q}: \text{each } q_i = \text{integer} \geq 0; \text{ at least one } q_i > 0\}, \tag{30.2-8}$$

where $\mathbf{x}^{\mathbf{q}}$ is an abbreviation for $x_1^{q_1} x_2^{q_2} \ldots x_K^{q_K}$, and where each coefficient $u_{\mathbf{q}}$ is an element of $\mathfrak{H}$.

According to (30.2-5, 6), the coefficients $u_{\mathbf{q}}$ of the linear terms in (30.2-7) are the first K eigenfunctions $\psi_1, \ldots, \psi_K$, while the remaining $u_{\mathbf{q}}$ are orthogonal to $\chi_1, \ldots, \chi_K$ in the sense of (30.2-6). However, the later coefficients $u_{\mathbf{q}}$ are not eigenfunctions, but satisfy certain inhomogeneous equations, described below; therefore, (30.2-7) is *not* an eigenfunction expansion.

We assume that the functions in (29.7-4), which describe the dynamical system in $\mathfrak{M}$, are also analytic and can be expanded in power series in $x_1, \ldots, x_K$, and we write

$$\dot{x}_j = \sum_{(\mathbf{p} \in \mathscr{L})} a_{j\mathbf{p}} \mathbf{x}^{\mathbf{p}} \qquad (j = 1, \ldots, K), \tag{30.2-9}$$

where the $a_{j\mathbf{p}}$ are numerical coefficients.

We compute $\dot{u} = du/dt$ from (30.2-7, 9), using the chain rule for differentiation. Then we substitute u and $\dot{u}$ into the equation of evolution (29.4-3) and require the result to be an identity in $x_1, \ldots, x_K$. That requirement suffices to determine the coefficients $u_{\mathbf{q}}$ and $a_{j\mathbf{p}}$.

To compute $\dot{u}$ from (30.2-7), we must differentiate $\mathbf{x}^{\mathbf{q}}$ with respect to x_j, for each j, and then use (30.2-9). We call $\mathbf{e}_j$ the K-component vector whose jth component is equal to 1 while the other components are equal to 0. Then,

$$\frac{\partial}{\partial x_j} \mathbf{x}^{\mathbf{q}} = q_j \mathbf{x}^{\mathbf{q} - \mathbf{e}_j}; \tag{30.2-10}$$

hence,

$$\dot{u} = \sum_{(\mathbf{p},\,\mathbf{q}\in\mathscr{L})} \sum_{j=1}^{K} q_j a_{j\mathbf{p}} \mathbf{x}^{\mathbf{p}+\mathbf{q}-\mathbf{e}_j} u_{\mathbf{q}}$$

$$= \sum_{(\mathbf{s}\in\mathscr{L})} \mathbf{x}^{\mathbf{s}} \sum_{j=1}^{K} \sum_{(\mathbf{q})}{}' q_j a_{j\mathbf{s}+\mathbf{e}_j-\mathbf{q}} u_{\mathbf{q}}, \qquad (30.2\text{-}11)$$

where $\sum'$ denotes the sum over all $\mathbf{q}$ in $\mathscr{L}$ such that $\mathbf{s} + \mathbf{e}_j - \mathbf{q}$ is also in $\mathscr{L}$; it is a finite sum, for given $\mathbf{s}$.

It is convenient to introduce a norm for vectors in the lattice $\mathscr{L}$: $|\mathbf{q}| = \sum_{j=1}^{K} q_j$; $|\mathbf{q}|$ is a positive integer. If $|\mathbf{q}| = 1$, $\mathbf{q}$ is one of the vectors $\mathbf{e}_j$.

We now substitute u from (30.2-7) and $\dot{u}$ from (30.2-11) into the equation of evolution (29.4-3) and equate the net coefficients of $\mathbf{x}^{\mathbf{s}}$ on the two sides, for each $\mathbf{s}$ in $\mathscr{L}$. First, if $\mathbf{s}$ is one of the vectors $\mathbf{e}_l$, the quadratic terms contribute nothing, and we have

$$\sum_{j=1}^{K} a_{j\mathbf{e}_l} M u_{\mathbf{e}_j} - L u_{\mathbf{e}_l} = 0 \qquad (l = 1, \ldots, K). \qquad (30.2\text{-}12)$$

But we have already seen that $u_{\mathbf{e}_j}$ is $= \psi_j$; hence [from (30.2-1)]

$$a_{j\mathbf{e}_l} = \begin{cases} \lambda_l & \text{for } j = l, \\ 0 & \text{for } j \neq l. \end{cases} \qquad (30.2\text{-}13)$$

To lowest order for small $x_1, \ldots, x_K$, (30.2-9) then gives $x_j = \text{const. } \exp\{\lambda_j t\}$, as expected.

Second, if $\mathbf{s}$ is not one of the $\mathbf{e}_l$, that is, if $|\mathbf{s}| > 1$, we find, using (30.2-13),

$$\left(\sum_{j=1}^{K} s_j \lambda_j M - L\right) u_{\mathbf{s}} = -\sum_{j=1}^{K} \sum_{(\mathbf{q})}{}'' q_j a_{j\mathbf{s}+\mathbf{e}_j-\mathbf{q}} M u_{\mathbf{q}} + \sum_{(\mathbf{q})}{}''' B(u_{\mathbf{q}}, u_{\mathbf{s}-\mathbf{q}}), \qquad (30.2\text{-}14)$$

where $\sum''$ denotes the sum over all $\mathbf{q}$ in $\mathscr{L}$ such that $\mathbf{s} + \mathbf{e}_j - \mathbf{q}$ is also in $\mathscr{L}$ and $\mathbf{q} \neq \mathbf{s}$, and $\sum'''$ denotes the sum over all $\mathbf{q}$ in $\mathscr{L}$ such that $\mathbf{s} - \mathbf{q}$ is also in $\mathscr{L}$; these are finite sums.

We now show that these equations, if taken in a suitable order for the various $\mathbf{s}$ in $\mathscr{L}$, determine the unknown functions $u_{\mathbf{s}}$ and unknown coefficients $a_{j\mathbf{s}}$ inductively. We assume the equations (30.2-14) so ordered that all equations with a given value of $|\mathbf{s}|$ appear earlier than the equations with one higher value of $|\mathbf{s}|$, and we assume that when any of the equations is encountered, all functions and coefficients appearing in previous equations have been determined. (The ordering of the equations with a given value of $|\mathbf{s}|$ is irrelevant.) We claim that then all the $u_{\mathbf{q}}$ appearing on the right of (30.2-14) are known. In fact, for most of the terms there, $|\mathbf{q}|$ is less than $|\mathbf{s}|$; the only exceptions are terms in which $\mathbf{q}$ is of the form $\mathbf{s} + \mathbf{e}_j - \mathbf{e}_l$, for some $l \neq j$ (recall that $\mathbf{q} \neq \mathbf{s}$); however, the term with that value of $\mathbf{q}$ contains the coefficient $q_j \mathbf{e}_l$, which is $= 0$ by (30.2-13); hence, all the $u_{\mathbf{q}}$ that appear on the right of the equation may be regarded as known. The unknowns in that equation are therefore the function $u_{\mathbf{s}}$ and the coefficients $a_{l\mathbf{s}}$ $(l = 1, \ldots, K)$.

To determine the coefficient a_{ls}, for any $l = 1, \ldots, K$, we take the inner product with χ_l throughout (30.2-14). The left member gives zero, and all the terms in the first sum on the right give zero, except when $\mathbf{q} = \mathbf{e}_l$ and $j = l$; hence,

$$a_{ls} = -\left(\chi_l, \sum_{(\mathbf{q})}''' B(u_{\mathbf{q}}, u_{s-\mathbf{q}})\right).$$

With these coefficients known for $l = 1, \ldots, K$, equation (30.2-14) can then be solved for u_s. In this way, the invariant K-dimensional manifold $\mathfrak{M}$ is determined in that neighborhood of the origin in $\mathfrak{H}$ in which the series (30.2-7, 9) converge.

This is the method of Davey, of Davey, DiPrima, and Stuart, and of Eagles. Although those authors don't describe it in those terms, its main feature is a calculation of the unstable manifold of the origin in $\mathfrak{H}$; then the equations (30.2-9) represent a finite-dimensional dynamical system in that manifold and can be studied by any of the standard methods, by analytical search for fixed points and cycles, or by calculations of orbits by numerical methods for ordinary differential equations.

An important feature of their method is the assumed orthogonality of all the $u_{\mathbf{q}}$, except for $|\mathbf{q}| = 1$, to the adjoint functions $\chi_1, \ldots, \chi_K$. We have presented that assumption by equations (30.2-5, 6) as simply a convenient way of prescribing the coordinates $x_1, \ldots, x_K$ in the unstable manifold, but it has a deeper significance, and is the fundamental idea that makes the method successful. Each u_s is determined in terms of known quantities by an inhomogeneous equation (30.2-14), in practice a differential equation with boundary conditions. For some of those equations, when the Reynolds number is close to one of the critical values, the operator on the left is nearly singular, and in fact is singular when the Reynolds number is equal to that value. According to the alternative theorem of linear algebra, which applies also to linear equations in a Hilbert space, singular inhomogeneous linear equations have no solution unless the right member is orthogonal to all solutions of the corresponding transposed homogeneous equation, which are just $\chi_1, \ldots, \chi_K$; if it is, the solutions are nonunique, but they can be made unique by requiring that they also be orthogonal to $\chi_1, \ldots, \chi_K$.

If this assumption, or something similar, is not made, some of the $u_{\mathbf{q}}$ can be unreasonably large, for Reynolds numbers of interest, presumably preventing the convergence of the series (30.2-7).

30.3 Cylindrical Coordinates

The Navier–Stokes equations (29.3-4 and 5) are

$$\frac{\partial \mathbf{u}}{\partial t} + (\mathring{\mathbf{u}} \cdot \nabla)\mathbf{u} + (\mathbf{u} \cdot \nabla)\mathring{\mathbf{u}} + (\mathbf{u} \cdot \nabla)\mathbf{u} + \nabla p - \nu\nabla^2\mathbf{u} = 0, \quad (30.3\text{-}1)$$

$$\nabla \cdot \mathbf{u} = 0, \quad (30.3\text{-}2)$$

where $\hat{u}(x)$ is the basic laminar flow (Couette flow for the Taylor Problem). We introduce cylindrical coordinates r, θ, z and corresponding velocity components u, v, w, so that

$$\mathbf{u} = u\mathbf{k}_r + v\mathbf{k}_\theta + w\mathbf{k}_z, \tag{30.3-3}$$

where $\mathbf{k}_r, \mathbf{k}_\theta, \mathbf{k}_z$ are unit vectors in the directions of increasing r, θ, and z, respectively. In the cylindrical coordinates, the operators ∇^2 and $\mathbf{u} \cdot \nabla$ have the forms

$$\nabla^2 = \frac{\partial^2}{\partial r^2} + \frac{1}{r}\frac{\partial}{\partial r} + \frac{1}{r^2}\frac{\partial^2}{\partial \theta^2} + \frac{\partial^2}{\partial z^2},$$

$$\tag{30.3-4}$$

$$\mathbf{u} \cdot \nabla = u\frac{\partial}{\partial r} + \frac{v}{r}\frac{\partial}{\partial \theta} + w\frac{\partial}{\partial z}.$$

When these operators are applied to a vector field in the form (30.3-3), the dependence of the unit vectors $\mathbf{k}_r$ and $\mathbf{k}_\theta$ on θ must be taken into account:

$$\frac{\partial}{\partial \theta}\mathbf{k}_r = \mathbf{k}_\theta, \qquad \frac{\partial}{\partial \theta}\mathbf{k}_\theta = -\mathbf{k}_r.$$

For the Taylor problem, the basic laminar flow is given, according to (30.1-1), by $\hat{u} = 0$, $\hat{w} = 0$, and

$$\hat{v} = \hat{v}(r) = Ar + B/r. \tag{30.3-5}$$

When the foregoing equations are combined, they give a simultaneous system of partial differential equations for the quantities u, v, w, and p, as functions of r, θ, z, and t. For the practical application of the method described in the preceding section, it is convenient to put those equations in the form introduced by Eagles 1971, by the introduction of six-component vectors $\mathbf{U}, \mathbf{V}$, etc., whose components are $p, \partial v/\partial r, \partial w/\partial r, u, v, w$. The system then takes the form

$$\frac{\partial}{\partial r}\mathbf{U} - A\mathbf{U} - M\frac{\partial}{\partial t}\mathbf{U} - K(\mathbf{U})\mathbf{U} = 0, \tag{30.3-6}$$

where A, M, and $K(\mathbf{U})$ are 6×6 operator-valued matrices containing $\partial/\partial \theta$ and $\partial/\partial z$, but not $\partial/\partial r$. [To achieve the last, the divergence equation $\nabla \cdot \mathbf{u} = 0$ has been used to eliminate explicit reference to $\partial u/\partial r$, except in the first term on the left of (30.3-6).] The matrices are written out in the Appendix to this chapter; they are essentially the same as in the paper of Eagles, but in a slightly different notation. All matrix elements of M and $K(\mathbf{U})$ are zero except in the upper right quarter of the matrices. To explain the notation $K(\mathbf{U})$, we may consider the expression $K(\mathbf{U})\mathbf{V}$; the matrix elements depend linearly on the components of $\mathbf{U}$ and act linearly on the components of $\mathbf{V}$.

30.4 The Hilbert Space

For the Taylor problem, theory and experiment agree in indicating that, once the Taylor vortices (or wavy vortices, or helical vortices) have been established, the entire flow is periodic in the z direction with a period of the order of twice the separation of the cylinders, at least when the cylinders are very long, and in the approximation in which end effects are neglected. Here, we shall simply assume such periodicity, and we shall assume that the wave number α is known, so that the period is $2\pi/\alpha$. We call $\mathscr{R}$ the region given by

$$\mathscr{R}: \quad r_1 \le r \le r_2, \quad 0 \le \theta \le 2\pi, \quad 0 \le z \le \frac{2\pi}{\alpha}, \quad (30.4\text{-}1)$$

and we take $\mathfrak{H}$ as the Hilbert space $L^2(\mathscr{R})^6$ with the inner product

$$(\mathbf{U}, \mathbf{V}) = \iiint_{\mathscr{R}} \bar{\mathbf{U}} \cdot \mathbf{V} \, dr \, d\theta \, dz. \quad (30.4\text{-}2)$$

With this choice of inner product, the eigenvalue problem adjoint to

$$\frac{\partial}{\partial r} \mathbf{U} - A\mathbf{U} - \lambda M\mathbf{U} = 0,$$

$$U_4 = U_5 = U_6 = 0 \quad \text{at} \quad r = r_1 \quad \text{and} \quad r = r_2 \quad (30.4\text{-}3)$$

is the problem

$$-\frac{\partial}{\partial r} \mathbf{U} - A^*\mathbf{U} - \bar{\lambda} M^*\mathbf{U} = 0,$$

$$U_1 = U_2 = U_3 = 0 \quad \text{at } r = r_1 \quad \text{and} \quad r = r_2, \quad (30.4\text{-}4)$$

where A^* and M^* are the transposes of the matrices A and M with $\partial/\partial\theta$ and $\partial/\partial z$ replaced by $-\partial/\partial\theta$ and $-\partial/\partial z$.

As in the preceding section, it is assumed that:

1. For a given value of λ, the problem (30.4-3) has a solution if and only if the problem (30.4-4) has one.
2. Each of the problems has a complete set of eigenfunctions, and for this purpose it is not necessary to include generalized eigenfunctions of higher order, e.g., solutions of $(\partial/\partial r - A - \lambda M)^2 \mathbf{U} = 0$, etc.

These assumptions have been confirmed as far as possible by numerical calculations involving the first ≈ 40 eigenfunctions. The eigenfunctions $\mathbf{U}^{(j)}$ of the direct problem and those $\mathbf{U}^{\dagger(j)}$ of the adjoint problem can be chosen to be biorthogonal in the sense that

$$(\mathbf{U}^{\dagger(j)}, M\mathbf{U}^{(l)}) = \delta_{jl}. \quad (30.4\text{-}5)$$

The choice (30.4-2) of inner product is somewhat arbitrary. For example, it might seem to be natural to have $r\,dr$ appear in place of dr. Then, the adjoint equation (30.4-4) would be somewhat different, and the adjoint functions $\mathbf{U}^{\dagger(j)}$ would be different, but the biorthogonality relations (30.4-5) would continue to hold. It is only those relations that are important; they are used to construct the projections that project the points of the unstable manifold $\mathfrak{M}$ onto the corresponding linear manifold $\mathfrak{M}_0$ tangent to $\mathfrak{M}$ at the origin. There is no choice of inner product that makes the operators self-adjoint.

30.5 Separation of Variables in Cylindrical Coordinates

The variables can be separated in the linearized problem, and in particular in the eigenvalue problems (30.4-3, 4), so that the eigenfunctions are of the form $\mathbf{U} = \mathbf{V}(r)e^{ipaz + im\theta}$, where p and m are integers; we write the kth eigenfunction as

$$\mathbf{V}_k(r)e^{ip_k z + im_k \theta}, \qquad k = 1, 2, 3, \ldots. \tag{30.5-1}$$

One of the advantages of the method of Davey, DiPrima, and Stuart, described in Section 30.2, is that, although the variables don't separate in the nonlinear problem, they do separate in each term of the expansion (30.2-7) of the unstable manifold in the coordinates $x_1, \ldots, x_K$. The coefficient $u_\mathbf{q}$ that appears there is an element of the Hilbert space, hence represents a vector function of r, θ, z in the region $\mathcal{R}$, and that function has just the form (30.5-1).

Specifically, those coefficients, which are now called $\mathbf{U}_\mathbf{q}$ [$\mathbf{q}$ being a point in the lattice $\mathcal{L}$ described by (30.2-8)] and the numerical coefficients $a_{j\mathbf{q}}$ that appear in (30.2-9) satisfy relations as follows:

Lemma.   *For general j and $\mathbf{q}$, $a_{j\mathbf{q}} = 0$ unless $p(\mathbf{q}) = p(\mathbf{e}_j)$ and $m(\mathbf{q}) = m(\mathbf{e}_j)$; furthermore, $\mathbf{U}_\mathbf{q}$ is of the form*

$$\mathbf{U}_\mathbf{q} = \mathbf{V}_\mathbf{q}(r)e^{ip(\mathbf{q})az + im(\mathbf{q})\theta}.$$

Here $p(\mathbf{q})$ and $m(\mathbf{q})$ are given by

$$p(\mathbf{q}) = \sum_{k=1}^{K} q_k p_k,$$

$$m(\mathbf{q}) = \sum_{k=1}^{K} q_k m_k,$$

and $\mathbf{e}_j$ is that vector in the lattice $\mathcal{L}$ having its jth component $= 1$ and its other components $= 0$.

The proof of the lemma is by an easy induction on the norm $|\mathbf{q}| = q_1 + \cdots + q_K$ introduced in Section 30.2, and is omitted.

The advantage of the separation of variables is that the linear equations that have to be solved for the $u_\mathbf{q}$ (30.2-14) are *ordinary* differential equations for functions of r in the interval $[r_1, r_2]$. They are sixth-order two-point boundary-value problems, with three boundary conditions at each end of the interval. There are generally many such equations per problem (for example, 800 of them when the number of dimensions K of the unstable manifold is $=14$ and the expansions (30.2-7, 9) are carried through fifth-degree terms), but the available numerical methods for such equations are faster and more accurate than methods for partial differential equations.

30.6 Results to Date for the Taylor Problem

We summarize here the general conclusions that come from the calculations of Davey, Di Prima, and Stuart 1968, those of Eagles 1971, and a few additional calculations that I have made recently on the Cray 1 Computer at NCAR (the National Center for Atmospheric Research, Boulder, Colorado, sponsored by the National Science Foundation).

The outer cylinder is taken to be at rest ($\Omega_2 = 0$). It is then customary to express the rate of rotation in terms of the Taylor number

$$T = \frac{2\Omega_1^2 r_1^2 (r_2 - r_1)^3}{v(r_1 + r_2)},$$

which is proportional to the square of the Reynolds number R_1 (R_2 is zero).

As T is increased, for fixed r_1/r_2, the first eigenvalue of the linearized problem to cross into the right half-plane is real and corresponds to $m_k = 0$ in the expression (30.5-1) for the eigenfunction. Next is a complex conjugate pair of eigenvalues corresponding to $m_k = 2$, and so on. Each eigenvalue is of multiplicity 2 (the degeneracy corresponds to the possibility of shifting the entire flow pattern in the z direction); hence the number of dimensions K of the unstable manifold takes on in succession the values 2, 6, 10, 14, ... Calculations have been made up to $K = 14$. (At still higher values of T, the values $m_k = 0, 1, \ldots$ come in again, but corresponding to eigenfunctions with more complicated radial dependence; no calculations have been made in this regime.)

The modes of motion (stable and unstable) that have been discovered so far are all periodic orbits of the dynamical system in the unstable manifold [see (30.2-9)], except for the basic laminar (Couette) flow and the Taylor vortices, which are fixed points of that system. The additional modes are as follows:

Helical vortices: These are similar to the Taylor vortices, except that if we follow one of them around in θ from 0 to 2π, we find that its end is not connected onto its own beginning, but onto the beginning of the second, or fourth, or sixth, ..., one above (or below) it. We define a corresponding integer $m = 1, 2, 3, \ldots$. (It cannot be connected onto the first or third, etc., because those vortices rotate in the opposite direction.) The entire pattern rotates about the axis with an angular velocity not far from the mean velocity $\Omega_1/2$ of the fluid.

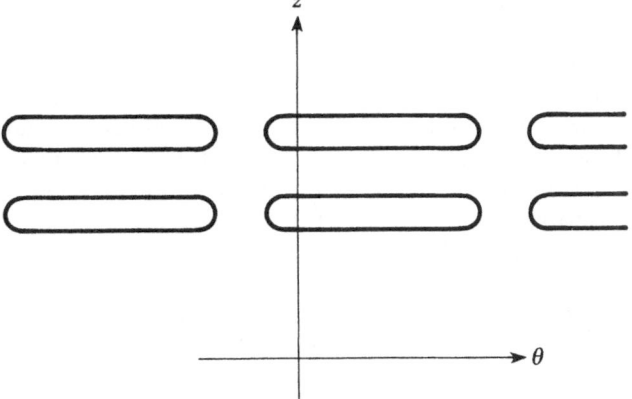

Figure 30.2

 Nonaxisymmetric simple mode (never stable): The vortex strength varies sinusoidally around the axis. In effect the vortex cores are connected as shown schematically in Figure 30.2. The pattern rotates about the axis.

Up-down wavy vortices: These are similar to the Taylor vortices, but the vortex cores are displaced alternately in the $+$ and $-$ z direction as θ goes from 0 to 2π. We call m $(=1, 2, \ldots)$ the number of such displacements in each direction.

In-out wavy vortices: Similar to the up-down wavy vortices, except that the cores are displaced alternately in and out in the radial direction. They are never stable. Both types of wavy vortices rotate about the axis with an angular velocity $\approx \Omega_1/2$.

In all cases it is found that as the Taylor number T is increased past a first critical value T_1, the laminar flow becomes unstable and is replaced by the Taylor vortices, which have a stength roughly proportional to $\sqrt{T - T_1}$, and which are then stable up to a second critical value T_2, where the waviness sets in with an amplitude roughly proportional to $\sqrt{T - T_2}$.

The helical vortices bifurcate from the basic laminar flow above T_1, i.e., after the basic flow has already become unstable; see Figure 30.3. The

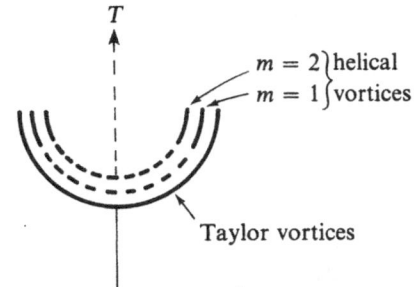

Figure 30.3   Inaccessible stable helical vortices.

helical vortices are unstable when they first appear, but then become stable at slightly higher values of T, as indicated by the solid curves in the figure. Then, they are stable modes that are inaccessible in the sense that they cannot be reached from the basic flow by a continuous sequence of stable modes.

The experimental work of Gollub and Swinney 1975 indicates that, at a Taylor number of the order of $200T_1$, a strange attractor should appear, because they observe a continuous power spectrum. It is not possible to extend the calculations to such high values of T, because the number of dimensions of the unstable manifold becomes unmanageably large. It is of course possible to continue calculating with the manifold of smaller dimension, determined by the eigenvalues farthest to the right in the complex plane. Such a manifold is invariant, but not attracting. Calculations of this sort suggest that the wavy vortices *may* be stable to quite high value of T, thus delaying the appearance of a stange attractor or even of further bifurcations.

Appendix to Chapter 30—The Matrices in Eagles' Formulation

The matrices that appear in equation (30.3-6) are as follows:

$$A =$$

$$
\begin{vmatrix}
0 & -\dfrac{\nu}{r}\partial_\theta & -\nu\partial_z & B & \dfrac{2V}{r}-\dfrac{\nu}{r^2}\partial_\theta & 0 \\[2ex]
\dfrac{1}{\nu r}\partial_\theta & -\dfrac{1}{r} & 0 & \dfrac{1}{\nu}\left(V'+\dfrac{V}{r}\right)-\dfrac{2}{r'}\partial_\theta & -\dfrac{1}{\nu}B+\dfrac{1}{r^2} & 0 \\[2ex]
\dfrac{1}{\nu}\partial_z & 0 & -\dfrac{1}{r} & 0 & 0 & -\dfrac{1}{\nu}B \\[2ex]
0 & 0 & 0 & -\dfrac{1}{r} & -\dfrac{1}{r}\partial_\theta & -\partial_z \\[2ex]
0 & 1 & 0 & 0 & 0 & 0 \\[2ex]
0 & 0 & 1 & 0 & 0 & 0
\end{vmatrix},
$$

where B is the operator

$$B = \nu\left(\frac{1}{r^2}\partial_\theta^2 + \partial_z^2\right) - \frac{V}{r}\partial_\theta;$$

$$
M =
\begin{vmatrix}
(0) & \begin{matrix} -1 & 0 & 0 \\ 0 & 1/\nu & 0 \\ 0 & 0 & 1/\nu \end{matrix} \\
(0) & (0)
\end{vmatrix},
$$

$$
K(U) = \begin{pmatrix}
(0) & \begin{array}{ccc}
\dfrac{u}{r} - \dfrac{v}{r}\partial_\theta - w\partial_z & \dfrac{v}{r} + \dfrac{u}{r}\partial_\theta & u\partial_z \\[2mm]
\dfrac{1}{v}\left(\dfrac{v}{r} + \dfrac{\partial v}{\partial r}\right) & \dfrac{1}{v}\left(\dfrac{v}{r}\partial_\theta + w\partial_z\right) & 0 \\[2mm]
\dfrac{1}{v}\dfrac{\partial w}{\partial r} & 0 & \dfrac{1}{v}\left(\dfrac{v}{r}\partial_\theta + w\partial_z\right)
\end{array} \\
\\
(0) & (0)
\end{pmatrix}.
$$

CHAPTER 31

The Early Onset of Turbulence

Periodic, quasi-periodic, almost periodic, and aperiodic motions; ω-limit set; attractors; power spectrum; Lyapounov stability; strange attractors; the Lorenz attractor; strongly generic, generic, nongeneric, and strongly nongeneric properties of systems.

Prerequisites: Chapter 29.

The word "turbulence," as used in practical fluid dynamics, refers to a flow whose chaotic aspects are so highly developed that statistical methods can be used for the study of at least many of its characteristics. It occurs at substantially higher Reynolds numbers than those we shall consider. For example, fully developed turbulence in air, in which the so-called inertial subrange is fully developed, requires more violent conditions than can be achieved in most wind tunnels, and is observed mainly in the free atmosphere.

Between laminar flow and turbulence there is an ill-defined regime called the transition to or onset of turbulence. We shall be concerned only with the part of that regime at quite low Reynolds numbers. The word "early" in the title refers to a physical arrangement in which the Reynolds number is increased very slowly with time. The flows we consider are simple and smooth, but, nevertheless, exhibit certain features of unpredictability and chaos at quite early stages.

31.1 The Landau–Hopf Model

A schematic model of the transition is described in *Fluid Mechanics*, Landau and Lifshitz 1959. It is assumed that, at least in some problems, there is a sequence of supercritical bifurcations forming schematically a tree, as in Figure 31.1. After the first bifurcation the motion is generally periodic; after the second it is generally quasi-periodic with two periods, and so on. A *quasi-periodic* function with m periods is a function of the form

$$f(t) = g(\omega_1 t, \omega_2 t, \ldots, \omega_m t), \qquad (31.1\text{-}1)$$

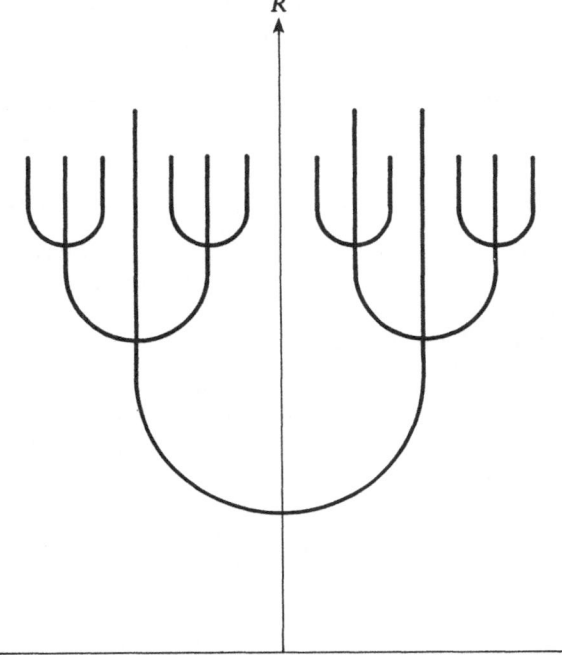

R

Figure 31.1   Bifurcation tree.

where $g(\cdot, \cdot, \ldots, \cdot)$ is periodic in each of its arguments with period 2π, and the frequencies ω_i are incommensurable, which means there is no vanishing linear combination $c_1\omega_1 + \cdots + c_m\omega_m$ with rational coefficients $c_1, \ldots, c_m$. If $\omega_1, \ldots, \omega_m$ are commensurable, the number of independent frequencies is less than m. Suppose, for example that $m = 2$ and $\omega_2/\omega_1 = p/q$, where p and q are integers. Then, for

$$t_0 = 2\pi\left(\frac{q}{\omega_1} + \frac{p}{\omega_2}\right), \tag{31.1-2}$$

we find that $\omega_1 t_0 = 4\pi q$ and $\omega_2 t_0 = 4\pi p$; hence $f(t)$ is periodic (not merely quasi-periodic) with period given by (31.1-2).

It was shown in the last section of the preceding chapter that if the first bifurcation leads to a closed orbit, the second can lead to an attracting invariant torus in the phase space $\mathfrak{H}$. If, furthermore, the motion is such that its orbit covers the torus densely, then a resulting function of time, such as one of the coordinates in the phase space, is quasi-periodic with two periods. Specifically, one can define two intrinsic angle-coordinates θ and φ on the torus such that $\theta = \omega_1 t + \text{const.}$, $\varphi = \omega_2 t + \text{const.}$, and the orbit is dense on the torus if and only if ω_1 and ω_2 are incommensurable. After the next bifurcation there may be motion on a 3-torus, and so on.

Exactly which branch of the tree in Figure 31.1 is followed depends on the structure of the infinitesimal perturbation that caused departure from the basic or laminar flow when the first critical value of the Reynolds number was reached. More generally, the phases associated with the various frequencies depend in a random manner on that perturbation, so that (31.1-1) might better be written as

$$f(t) = g(\omega_1 t + \beta_1, \ldots, \omega_m t + \beta_m). \qquad (31.1\text{-}3)$$

The idea behind the Landau–Hopf model was that as soon as there are many independent frequencies, the motion is so irregular in appearance that it must be regarded for practical purposes as chaotic.

There are various ways in which this model can be inappropriate:

1.   One of the bifurcations in the tree of Figure 31.1 may be subcritical; then, as soon as the corresponding critical value of the Reynolds number is exceeded, there is no nearby stable motion for the system to follow, and there is a so-called explosive transition to a motion involving more or less remote parts of the phase space.

2.   In some problems, such as flow in a circular pipe, the basic flow is stable to infinitesimal disturbances at all Reynolds numbers, but is unstable to finite disturbance of rather small amplitude, and the critical amplitude for instability decreases toward zero as the Reynolds number increases, so that stable flow cannot be achieved in practice at high Reynolds number, owing to the presence of small but finite disturbances.

3.   Although an invariant torus generally appears at the second bifurcation, the orbit need not be dense on it; it may return to its starting point after winding finitely many times around; then the orbit is closed and the motion is periodic, as mentioned in Section 29.11. In fact it is now believed, on the basis of Peixoto's theorem (see Appendix) that closed orbits on the torus are more likely than dense ones. This may lead to the Feigenbaum model—see Section 31.19.

4.   A possibility discussed by Ruelle and Takens 1971 is that, after a few bifurcations, there appears an invariant point set in the phase space, which is not a torus but a so-called strange attractor; then, as explained below, the motion is not quasi-periodic, but aperiodic.

31.2 The Hopf Example

In 1948 Eberhard Hopf gave an example of a simple dynamical system that has an infinite sequence of bifurcations, each leading to an attracting torus of one higher dimension than the preceding one. The functions $u(x, t)$ and $z(x, t)$ are generally complex-valued functions of real variables x and t and are periodic in x with period 2π. The equations are

$$\frac{\partial u}{\partial t} = -z \circ z - u \circ 1 + \mu \frac{\partial^2 u}{\partial x^2},$$

$$\frac{\partial z}{\partial t} = z \circ u + z \circ F + \mu \frac{\partial^2 z}{\partial x^2}. \qquad (31.2\text{-}1)$$

The small circle denotes convolution; generally,

$$f \circ g = (f \circ g)(x) = \frac{1}{2\pi} \int_0^{2\pi} f(x + y)\overline{g(y)}dy;$$

$u \circ 1$ is just the average value of u; μ is a positive constant, and $F = F(x)$ is a given complex-valued even periodic function.

The system may be regarded as a simple analogue of the Navier–Stokes equations on a compact manifold (the circle), with the nonlinearities written as convolutions rather than advection terms. $F(x)$ is a forcing term, and μ, the viscosity, is a parameter that can be varied.

Solutions can be found by expanding all functions in Fourier series with respect to x, for example $u(x, t) = \sum_{-\infty}^{\infty} u_n(t)e^{inx}$, with similar notations for the other functions. Since the Fourier transform of a convolution is the product of the Fourier transforms, the terms with given n are not coupled to terms with a different n. In that respect the system fails to model the Navier–Stokes equations, but its solutions can be found explicitly.

The only restrictions put on the forcing function $F(x)$ are to prevent the system's being too special, i.e., nongeneric. Namely, if we write

$$F(x) = \sum_{-\infty}^{\infty} (a_n + ib_n)e^{inx}, \tag{31.2-2}$$

then it is assumed that infinitely many of the a_n are positive, that the b_n are not rationally related (any finite set of them is linearly independent over the rationals), and that no two of the quantities a_n/n^2 are equal.

The critical values of μ are the numbers a_n/n^2; they can be arranged in a sequence $\mu_1 > \mu_2 > \mu_3 > \cdots \to 0$. The general solution represents a moving point in an infinite-dimensional space Ω with coordinates $u_n(t)$, $z_n(t)$, $n = 0, \pm 1, \pm 2, \ldots$. Hopf proved that, for $\mu > \mu_1$, the fixed point at the origin of Ω attracts all other solutions, so that $u_n \to 0, z_n \to 0$, as $t \to \infty$; as μ decreases past μ_1, that solution becomes unstable, and there is a bifurcation to an attracting periodic orbit that grows out of the origin; when μ decreases past μ_2, that orbit also becomes unstable, and there is a bifurcation to an attracting torus (2-torus) that grows out of the orbit, and so on. After the kth bifurcation there is an attracting k-dimensional torus, and the orbits on the torus are dense on it.

31.3 The Ruelle–Takens Model

In the model of the early onset of turbulence proposed by D. Ruelle and F. Takens 1971, the first four bifurcations are assumed, as in the Landau-Hopf model, to be supercritical and to lead to invariant tori T^k, $k = 1, 2, 3, 4$, each of which is attracting between its appearance and the next bifurcation. Concerning the existence of these tori, see the discussion of the Feigenbaum model in Section 31.19. Ruelle and Takens prove that, on T^4, motion on a particular kind of strange attractor contained in T^4 is rather likely. The attractor is locally the Cartesian product of a two-dimensional Cantor set and a two-dimensional surface.

Their theorem can be paraphrased as follows: Consider a Banach space $\mathfrak{B}$, each point of which represents a vector field on the torus T^4, with a norm containing the magnitudes of the components of the vector field and their derivatives of order ≤ 3. Two points in $\mathfrak{B}$ differing by a small amount in norm may be regarded as two physical systems, of which one can be obtained from the other by a small perturbation of the vector field. Then, given any constant vector field on T^4 (one for which the angle variables on the torus vary linearly with time), and given any $\varepsilon_1 > 0$, there is a perturbation of that field of magnitude less than ε_1 which yields a strange attractor of the kind they describe. Then there is another number $\varepsilon_2 > 0$ (possibly much smaller than ε_1) such that the strange attractor persists under any further perturbation of magnitude $< \varepsilon_2$. Hence, the vector fields that yield the strange attractor cannot be dismissed as unlikely.

Their particular choice of strange attractor is somewhat arbitrary; one can imagine many variations of it, each having the property stated.

The strange attractor discovered earlier by E. N. Lorenz (see Sections 31.9–31.17 below) arises by a somewhat different mechanism. Apparently, no one has found a specific vector field on a specific manifold that leads to a strange attractor precisely according to the Ruelle–Takens model. The important idea in their paper is that motions on strange attractors are in some sense likely, or at least not unlikely and are possibly even generic in certain circumstances. Their theorem does not say that the existence of a strange attractor is a generic property of vector fields on T^4 (see the Appendix to this chapter). The corresponding subset of the Banach space $\mathfrak{B}$ is open but not necessarily dense; it merely comes arbitrarily close to every point of $\mathfrak{B}$ that represents a constant vector field on T^4. It says simply that motion on a strange attractor is more likely, once an invariant T^4 has been established, than the quasi-periodic motion on T^4.

31.4   The ω-Limit Set of a Motion

Following the considerations of Section 29.7, we assume that we are dealing with a dynamical system

$$\dot{\mathbf{x}} = \mathbf{F}(\mathbf{x}) \tag{31.4-1}$$

on a finite-dimensional manifold $\mathfrak{M}$, where $\mathbf{F}(\cdot)$ is a smooth vector field on $\mathfrak{M}$. A solution $\mathbf{x}(t)$ of this equation is called a *motion* in $\mathfrak{M}$, and the point set in $\mathfrak{M}$ given by

$$\gamma = \{\mathbf{x}(t): \text{all } t\} \tag{31.4-2}$$

is called the *orbit* or *trajectory*) of the motion. We assume that the initial-value problem of (31.4-1) is well posed, and we denote by $\boldsymbol{\varphi}(\mathbf{x}_0, t)$ the solution that starts at $\mathbf{x}_0$, for any $\mathbf{x}_0$ in $\mathfrak{M}$ and all $t \geq 0$; that is, if $\mathbf{x}(t)$ is any solution, then

$$\mathbf{x}(t) = \boldsymbol{\varphi}(\mathbf{x}(0), t). \tag{31.4-3}$$

For fixed $\mathbf{x}$, $\varphi(\mathbf{x}, t)$ is a motion, while for fixed $t \geq 0$, a one-to-one mapping of $\mathfrak{M}$ into itself is given by $\mathbf{x} \to \varphi(\mathbf{x}, t)$. The function $\varphi(\mathbf{x}, t)$ is called a *semi-flow*; it obviously has the semigroup property that if t and s are ≥ 0, then

$$\varphi(\varphi(\mathbf{x}, t), s) = \varphi(\mathbf{x}, t + s). \tag{31.4-4}$$

For $t = 0$, φ is the identity mapping: $\varphi(\mathbf{x}, 0) \equiv \mathbf{x}$. We assume that φ is continuous in $\mathbf{x}$ and t.

A point ξ of $\mathfrak{M}$ is called an *ω-limit point* of a motion $\mathbf{x}(t)$ if $\mathbf{x}(t)$ comes arbitrarily close to ξ at times arbitrarily far in the future, that is, if there is a sequence $\{t_n\}_1^\infty$ such that

$$\left.\begin{array}{r} t_n \to \infty \\ |\mathbf{x}(t_n) - \xi| \to 0 \end{array}\right\} \quad \text{as } n \to \infty. \tag{31.4-5}$$

The set of all ω-limit points of a motion is called its *ω-limit set* and is denoted by $\Omega_\mathbf{x}$, where $\mathbf{x}$ is the initial point of the motion; it is a closed point set. If $\mathbf{y}$ is any other point of the same orbit, then $\Omega_\mathbf{y} = \Omega_\mathbf{x}$.

As an example, if a motion tends to a fixed point, as $t \to \infty$, then that point is the ω-limit set of the motion. If a motion in a plane spirals outward toward a closed curve, then that curve is the ω-limit set of the motion. If the orbit of a motion on a torus is dense on the torus, then the entire torus is the ω-limit set of every point of that orbit.

The symbol ω refers to future time. When a motion exists for all t, α-limit points and α-limit sets are similarly defined by letting $t \to -\infty$.

If a motion $\mathbf{x}(t)$ lies in a bounded region of $\mathfrak{M}$, for $t \geq 0$, then its ω-limit set Ω is nonempty, and $\mathbf{x}(t)$ gets closer to Ω as times goes on; that is,

$$\text{distance } \{\mathbf{x}(t), \Omega\} \to 0, \quad \text{as } t \to \infty. \tag{31.4-6}$$

(See Nemytskii and Stepanov 1960, Chapter V, Section 3.) However, Ω need not be stable or attracting; other nearby motions may move away from it and never return.

An important property of ω-limit sets has to do with the reversibility of the motion. A solution of (31.4-1) cannot generally be continued to all negative t. For example, the solutions of $\dot{x} = -x^3$ on $\mathbb{R}$ cannot be [except the solution $x(t) \equiv 0$], and solutions of the Navier–Stokes equations generally cannot be, because of the parabolic nature of the equations. However, certain special motions can be.

From the continuity and semigroup property of φ, it follows that if the initial point $\mathbf{x}(0)$ is an ω-limit point of $\mathbf{x}(t)$, then every subsequent point $\mathbf{x}(t_0)$ is also an ω-limit point of $\mathbf{x}(t)$. In other words, if a motion starts in its own ω-limit set, it stays there. It can be proved that such motions can also be continued indefinitely backward in time, and hence lie in $\Omega_{\mathbf{x}(0)}$ for all t. See Sell 1971, Theorem II.8. This last result may seem paradoxical from the point of view of physical observation or numerical simulation, where there is only a finite accuracy. According to (31.4-6), any bounded motion $\mathbf{x}(t)$ lies for practical purposes in its own ω-limit set after a finite lapse of time. That doesn't imply, of course, that finite-difference methods can be used to follow such an orbit indefinitely backward without escaping from $\Omega_{\mathbf{x}(0)}$; it means merely that if a motion has been somehow followed forward for a sufficiently long time interval, after transients have disappeared, the result may be regarded as a rather accurate sampling of the kind of motion which ideally is confined to $\Omega_{\mathbf{x}(0)}$ for all t.

31.5 Attractors

An attractor is roughly a set in $\mathfrak{M}$ such that any sufficiently nearby motion gets closer and closer to it as time goes on. Specifically, we shall call a connected closed bounded set S in $\mathfrak{M}$ an *attractor* if

1. S is contained in an open set $\mathscr{R}_0$ such that for any $\mathbf{x}$ in $\mathscr{R}_0$ the motion $\boldsymbol{\varphi}(\mathbf{x}, t)$ is in $\mathscr{R}_0$ for all $t > 0$;
2. If $\mathscr{R}$ is any open set containing S (see Figure 31.2), then for any $\mathbf{x}$ in $\mathscr{R}_0$ there is a time τ such that $\boldsymbol{\varphi}(\mathbf{x}, t)$ is in $\mathscr{R}$ for all $t > \tau$.
3. For the given region $\mathscr{R}_0$, S is the smallest set having the above properties in the sense that if $\mathscr{R}(t)$ is the image of $\mathscr{R}_0$ under the flow $\boldsymbol{\varphi}$, then, as $t \to \infty$, $\mathscr{R}(t)$ shrinks down onto S but no further:

$$S = \bigcap_{t \geq 0} \mathscr{R}(t).$$

For a given attractor S, the largest open set $\mathscr{R}_0$ having the properties stated is called the *region of attraction* of S. We require S to be connected, for if it consisted of two disconnected pieces S_1 and S_2, then each of these could be enclosed in a suitable region $\mathscr{R}_1$ or $\mathscr{R}_2$ and be an attractor in its own right.

Smale 1967 imposes the further requirement that there be an orbit dense in S. An example in which that requirement is not met, while the others are, is provided by the Taylor problem of flow between rotating cylinders. After the first bifurcation, there is a closed curve (in fact a circle) in phase space (Hilbert space) consisting of fixed points, and any nearby motion comes

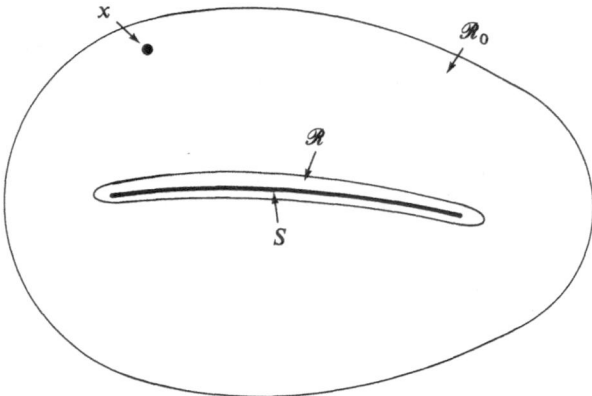

Figure 31.2

asymptotically to rest at one of those fixed points. However, this example is nontypical in the sense that an arbitrarily small perturbation (for example, subjecting the entire apparatus to a small velocity in the axial or z direction) can make points march around the circle, whereupon the circle becomes an orbit.

The ω-limit set of a motion is an attractor if it also attracts all other nearby motions. More generally, an attractor S is the union of many ω-limit sets, namely the ω-limit sets of all motions that start in $\mathcal{R}$. In particular, S may contain fixed points and closed orbits, which are of course their own ω-limit sets.

Attracting fixed points, attracting closed orbits, and attracting invariant tori of various dimensions are examples of attractors. Other attractors often have a more complicated geometric structure, for example involving Cantor sets. A *strange* attractor is one on which the motions are unstable in the sense of Lyapunov and hence are characterized by a continuous power spectrum, as explained in the rest of this chapter.

31.6 The Power Spectrum for Motions in $\mathbb{R}^n$

The theory of the power spectrum was discussed in Section 4.6 of Volume I. It concerns a bounded continuous function $f(t)$, which oscillates more or less irregularly for all t from $-\infty$ to $+\infty$. Such a function cannot be expressed as a classical Fourier series, because it is not periodic, nor as a classical Fourier integral, because it is not in L^2; but it is a tempered distribution, and hence has a Fourier transform in the sense of distribution theory. The power spectrum is a function $S(\omega)$, which describes the distribution of the power associated with $f(t)$ over the frequencies of the Fourier components, ignoring their phases. Namely, $S(\omega)$ is a nondecreasing real function, and $S(\omega_2) - S(\omega_1)$ is the power contained in the Fourier components with frequencies in the interval (ω_1, ω_2).

We summarize the results here, as applied to a *vector*-valued function $\mathbf{x}(t)$ in $\mathbb{R}^n$, in place of $f(t)$.

The *autocovariance* of $\mathbf{x}(t)$ is given by

$$R(\tau) = \lim_{T \to \infty} \frac{1}{2T} \int_{-T}^{T} \overline{\mathbf{x}(t + \tau)} \cdot \mathbf{x}(t)dt, \qquad (31.6\text{-}1)$$

and then the power spectrum is given by

$$S(\omega) = \int_{-\infty}^{\infty} R(\tau) \frac{e^{i\omega\tau} - 1}{2\pi i\tau} d\tau. \qquad (31.6\text{-}2)$$

Comments

1.   It has been tacitly assumed that all components of the vector $\mathbf{x}(t)$ contribute equally to the energy of the motion. A generalization is to replace the scalar product by

$$\mathbf{x}(t + \tau) \cdot B\mathbf{x}(t), \qquad (31.6\text{-}3)$$

where B is a positive definite matrix, to be determined by physical considerations so that (31.6-3) represents the energy.

2.   If the spectrum is "continuous," i.e., if $S(\omega)$ is absolutely continuous, which happens if $R(\tau)$ tends to zero sufficiently rapidly as $\tau \to +\infty$, then the spectral density is given by an ordinary Fourier transform

$$S'(\omega) = \frac{1}{2\pi} \int_{-\infty}^{\infty} R(\tau)e^{i\omega\tau}\, d\tau. \tag{31.6-4}$$

In numerical calculations, $R(\cdot)$ is given only at discrete values of τ and $S(\cdot)$ or $S'(\cdot)$ only at discrete values of ω. Then, $S'(\cdot)$ is usually obtained from $R(\cdot)$ by the fast Fourier transform algorithm, and if the spectrum contains lines, they show up not as δ-function contributions to $S'(\cdot)$ but as single values of $S'(\cdot)$ that are much larger than the surrounding values.

3.   If $\mathbf{x}(t)$ comes from experiment or numerical simulation, we do not take the limit in (31.6-1), but write instead

$$R(\tau) \approx \frac{1}{b-a} \int_{a}^{b} \overline{\mathbf{x}(t+\tau)} \cdot \mathbf{x}(t) dt, \tag{31.6-5}$$

where (a, b) is a long interval. As noted at the end of Section 31.4, although $\mathbf{x}(t)$ is regarded as an approximation to a motion on its own ω-limit set, hence defined for all t, it is known in practice only for $t \geq 0$; hence in the above expression, we have $0 < a < b$. Still, that expression should be a good approximation if all transients have decayed sufficiently by the time $t = a$.

For further details and examples, see Section 4.6 of Volume I.

31.7   Almost Periodic and Aperiodic Motions

The quasi-periodic motions predicted by the Landau–Hopf model are a special case of almost periodic motions, and hence have a pure line spectrum.

If the m-tuple periodic function g in (31.1-1) is expanded in an m-tuple Fourier series,

$$g(z_1, \ldots, z_m) = \sum_{-\infty(k_1, \ldots, k_m)}^{\infty} c(k_1, \ldots, k_m)e^{i(k_1 z_1 + \cdots + k_m z_m)}, \tag{31.7-1}$$

then, after suitable renumbering of the terms, the quasi-periodic function $f(t)$ given by (31.1-1) can be written as

$$f(t) = \sum_{j=1}^{\infty} c_j e^{i\tilde{\omega}_j t}, \tag{31.7-2}$$

where each $\tilde{\omega}_j$ is a linear combination of $\omega_1, \ldots, \omega_m$ with integer coefficients.

If there were a value of $\tau > 0$ such that the quantities $\omega_k \tau$ $(k = 1, \ldots, m)$ were all simultaneously integer multiples of 2π, then $f(t + \tau)$ would be $\equiv f(t)$, and $f(\cdot)$ would be periodic. Since the ω_k's are incommensurable, there is no such τ. However, it can be proved that τ can be so chosen that the difference $f(t + \tau) - f(t)$ is arbitrarily small. Not only that, but given $\varepsilon > 0$, there is a $T = T(\varepsilon) > 0$ such that every t-interval of length T contains at least one value τ such that

$$|f(t + \tau) - f(t)| < \varepsilon \quad \text{for all } t.$$

Any continuous function having this last property is called *almost periodic* in the sense of H. Bohr and can be expanded in a series of the form (31.7-2), which then converges in a certain L^2 sense—see Riesz and Sz. Nagy 1953, Chapter VI. A quasi-periodic function, as we have defined it, is an almost periodic function in which only a finite number of the frequencies $\tilde{\omega}_j$ are linearly independent over the rationals.

A vector-valued almost periodic function $\mathbf{x}(t)$ is similarly defined.

As shown earlier (Section 4.6 of Volume I), the power spectrum of an almost-periodic function is a pure line spectrum. Hence, in the Landau–Hopf model of the transition to turbulence, the power spectrum is a pure line spectrum after any finite number of bifurcations, although the number of lines in a given frequency interval might increase considerably as the Reynolds number increases.

The Ruelle–Takens model, as we shall see, predicts that a continuous spectrum will appear after a rather small number of bifurcations.

The almost periodic character of the motion in the Landau–Hopf model seems implausible on intuitive grounds. If the motion is in any sense random, it should not be able to "remember" its past behavior so precisely as to reproduce that behavior, to any desired accuracy, at times arbitrarily far in the future, as would be the case if it were almost periodic. This can also be expressed in terms of the autocovariance $R(\tau)$, because $R(\tau)/R(0)$ is the autocorrelation, that is, the correlation of the functions $f(t)$ and $f(t + \tau)$. A theorem on almost periodic functions says that the convolution of two such functions, in the sense of (31.6-1), is also almost periodic (see Riesz–Nagy). Hence $R(\tau)$ would be almost periodic if $f(t)$ were, and the correlation coefficient of $f(t)$ and $f(t + \tau)$ would come arbitrarily close to 1.0 repeatedly for certain arbitrarily large values of τ.

By contrast, for typical motion on a strange attractor, $R(\tau)$ decreases rapidly to zero as $t \to \pm \infty$, and then the power spectrum is purely continuous and is given by (31.6-4). For the case of the Lorentz attractor, see Section 31.11.

31.8 Lyapunov Stability

A motion $\mathbf{x}(t)$ [a solution of (31.4-1)] is *stable in the sense of Lyapunov* if any other motion $\bar{\mathbf{x}}(t)$ [another solution of (31.4-1)] that is sufficiently close to it initially stays close at later times, specifically if for any $\varepsilon > 0$ there is a $\delta = \delta(\varepsilon) > 0$ such that

$$\text{if } |\bar{\mathbf{x}}(0) - \mathbf{x}(0)| < \delta,$$

$$\text{then } |\bar{\mathbf{x}}(t) - \mathbf{x}(t)| < \varepsilon \quad \text{for all } t > 0.$$

If the motion is not stable in this sense, there is a positive ε (not necessarily very small) such that no matter how small δ is, there is a perturbation of initial size $<\delta$ that grows to a size $\geq \varepsilon$ at some later time. One of the motivations for Lorenz's work, described below, was to show that a simple prototype of atmospheric motion is Lyapunov unstable, with obvious implication for the problem of weather forecasting.

31.9 The Lorenz System; the Bifurcations

The first strange attractor in a problem arising from fluid dynamics was discovered by E. N. Lorenz in 1963. Lorenz expanded the Bénard equations of thermal convection for a horizontal layer of fluid heated from below in a triple Fourier series with respect to the space variables, then truncated the resulting system of ordinary differential equations for the time dependence of the Fourier coefficients to three equations. If the Fourier coefficients in those equations are denoted by $X(t)$, $Y(t)$, and $Z(t)$, the equations are

$$\dot{X} = -\sigma X + \sigma Y,$$
$$\dot{Y} = rX - Y - XZ, \tag{31.9-1}$$
$$\dot{Z} = -bZ + XY,$$

or, more briefly,

$$\dot{\mathbf{X}} = \mathbf{F}(\mathbf{X}). \tag{31.9-2}$$

The constants σ, r, and b are dimensionless; for the physical system considered by Lorenz, they have the values

$$\sigma = 10, \qquad b = \tfrac{8}{3}, \qquad 0 < r < \infty; \tag{31.9-3}$$

r is proportional to the Rayleigh number and is a measure of the intensity of the heating.

Lorenz was interested in exhibiting the general kind of instability found in atmospheric physics and did not intend the above system to be a realistic model of the atmosphere or of thermal convection. Later Curry 1978 studied a more realistic model of the Bénard convection equations by truncating to 14 rather than three equations. He found a more complicated sequence of bifurcations, as the Rayleigh number r is increased, but the strange attractor was still present for certain values of r.

Lorenz showed there is a constant R, depending on σ, r, and b, such that any solution $\mathbf{X}(t)$ of (31.9-1) is eventually trapped in the ball

$$X^2 + Y^2 + Z^2 < R^2. \tag{31.9-4}$$

Furthermore, it follows from (31.9-1) that the divergence of the vector field $\mathbf{F}(\mathbf{X})$ has the constant value

$$\nabla \cdot \mathbf{F} = -(\sigma + b + 1) = -13\tfrac{2}{3}, \tag{31.9-5}$$

so that the volume of a region carried along by the flow (31.9-1) in $\mathbb{R}^3$ decreases with time as $\exp(-13.67t)$. Therefore there is at least one attractor in the ball (31.9-4), and any such attractor occupies zero volume in $\mathbb{R}^3$.

The fixed points or stationary solutions of the system (31.9-1) are as follows:

1.   For any r, the origin, $X = Y = Z = 0$ is a fixed point. For $0 < r < 1$ it is stable (in fact attracting). For $r > 1$ it is unstable; the linearized problem has one positive eigenvalue and two negative ones. There is a one-dimensional unstable manifold with a horizontal tangent vector at the origin (a vector parallel to the plane $Z = 0$) and a two-dimensional stable manifold with a vertical tangent plane.

2.   For any $r > 1$ there are two more fixed points, called P_1 and P_2:

$$P_2 : X = Y = \sqrt{b(r-1)}, \qquad Z = r - 1,$$
$$P_1 : X = Y = -\sqrt{b(r-1)}, \qquad Z = r - 1. \tag{31.9-6}$$

Hence, there is a first bifurcation, of the type discussed in Section 29.9, at $r = 1$.

To determine the stability of the new fixed points, we write $\mathbf{X} = \mathbf{X}_0 + \mathbf{X}_1$, where $\mathbf{X}_0$ is given by (31.9-6), and we linearize with respect to $\mathbf{X}_1$; we find

$$\begin{pmatrix} \dot{X}_1 \\ \dot{Y}_1 \\ \dot{Z}_1 \end{pmatrix} = \begin{pmatrix} -\sigma & \sigma & 0 \\ r - Z_0 & -1 & -X_0 \\ Y_0 & X_0 & -b \end{pmatrix} \begin{pmatrix} X_1 \\ Y_1 \\ Z_1 \end{pmatrix}. \tag{31.9-7}$$

When X_0, Y_0, and Z_0 are substituted from (31.9-6), the matrix of this system becomes

$$\begin{pmatrix} -\sigma & \sigma & 0 \\ 1 & -1 & -\sqrt{b(r-1)} \\ \sqrt{b(r-1)} & \sqrt{b(r-1)} & -b \end{pmatrix};$$

It has one negative real eigenvalue and two complex conjugate ones. The complex eigenvalues are in the left half-plane (hence the new fixed points are stable) if $r < r_0$, where

$$r_0 = \sigma(\sigma + b + 3)/(\sigma - b - 1) = 24.74. \tag{31.9-8}$$

Hence there is a second bifurcation at $r = r_0$, and this one is of the kind discussed in Section 29.10, leading to periodic solutions. However, this bifurcation is subcritical, as shown by the calculations of Marsden and McCracken 1976. Hence the periodic solutions are present only for $r < r_0$ and are unstable, while for $r > r_0$ an explosive transition to something else must be expected. It turns out, as discussed below, that the transition is not really "explosive," because of the presence of another attractor (in fact a strange attractor) in the near vicinity in $\mathbb{R}^3$; see Section 31.17.

With each of the points P_1 and P_2 is associated, for $r > r_0$, a one-dimensional stable manifold and a two-dimensional unstable one. In the latter, solutions spiral outward from the fixed point. Nearby solutions also spiral outward and, at the same time, are drawn rapidly toward the unstable manifold, because of the large negative eigenvalue associated with the stable one.

31.10 The Lorenz Attractor; General Description

To investigate the behavior of the system after the second bifurcation, at
$r = r_0 = 24.74$, Lorenz calculated solutions of the system (31.9-1) numerically, for $r = 28$. He found that, after transients have decayed, the orbits
appear to lie, as accurately as one can tell from the calculations, on a branched
surface L_0, shown schematically in Figure 31.3a, where the directions of
motion are also indicated. It is roughly heart-shaped and lies roughly in the
vertical plane $X = Y$ [which contains the fixed points P_1 and P_2 given by
(31.9-6)]. It is symmetric with respect to reflection about the Z axis and has
two holes, one surrounding each of the fixed points P_1 and P_2. Below a
branch line BB, which joins the two holes and dips down somewhat below
the horizontal, there is a single sheet. Above it, there are two sheets, one a
little to the left and behind, the other a little to the right and in front. They
are joined along the heavy part of the branch line BB.

L_0 is bounded by part of the unstable manifold of the origin, which consists of two orbits W_l^u and W_r^u starting out from the origin in opposite directions. These orbits later continue into the interior of the surface L_0 after going
round the holes shown and then crossing the branch line BB at the ends
of the heavy part. The fixed points P_1 and P_2 lie inside the holes, and if the
surface L_0 were continued into the holes, orbits would spiral outward in
them; it is seen from the figure, however, that an orbit once in L_0 can never
subsequently get into the holes.

Every orbit crosses the branch line BB downward; if it crosses at the
center of BB, it comes asymptotically to rest at the origin 0; otherwise, it
encircles one of the holes and goes down across BB again, and so on. We
define a Poincaré map $s \to \psi_0(s)$ of BB as follows: Let s be a parameter
(possibly arclength) along BB, with $s = 0$ at the center. If an orbit crosses
BB at $s = s_1 \neq 0$, then it next crosses BB at $s = \psi_0(s_1)$; $\psi_0(0)$ is undefined.

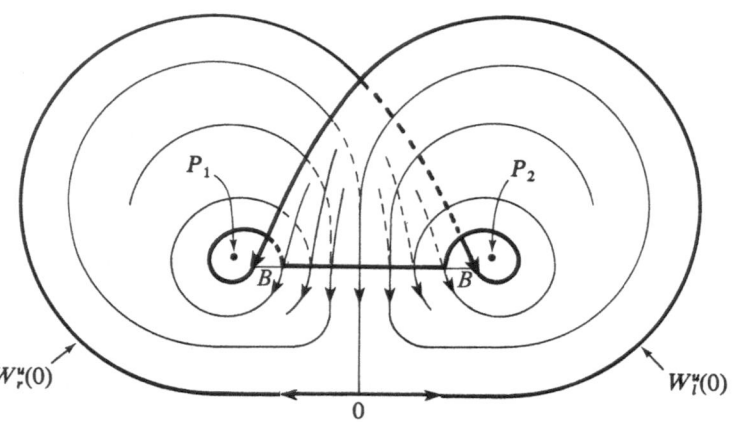

Figure 31.3a   The branched surface.

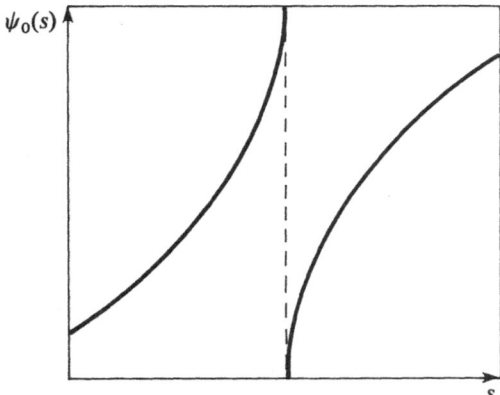

$\psi_0(s)$

s

Figure 31.3b   The Poincaré mapping.

See Figure 31.3b. Numerical orbit calculations show that this mapping has the property called *locally eventually onto* by Williams 1977: If I is any interval $s_0 < s < s_0 + \varepsilon$, no matter how narrow, then for some n, the n-times iterated mapping of I is all of BB. In other words, no matter how close together two orbits on L_0 are initially, they eventually become completely separated as time goes on. This follows, for example, from showing that if the parameter s is suitably chosen, then $\psi_0'(s) \geq \text{const.} > 1$ for all s.

An orbit that goes down across the branch line BB to left of center encircles P_1 clockwise before returning to BB, and one that goes down to the right of center encircles P_2 counterclockwise. The number of times an orbit encircles one of the points P_1 or P_2 before moving to the other depends on how rapidly it spirals outward from that point and also in a critical way or how far from the center it first cut BB after coming from the other side. It is the essence of Lorenz's discovery that the successive numbers of circuits round those points vary in a pseudorandom way so that the motion is aperiodic.

As Lorenz pointed out, the picture based on the branched surface L_0 cannot be precise, because two orbits that go down across the branch line BB at the same point of BB would then coincide subsequently, and that would contradict the unique reversibility of the orbits. Hence, the two sheets of the surface can't merge into a single sheet, but must remain separated by a possibly very small distance. If we follow the orbits round again, the two sheets become four, and so on. We conclude that the attractor must contain infinitely many sheets, probably shomehow connected together into a single structure in $\mathbb{R}^3$, which will be called the *Lorenz attractor* and denoted by L. However, at least for the parameter values studied by Lorenz ($\sigma = 10$, $b = \frac{8}{3}$, $r = 28$), the fine structure just described is really quite fine; the separation of the sheets is very small, so that, to something like four-decimal accuracy, the branched surface L_0, with the motions on it as sketched, describes the attractor fully.

31.11 The Lorenz Attractor; Aperiodic Motions

Lorenz observed that the randomness of the motion could be analyzed by considering the successive maxima Z_n, $n = 0, 1, 2, \ldots$, of $Z(t)$ on the computed orbit. [From the third equation of (31.9-1), we see that these maxima occur at intersections of the orbit with the hyperboloid given by $Z = (1/b)XY$.] Lorenz found that if Z_{n+1} is plotted against Z_n, the points lie quite accurately on a curve $Z_{n+1} = f(Z_n)$, which is smooth except for a central cusp, as shown in Figure 31.4.

As Lorenz pointed out, Z_{n+1} cannot be exactly a single-valued function of Z_n, since its value depends generally also on the values of $X(t)$ and $Y(t)$ at the instant when $Z(t) = Z_n$. Hence the Lorenz graph ought to be slightly fuzzy and ought to have some transverse structure, although possibly of a very narrow extent. This structure is related to the fine structure of the attractor L, and one can begin to see it if the deviations of the points on the graph from a smooth curve are magnified by about 10^3 (see Richtmyer 1981).

If we ignore the fine structure of the Lorenz graph, the sequence $\{Z_n\}_0^\infty$ of maxima of $Z(t)$ may be regarded as obtained from an initial value Z_0 by iteration of the mapping $f: Z \to f(Z)$, so that $Z_{n+1} = f(Z_n)$.

In order to study the iterated mapping statistically, we consider the problem of transforming from Z to another variable W by equations

$$Z = \zeta(W), \qquad W = \zeta^{-1}(Z), \tag{31.11-1}$$

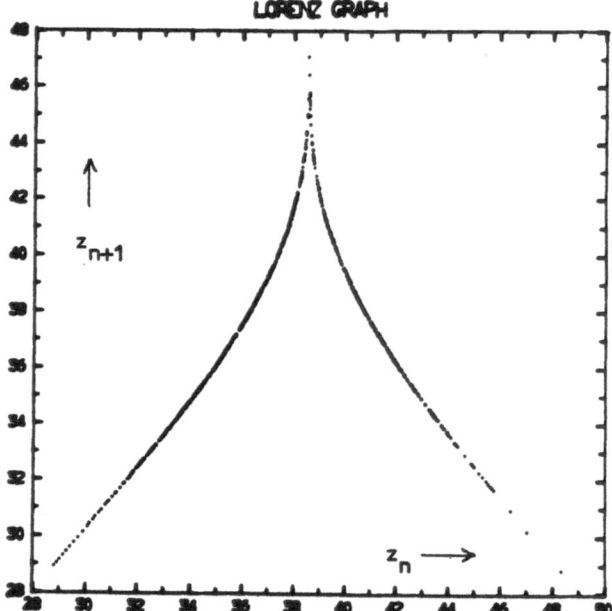

Figure 31.4 The Lorenz graph.

such that the resulting mapping $g: W \to g(W)$ takes an especially simple form. In particular, we wish to choose (31.11-1) so that $g(W)$ is the triangular function

$$g(W) = \begin{cases} 2W, & \text{if } 0 \le W \le \frac{1}{2}, \\ 2 - 2W, & \text{if } \frac{1}{2} \le W \le 1. \end{cases} \tag{31.11-2}$$

Lorenz considered the mapping g based on this function as a sort of model for the mapping f that arose in his calculations, in order to predict qualitatively the statistical properties of the motion. We show how to compute the transformation (31.11-1) under certain assumptions about the function $f(Z)$. First, we assume that a linear change of the coordinate Z has been made so that the least and greatest possible values of Z_n are 0 and 1. Then the function $f(Z)$ maps the interval $[0, 1]$ onto itself. We assume that $f(0) = f(1) = 0$. [Actually, $f(0)$ is ≈ 0.0035, but we shall ignore this difference along with the fine structure of the Lorenz graph.] We also assume that $f(Z)$ is differentiable, except at the cusp, and that $|f'(Z)|$ is greater than some constant $\alpha > 1$ for all Z. Then, it can be shown—see Rüssmann and Zehnder 1980 or Richtmyer 1981—there is a unique continuous increasing function $\zeta(W)$ that transforms the mapping $Z \to f(Z)$ into the mapping $W \to g(W)$.

To calculate $\zeta(W)$, we denote by $\varphi_1(Z)$ and $\varphi_2(Z)$ the inverses of the rising and falling parts of $f(Z)$, as shown in Figure 31.5. Then, it is seen from (31.11-2) that $\zeta(W)$ must satisfy the equations

$$\zeta(W) = \varphi_1(\zeta(2W)) \quad \text{for } 0 \le W \le \frac{1}{2},$$

$$\zeta(W) = \varphi_2(\zeta(2 - 2W)) \quad \text{for } \frac{1}{2} \le W \le 1.$$

From these equations, $\zeta(W)$ is calculated in succession for dyadic values of W in the order $W = 0, 1, \frac{1}{2}, \frac{1}{4}, \frac{3}{4}, \frac{1}{8}, \ldots$, starting with $\zeta(0) = 0$ and $\zeta(1) = 1$, and then for other values by the requirement of continuity. Figure 31.6a shows the result, and Figure 31.6b shows the result of applying the transformation $\zeta(W)$ to the Lorenz graph.

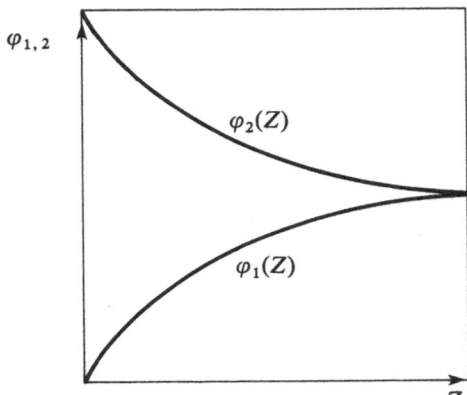

$\varphi_{1,2}$

$\varphi_2(Z)$

$\varphi_1(Z)$

Z **Figure 31.5**

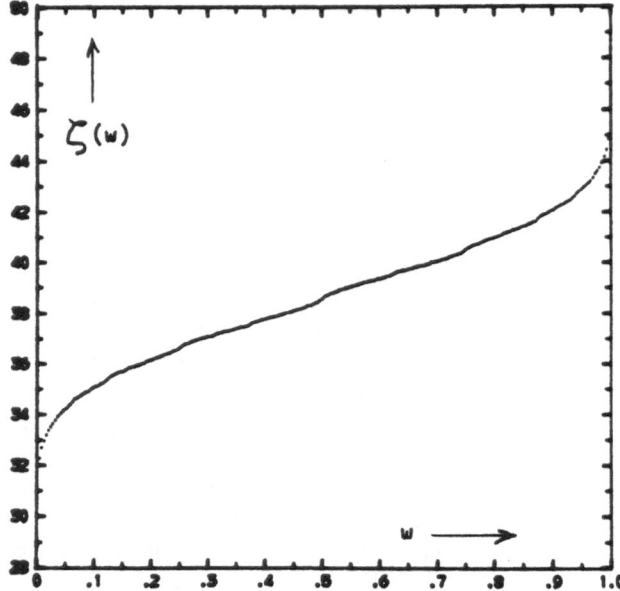

Figure 31.6a

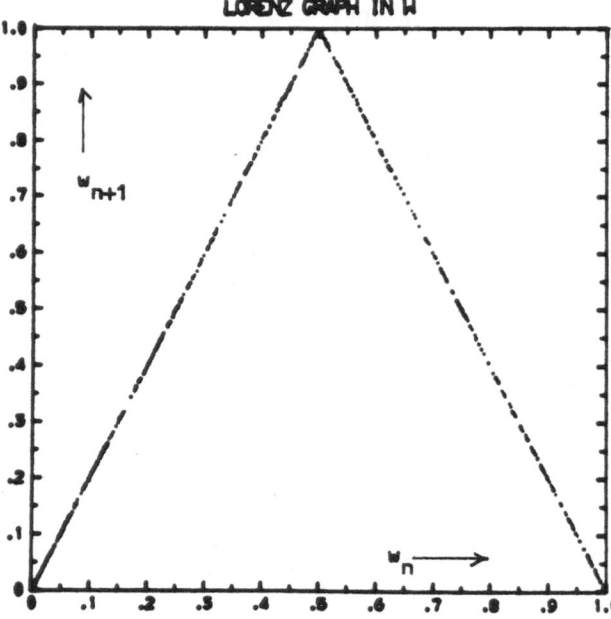

Figure 31.6b

The function $\zeta(W)$ is Hölder continuous with Hölder exponent $\log_2 \alpha$, where α is the greatest lower bound of $|f'(Z)|$, as above, but it is not absolutely continuous; hence what is true of the mapping $g: W \to g(W)$ for almost all W is not necessarily true of the mapping $f: Z \to f(Z)$ for almost all Z.

A consequence of the continuity of $\zeta(W)$ (which does not require absolute continuity) is that the mapping f is *locally eventually onto*, in the terminology of Williams, referred to in Section 31.10; if I is any interval $a < Z < a + \varepsilon$, then for some finite number n of iterations, $f^{(n)}(I)$ is the entire interval $[0, 1]$. Clearly f has that property if and only if g also has it, but for g it is nearly obvious. Namely, the length of any interval I in W is doubled under g unless $g(I)$ contains the point $W = \frac{1}{2}$, in which case the length is at least not decreased. Hence, under each pair of iterations, the length is at least doubled, unless, for some I, $g(I)$ and $g(g(I))$ both contain the point $W = \frac{1}{2}$, but in that case $g(g(I))$ contains all of $[\frac{1}{2}, 1]$, so that $g(g(g(I)))$ is $[0, 1]$.

It follows that iteration of the mapping f is unstable in the sense of Lyapounov, for no matter how close together two points are initially, they will eventually be separated by a finite amount (for example at least $\frac{1}{2}$). Hence, if we can neglect the fine structure of the Lorenz graph, the motion on the Lorenz attractor is also Lyapounov unstable, and hence has a purely continuous power spectrum.

31.12 Statistics of the Mappings f and g

The two mappings have slightly different statistical properties; we discuss g first. Because of the special character of the function $g(W)$, as given by (31.11-2), we represent W in binary form as $W = .a_0 a_1 a_2 \ldots$, where each a_i is 0 or 1. Then $g(W)$ is simply

$$g(.a_0 a_1 a_2 \ldots) = \begin{cases} .a_1 a_2 \ldots, & \text{if } a_0 = 0, \\ .\bar{a}_1 \bar{a}_2 \ldots, & \text{if } a_0 = 1. \end{cases}$$

where the overbar denotes complementation, i.e., replacement of 0 by 1 and 1 by 0. Now suppose the initial element W_0 of the sequence $\{W_n\}_{n=0}^{\infty}$, obtained by iterating g, is chosen at random from a uniform distribution in $[0, 1]$. Then each binary digit of W_0 is equally likely to be 0 or 1, independently of the choice of the others, and furthermore W_n is $< \frac{1}{2}$ or $\geq \frac{1}{2}$ according to the choice of the nth digit of W_0. We conclude that each W_n is equally likely to be $< \frac{1}{2}$ or $\geq \frac{1}{2}$, and there is no correlation in this matter between successive members of the sequence.

To find the corresponding properties of f, a sequence $\{Z_n\}$ of length 200,000 was generated by iterating f numerically (Richtmyer 1981). To analyze the result, s_n was defined as

$$s_n = \begin{cases} -1, & \text{if } Z_n < \frac{1}{2}, \\ +1, & \text{if } Z_n \geq \frac{1}{2}. \end{cases}$$

The average of s_n was found to be -0.1416 (about 64 standard deviations) and significant positive correlations were found of s_n with s_{n+1}, with s_{n+2}, and with s_{n+3}. The correlation of s_n with s_{n+k}, for $k \geq 4$, was not significant in the sample size available.

The difference between f and g reflects the nonabsolute continuity of the functions $\zeta(W)$ that connects Z and W. Clearly, W is $< \frac{1}{2}$ if and only if $Z = \zeta(W)$ is $< \frac{1}{2}$ (because the Lorenz graph, Figure 31.4, is symmetric with high accuracy), but the behavior for almost all W is not the same as the behavior for almost all Z. There is a set of measure zero in $[0, 1]$ that is transformed into a set of positive measure, in fact measure $= 1$, under ζ, and there are similar sets for ζ^{-1}. Nevertheless, the two mappings behave similarly; hence the qualitative statistical properties of the Lorenz system can be inferred by study of f, as was done by Lorenz.

31.13 The Lorenz Attractor; Detailed Structure I

According to Section 31.10, the attractor L contains infinitely many sheets (in fact, uncountably many), all lying close to the idealized branched surface L_0, and somehow connected together to form a many-sheeted structure in $\mathbb{R}^3$. The structure of L was investigated by R. F. Williams 1977, and we shall describe his main results in somewhat intuitive geometrical terms, ignoring certain topological difficulties such as arise from the presence of infinitely many vertices of the cell complex F in a bounded region of the strip S (see below).

The attractor is of course completely determined by the differential equations (31.9-1). At present, however, there is no known method for determining it precisely, starting from those equations, or even for determining its general topological properties. Furthermore, the equations are somewhat specialized and artificial; hence there is more interest in the general kinds of attractor that can result from equations generally similar to (31.9-1). The objective of the work of Williams was to find all attractors that are related to a branched manifold L_0 of the kind described above in the way the attractor L appears to be. He calls them *Lorenz attractors* generally.

Williams used an abstract topological construction known as the inverse limit. The topological properties of the resulting attractor and the flow on it are completely determined by the given semiflow on L_0, and in fact by the so-called kneading sequences of the orbits W_l^u and W_r^u. Williams showed that there are uncountably many different such attractors, i.e., topologically different ones.

The attractors found by Williams can be embedded in $\mathbb{R}^3$, but the question which of them would result from a given differential equation system of the type (31.9-1) remains open.

To investigate the structure, following Williams, we consider a piece of the surface and then continue it by following the orbits in it both forward and backward in time. If an orbit encircles P_1 it moves to a sheet of L behind the one it was on before (from the point of view of an observer looking at Figure 31.3), whereas, if it encircles P_2, it moves to a sheet in front. If it encircles first one, then the other, we don't know whether it is then in front of the starting point or behind; it might even return to its starting point, and then we should have a periodic orbit. One of Williams's results is that the periodic orbits are dense in L.

The unstable manifold $W^u(0)$ plays a role. As stated in Section 31.10, it consists of two orbits that start horizontally in two opposed directions from 0, form the boundary of L_0, and then continue into the interior. Conceivably either or both of them might eventually hit the exact center of BB and then come to rest asymptotically at 0. We shall ignore that possibility as being unlikely and assume that they continue winding indefinitely round in L_0. Following Williams, we denote by $W_l^u(0)$ the orbit that leaves 0 to the right, hence first meets BB at its *left* end and by $W_r^u(0)$ the orbit that leaves 0 to the left, hence first meets BB at its *right* end. We define so-called *kneading sequences* for these orbits as $z_1 z_2 z_3 \ldots$, where z_k is $= 1$ or $= 2$ according as the kth circuit of the orbit is around P_1 or P_2. Below, it will be assumed that these sequences start as $2111\ldots$ and $1222\ldots$, respectively, for $W_l^u(0)$ and $W_r^u(0)$. It will be seen that they completely determine the topology of the attractor L.

In order to carry out the program, we have to make the following assumptions. [That is, we have to conjecture that these things are consequences of the differential equations (31.9-1)—numerical calculations give considerable support for most of them.]

1.   The branched surface L_0 and the semiflow on it are as described in Section 31.10. [We say "semiflow," not "flow," because an orbit $\mathbf{x}(t)$ in L_0, in contrast with those in L, cannot be uniquely determined, for $t < 0$, from $\mathbf{x}(0)$, owing to the branching.]

2.   The Poincaré map ψ_0 of BB is locally eventually onto.

3.   There is a continuous mapping $\mathbf{p}$ (projection) of L onto L_0 which carries orbits of L onto orbits of L_0; if $\mathbf{x}(t)$ is a motion in L, then $\mathbf{p}(\mathbf{x}(t))$ is a motion in L_0.

4.   Every motion in L_0 is the image under $\mathbf{p}$ of a unique motion in L. (It would seem to be difficult to get supporting evidence for this assumption from the numerical work.) Furthermore, the motion in L depends continuously on the motion in L_0; if orbits $\mathbf{x}_0(t)$ and $\mathbf{x}_0'(t)$ are close together in L_0 for a long interval $(-T, T)$, then they are the projections of orbits $\mathbf{x}(t)$ and $\mathbf{x}'(t)$ that are close together in space (i.e., in L) for a long interval.

5.   The kneading sequences of $W_l^u(0)$ and $W_r^u(0)$ start as $2111\ldots$ and $1222\ldots$, respectively. That is, $W_l^u(0)$, after initially encircling P_2, then encircles P_1 at least three times (in Lorenz's system it does so 25 times) before again encircling P_2, and similarly for $W_r^u(0)$.

To follow the orbits properly in three dimensions, we replace the branch line BB by a strip S transverse to L_0 and intersecting L_0 on BB. It turns out to be convenient to pull S and BB both downward until they pass through the origin 0, as shown in Figure 31.7. We shall develop a detailed description of the intersection $S \cap L \stackrel{\text{def}}{=} F$ of the strip S and the attractor L. F will be

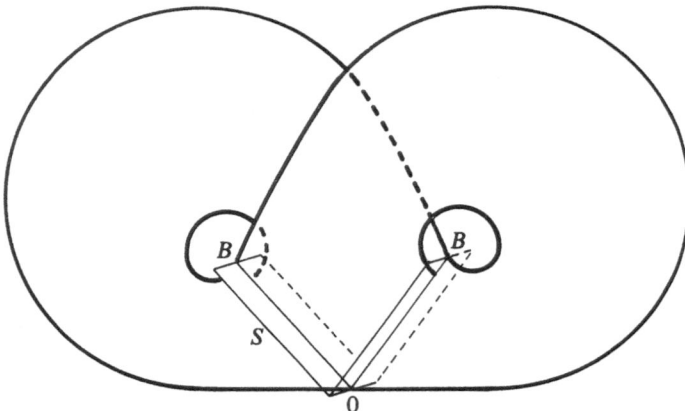

Figure 31.7   The strip S.

described as a so-called cell complex, consisting of points called vertices and curves called 1-cells, each of which connects one vertex to another. We shall find that the vertices are the origin 0 and the intersections of $W^u(0)$ with S, and the 1-cells are the intersections of the sheets that make up L with S. F consists of the origin 0 and two parts, one lying on either side of a line crossing the strip S at 0. That line lies in the stable manifold of 0 and serves to separate the orbits that will next encircle P_1 from those that will next encircle P_2.

It will be seen that $W^u(0)$ is throughout its length the boundary of infinitely many sheets of L, which come together at that boundary so as to form what Williams calls the "spine" of a "Cantor book"—see Figure 31.9. A line transverse to L_0 just inside the boundary intersects those sheets on a Cantor set, as explained below.

We denote by $l_1, l_2, \ldots$ the successive intersections of $W^u_l(0)$ with S (they can lie on either side of 0) and by $r_1, r_2, \ldots$ the intersections of $W^u_r(0)$ with S; they are points in S. When projected onto L_0, l_1 is at the extreme left of the branch line BB, and r_1 at the extreme right; the others lie between.

Each sheet of L intersects the strip S on a curve $\mathscr{C}$, a 1-cell of F, and if we let $\mathscr{C}$ be carried round by the flow until its points again meet S in another curve $\mathscr{C}'$, it sweeps out a surface element Σ in $\mathbb{R}^3$, as shown in Figure 31.8, and we shall construct L be piecing together such surface elements; $\mathscr{C}$ and $\mathscr{C}'$ are respectively the *initial* and *final* curves of Σ. We assume tentatively that there is a curve connecting l_1 to r_1 in S and consisting of points of the attractor L. (This assumption is justified in Section 31.17 below.) That curve goes through 0, hence consists of two 1-cells of F, and we have to consider the two separately, so let $\mathscr{C}$ be the 1-cell that connects 0 to r_1. As $\mathscr{C}$ is carried round by the flow to sweep out a surface Σ, we let its left end, which starts at 0, be carried to the right from 0 along $W^u_r(0)$, so that it, like all other points of $\mathscr{C}$, encircles the point P_2 once counterclockwise; it then

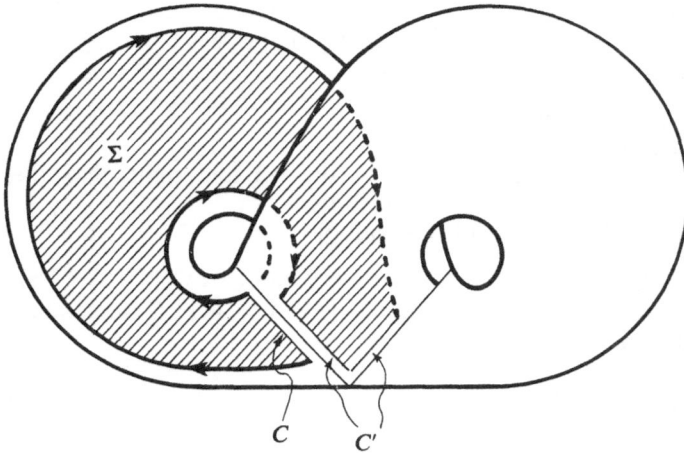

Figure 31.8   A surface element in the Lorenz attractor.

meets S at the point l_1; hence the final curve $\mathscr{C}'$ of the surface element Σ connects l_1 to r_2 in S, a little in front of $\mathscr{C}$ (toward the viewer).

Before we can carry the points of $\mathscr{C}'$ round again, we must note that since l_1 and r_2 are on opposite sides of 0 by assumptions 5 above, we must divide $\mathscr{C}''$ into two 1-cells, say $\mathscr{C}'_1$ and $\mathscr{C}'_2$, one connecting l_1 to 0 and the second connecting 0 to r_2. Then, $\mathscr{C}'_1$ will be carried clockwise round P_1 and $\mathscr{C}'_2$ counterclockwise round P_2, thus creating two new surface elements within L, say Σ'_1 and Σ'_2. This process can be continued indefinitely.

Generally, given two points l_i and r_j of the intersection of $W^u_l(0)$ and $W^u_r(0)$, respectively, with S, then, if they lie on the same side of 0, there may (or may not) be a 1-cell of F connecting l_i to r_j (i.e., a curve in S connecting l_i to r_j and consisting of points of L—if there is one, there are infinitely many). If that curve is carried along by the flow, as described above, it sweeps out a surface element Σ of L.

The origin 0 is also denoted by l_0 and r_0. We denote it by r_0 if it is then to be carried along $W^u_r(0)$ to r_1 and by l_0 if it is then to be carried along $W^u_l(0)$ to l_1. The general surface element Σ of L is then obtained by carrying a 1-cell connecting l_i to r_j in S (where now we permit $i = 0$ or $j = 0$) round P_1 or round P_2 to a curve connecting l_{i+1} to r_{j+1}. (Then, if l_{i+1} and r_{j+1} lie on opposite sides of 0, it is necessary to divide the new curve into two 1-cells, as before.)

31.14   The Symbols $[i, j]$ of Williams

For given i and j, if l_i and r_j are connected by a 1-cell of F (hence, by infinitely many, as we shall see), we denote that fact by saying that a symbol $[i, j]$ is defined or exists. We now state the rules for the existence of these symbols, following Williams. Our tentative assumption implied that $[0, 1]$ and $[1, 0]$ are symbols (and we concluded, assuming that $\mathbf{r}_2$ lies to the right of 0, that $[0, 2]$ is also a symbol). The rules are:

1. $[0, 1]$ and $[1, 0]$ are symbols.
2. If $[i, j]$ is a symbol, and
    (a) $\mathbf{l}_{i+1}$ and $\mathbf{r}_{j+1}$ lie respectively to the left and right of 0. then $[i + 1, 0]$ and $[0, j + 1]$ are symbols;
    (b) otherwise, $[i + 1, j + 1]$ is a symbol.

In case 2(a) we say that $[i, j]$ *precedes* (or *is followed by*) $[i + 1, 0]$ and $[0, j + 1]$, and in case 2(b) that it precedes $[i + 1, j + 1]$ ("Immediately precedes" might be a better wording.)

(Note. In contrast with Williams, we are assuming, as said earlier, that neither $\mathbf{l}_i$ or $\mathbf{r}_j$ ever falls exactly on 0 for any $i > 0$ or $j > 0$.)

The rules are so chosen that $[i, j]$ precedes $[i', j']$ if and only if there is a surface element Σ whose initial curve connects $\mathbf{l}_i$ to $\mathbf{r}_j$ and whose final curve connects $\mathbf{l}_{i'}$ to $\mathbf{r}_{j'}$. [This describes only a part of the final curve in case 2(a).]

The following are immediate consequences of the rules (some of them require an induction in the proof):

1. If $[i, j]$ is a symbol, then $i \neq j$.
2. Each symbol has either one or two successors, depending on whether case (a) or case (b) holds.
3. A symbol $[i, j]$ with i and j both > 0 has exactly one predecessor, namely $[i - 1, j - 1]$.
4. If $[i, j]$ is a symbol, then either $\mathbf{l}_i < \mathbf{r}_j \leq 0$ or $0 \leq \mathbf{l}_i < \mathbf{r}_j$, where $<$ denotes the ordering obtained by projecting the strip S onto the branch line BB of L_0.
5. Since, by assumption 5 of the preceding section, the orbit $W_1''(0)$, after coming to $\mathbf{l}_1$, goes at least twice more round the fixed point P_1 before crossing over the center into the right half of the strip S, we see that the first few symbols, starting with $[1, 0]$ are

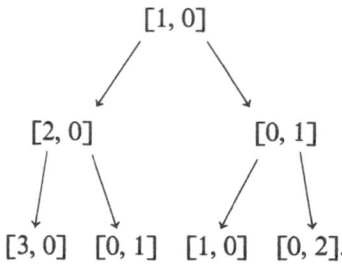

In particular $[0, 1]$ has at least two different predecessors, $[1, 0]$ and $[2, 0]$.

6. Each symbol is an ultimate predecessor of any other, in the sense that if $\sigma = [i, j]$ and $\tau = [i', j']$ are given, then there is a finite sequence $\sigma_0 = \sigma$, $\sigma_1, \sigma_2, \ldots, \sigma_n = \tau$ such that σ_k always precedes σ_{k+1}. That is clear if $\sigma_0 = [1, 0]$

(or [0, 1]), since, by the rules, all other symbols follow from [1, 0] or [0, 1], and we have just seen that [1, 0] and [0, 1] follow from each other. On the other hand, if $\tau = [1, 0]$ (or [0, 1]), and σ is arbitrary, the sequence can be found by appeal to what is essentially the locally eventually onto property of the Poincaré mapping $\psi_0(s)$ of the branch line BB, which says roughly that the sequence can be so chosen that the width of the interval (l_i, r_j) constantly increases. Williams proved that [1, 0] (hence also [0, 1]) can be reached from arbitrary σ in a finite number of steps if the derivative $\psi_0'(s)$ is $> \sqrt{2}$ for all s. In the case studied by Lorenz, the minimum of $\psi_0'(s)$ is more like 1.05; however, we can replace the archlength s by a new parameter on BB, by the method discussed in Section 31.11, so as to convert the graph of $\psi_0(s)$ into two straight lines, and then $\psi_0'(s)$ is indeed $> \sqrt{2}$ (in fact about 1.9).

7. For each $i > 0$ (and each $j > 0$) there is at least one j (one i) such that $[i, j]$ is a symbol.

31.15 Prehistories

According to Section 31.4, one of the characterizing features of an attractor L is that a point $\mathbf{x}$ belongs to L if and only if the orbit $\mathbf{x}(t)$ for which $\mathbf{x}(0) = \mathbf{x}$ lies in L for all $t < 0$ as well as all $t \geq 0$. As $t \to -\infty$, $\mathbf{x}(t)$ passes through a unique sequence of surface elements $\Sigma, \Sigma', \Sigma'', \ldots$ of the kind described in Section 31.13. If the orbits in Σ are followed backward in time, they lie in Σ', and so on. Hence, the initial curve of Σ is at least a part of the final curve of Σ'. If the initial curve of Σ connects l_i to r_j and that of Σ' connects $l_{i'}$ to $r_{j'}$, then the symbol $[i', j']$ precedes the symbol $[i, j]$. Generally, $[i, j]$ may have many predecessors, but as soon as a particular 1-cell connecting l_i to r_j is chosen, a unique predecessor is thereby singled out, because the initial curve of the surface element Σ' is uniquely determined by its final curve.

It follows that if $[i, j]$ is a symbol, each 1-cell connecting l_i to r_j is characterized by a unique infinite sequence of symbols $\ldots, \sigma_{-2}, \sigma_{-1}, \sigma_0 = [i, j]$ such that σ_{-k-1} precedes σ_{-k}, for each k.

It can be shown that if we extend the sequence also indefinitely in the other direction,

$$\ldots, \sigma_{-2}, \sigma_{-1}, \sigma_0, \sigma_1, \sigma_2, \ldots, \tag{31.15-1}$$

we single out not merely a particular 1-cell connecting l_i to r_j, but a single point of that 1-cell, so that each sequence (31.15-1), where σ_k always precedes σ_{k+1}, corresponds to a unique orbit on L. That follows from the locally eventually onto character of the Poincaré map in BB. If we follow two orbits forward in time, then no matter how close together they are at $t = 0$, they will eventually separate so as to lie in different surface elements Σ at some time $t > 0$.

31.16 The Lorenz Attractor; Detailed Structure II

We show first that if $[i, j]$ is a symbol, there are uncountably many 1-cells in F connecting $\mathbf{l}_i$ to $\mathbf{r}_j$. According to the preceding section, any such 1-cell corresponds to a unique sequence of symbols

$$\ldots, \sigma_{-2}, \sigma_{-1}, \sigma_0 = [i, j], \tag{31.16-1}$$

where each σ_{-k-1} is a predecessor of σ_{-k}. Conversely (see next section), any such sequence determines a 1-cell, and we shall see that there are uncountably many choices for the sequence, for given σ_0. Once any σ_{-k} has been chosen, it is possible, according to (6) of Section 31.4, to choose $\sigma_{-k-1}, \sigma_{-k-2}, \ldots$, so as to reach $[0, 1]$ in a finite number of steps, say $\sigma_{-k-l} = [0, 1]$. Then, according to (5) of that section, there are at least two choices of σ_{-k-l-1}. Hence, there is the possibility of a twofold choice infinitely many times in choosing the sequence. If we represent the nth choice by a binary digit a_n, we see that there are at least as many sequences (31.16-1), for given σ_0, as there are real numbers $.a_0 a_1 a_2 \ldots$ in the unit interval, namely uncountably many.

It follows that a line transverse to the branched surface L_0 intersects uncountably many sheets of the attractor L. Let M denote the point set, on that line, of intersections with L. Since L is a closed set in $\mathbb{R}^3$, M is a closed set on that line. As noted in Section 31.9, L has zero Lebesgue measure in $\mathbb{R}^3$; hence M has zero measure on the line, for otherwise the Cartesian product of M with a piece of surface of one of the sheets of L would have positive 3-dimensional measure. Hence M is an uncountable closed set of measure zero. Lastly, M has no isolated points, for the topological arguments of Williams show that if a 1-cell is determined by a sequence (31.17-1), there are other 1-cells arbitrarily close to it obtained from sequences that agree with (31.16-1) sufficiently far back. Hence, M is a Cantor set.

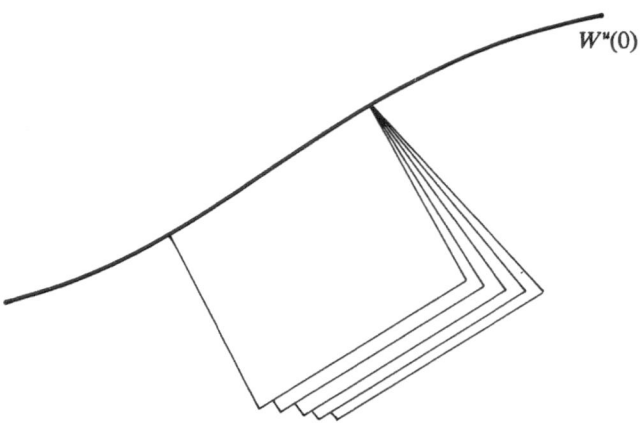

*W*ᵘ(0)

Figure 31.9   A "Cantor book."

According to (7) of Section 31.14, each point l_i and each point r_j is the terminus of uncountably many 1-cells in F. If we move those 1-cells along by the flow, we see that the unstable manifold $W^u(0)$ is, throughout its entire length, the spine of a Cantor book, in the terminology of Williams (see Figure 31.9).

A sequence (31.15-1), if periodic, corresponds to a periodic orbit on L. Given any sequence (31.15-1), we can clearly find a periodic sequence that agrees with the given one for say $-K < k < K$, where K is large. Hence, given any orbit, we can find a periodic orbit arbitrarily close to it; the periodic orbits are dense in L. (In a physical realization or numerical simulation, of course, the notation of a strictly periodic orbit is an empty concept, owing to the finite accuracy and the Lyapuonov instability.)

31.17 Existence of 1-Cells in F

In Section 31.13 we found it necessary to assume in advance that there is at least one 1-cell in F, for example one connecting 0 to r_1. (We then deduced the existence of other 1-cells by letting the first be carried round by the flow until it intersects the strip S again, and so on.) That is, we assumed that the attractor L contains a curve lying in S and connecting 0 to r_1. The projection of F onto the surface L_0 is the branch line BB, which is a curve in space. That much is supported by the numerical evidence. For the detailed structure of F itself, we must appeal to the assumptions 1–5 that were made in Section 31.13 about the attractor L. Here we shall indicate how those assumptions imply the existence of a 1-cell connecting 0 to r_1 (or, by the same argument, l_1 to 0).

We shall consider also the following problem. It was shown that if $\sigma_0 = [i, j]$ is a symbol, then any 1-cell connecting l_i to r_j determines a unique sequence $\ldots, \sigma_{-2}, \sigma_{-1}, \sigma_0$ such that σ_{-k-1} always precedes σ_{-k}. The problem is to show that conversely any such sequence determines a 1-cell connecting l_i to r_j.

Williams's solution of the first problem starts by letting x_{00} be any point of the branch line BB between 0 and the right end r_j, constructing an orbit $x_0(t)$ in the branched surface L_0, such that $x_0(0)$ is $= x_{00}$, by specifying its prehistory, then calling $x(t)$ the orbit in L whose projection onto L_0, according to assumption 4 in Section 31.13, is $x_0(t)$. The construction is such that the orbit $x(t)$ depends continuously on the position of x_{00} on BB; hence the point $x(0)$ of the orbit depends continuously on x_{00}, hence traces of a curve in F, as x_{00} is varied, and furthermore that curve touches 0 and r_1 at its two ends.

There are presumably many ways of choosing the prehistory of $x_0(t)$ that would achieve the desired effect, for there are infinitely many 1-cells connecting 0 to r_1 in F. The one selected by Williams is to make the prehistory alternate between the two halves of BB; namely, if for some t_1, the point $x_0(t_1)$ of the prehistory is on BB (but not at 0 or either end), the orbit is continued backward (upward) in the sheet of L_0 behind if $x_0(t_1)$ lies to the right of 0 and in the sheet in front if it lies to the left. It is easily verified by reference to Figure 30.3a that that is always possible and never causes the prehistory to hit 0 or r_1 or l_1. To show the continuous dependence on x_{00}, we observe that if x_{00} and x'_{00} are close together on the branch line BB, the resulting orbits $x_0(t)$ and $x'_0(t)$ in L_0 are close together for a long interval, and then the same follows for the orbits $x(t)$ and $x'(t)$ in L by assumption 4 of Section 31.13. To show that the resulting curve in F has the correct end-points, note first that if x_{00} is very close to the right end of BB, the prehistory is very close to the outside edge of that sheet of L_0 that lies behind; hence, in the past, $x(t)$ must have spent a long time near the stagnation point 0, and hence was close for a long time to the orbit $W''_r(0)$ that goes from 0 to r_1. Lastly, if x_{00} is very near the center of the branch line BB, the orbit is already very near the stagnation point, and hence was close, for a long past interval, to the constant orbit $x(t) \equiv 0$.

To show similarly that any sequence

$$\ldots, \sigma_{-2}, \sigma_{-1}, \sigma_0 = [i, j], \qquad (31.17\text{-}1)$$

where σ_{-k-1} is a precedessor of σ_{-k}, determines a 1-cell in F is more complicated, and we refer the reader to the paper of Williams, where the point set F is approximated by so-called retractions. The idea is that if x_{00} is a point of BB between the projections onto BB of l_i and r_j, we let the sequence (31.17-1) dictate which sheet of L_0 the prehistory $x_0(t)$ follows each time the branch line is encountered, as we follow the orbit $x_0(t)$ backward (upward) across BB. For technical reasons, it is expedient to follow the sequence (31.17-1) in this way for only a finite number, say n, of steps and to decree that prior to that the prehistory alternated between the two halves of BB as in the above discussion. The resulting orbit $x(t)$ is then not quite the one we want, but is close to the one we want and gets closer, as $n \to \infty$.

The conclusion that each sequence (31.17-1) determines a unique 1-cell in F has already been used in the preceding section, to show that the sheets of L are uncountable.

31.18 Bifurcation to a Strange Attractor

We noted in Section 31.9 that the bifurcation in the Lorenz system at $r = r_0 = 24.74$ is subcritical, and we mentioned the possibility of an explosive transition when the fixed points P_1 and P_2 lose stability. However, as r is slowly increased past r_0, the motion on the attractor L takes over without any explosive transition, except insofar as the sudden appearance of the motion on L may be regarded as an explosion.

The attractor has been described for $r = 28$. If we let r decrease down to r_0, the two holes in the branched surface L_0 close up by contracting onto the fixed points P_1 and P_2. For still smaller values of r, down to 13.96, L continues to exist as an invariant set for the motion (though not an attractor), but it is in contact with the fixed points P_1 and P_2 only at $r = r_0$. For r only slightly above r_0, an orbit emerging from P_1 or P_2 is immediately an orbit on L. In this sense the bifurcation at r_0 results in an abrupt transition from a stationary orbit at P_1 or P_2 to motion on the attractor L.

31.19 The Feigenbaum Model

While the Lorenz attractor appears in connection with a subcritical Hopf bifurcation, the Landau–Hopf model and the Ruelle–Takens model both require a sequence of supercritical bifurcations leading to invariant tori of successively higher dimension, arbitrarily high in the former model and of dimension at least 4 in the latter. However, such a sequence is unlikely, according to Peixoto's theorem. As was pointed out at the end of Section 29.10, the appearance of an invariant 2-torus at a bifurcation from a periodic orbit does not imply the appearance of orbits dense on that torus. Instead, the appearance of finitely many periodic orbits and fixed points is generic. Other orbits tend asymptotically to those periodic orbits and fixed points. The appearance of an invariant 3-torus at the next bifurcation depends, at least in the theory of Chenciner and Jooss 1976, on the existence of an orbit dense on the 2-torus. Hence, the bifurcation to an invariant 3-torus seems unlikely.

If a periodic orbit on the 2-torus goes round the long way n times before closing, then the bifurcation is subharmonic with a sudden n-folding of the period at the bifurcation (see Section 29.11). Recently, M. Feigenbaum has developed a model based on a sequence of subharmonic bifurcations with period doublings. (See Feigenbaum 1980 and the references given there.) It turns out that such doublings occur in many examples of iterated mappings and simple dynamical systems. Furthermore, as the number n of doublings increases, the behavior of the system is governed by certain asymptotic laws that involve universal constants and functions, independent of the system under study. In addition, the asymptotic laws appear to hold quite accurately for rather small values of n. In particular, the values μ_n of the dimensionless parameter μ at which the bifurcations (doublings) take place converge to a value μ_∞ geometrically, with

$$\frac{\mu_{n+1} - \mu_n}{\mu_n - \mu_{n-1}} \approx 0.21416938$$

for large n. As $n \to \infty$, at least in the cases studied, the power spectrum of the motion approaches a continuous spectrum with certain universal features. At $\mu = \mu_\infty$, the motion is presumably aperiodic on a strange attractor.

There is evidence (Lorenz 1981) for an example of this behavior in the Lorenz system at considerably higher values of the dimensionless parameter r than values studied by Lorenz. Namely, the strange attractor that appears at $r = 24.74$ persists up to a value $r = r^*$ (≈ 250). For r considerably greater than r^*, there is a periodic orbit, and as r is *decreased* toward r^*, there is a sequence of doublings at values r_n of r that converge to r^* from above, with

$$\frac{r_{n+1} - r_n}{r_n - r_{n-1}} \approx 0.214.$$

Appendix to Chapter 31—Generic Properties of Systems

In this appendix, generic and nongeneric properties of systems are explained and explored. It might be thought that these notions will in a sense replace probability in some parts of physics. We shall conclude that that is not so, but that they may be an important guide to the further development of our ideas of probability in physics.

31.A Spaces of Systems

Because a physical system cannot be specified exactly, and for other reasons, it is often desirable to consider not a single system but a large family of them. If each system is characterized by the values of n parameters $\alpha_1, \ldots, \alpha_n$, then each system corresponds to a point in the space $\mathbb{R}^n$. Conversely, each point in $\mathbb{R}^n$ or in some region $\mathcal{R} \subset \mathbb{R}^n$, may correspond to a unique system of the family.

More generally, the systems of a family may be represented by the points in a Banach or Hilbert space, or in some more general metric space, or in some still more general topological space, which we shall call *the space of systems*. For example, each system might be a dynamical system in the plane:

$$\dot{x} = X(x, y), \qquad \dot{y} = Y(x, y),$$

where X and Y are the components of a given smooth vector field $\mathbf{X}(\mathbf{x})$. Then each such vector field determines a system and may be represented by a point in a Banach space $\mathfrak{B}$ with a suitably chosen norm $\|\mathbf{X}(\cdot)\|$. Clearly $\mathfrak{B}$ is infinite-dimensional, because no finite number of parameters can specify a vector field completely.

We shall suppose generally that the space at least has a metric. Then if the distance $d(X_1, X_2)$ between two systems is small, we can think of either one as obtainable from the other by a small perturbation. We shall even usually assume that the space has a norm, so that $d(X_1, X_2) = \|X_1 - X_2\|$.

31.B Absence of Lebesgue Measure in a Hilbert Space

If there are finitely many parameters $\alpha_1, \ldots, \alpha_n$ and if they are distributed according to a continuous probability law in $\mathbb{R}^n$, and if a property holds in all $\mathbb{R}^n$ except for a set of Lebesgue measure zero, we say that it holds for almost all systems, or that the probability of its not occurring is zero.

If we let $n \to \infty$, so that $\mathbb{R}^n$ is replaced by an infinite dimensional Hilbert space, then, as shown in Section 13.11 of Volume I, there is no Lebesgue measure; hence, probability statements of the above kind cannot be made. (Non-Lebesgue probability measures are discussed in Section 31.H at the end of this appendix.)

31.C Generic Properties of Systems

In any space of systems, whether it has Lebesgue measure or not, assuming it at least has a topology, one can define what is meant by generic and nongeneric properties of the systems. This is often done with the implied suggestion that nongeneric properties can be ignored from some points of view.

A subset of the space is called a *Baire set* if it is the intersection of countably many dense open subsets. The complement of a Baire set is called a *meager set*; it is the union of countably many nowhere dense sets.

A property of a system is called *generic* if it occurs on a Baire set in the space of systems. A property is called *nongeneric* if it occurs on a meager set. Note that "nongeneric" does not mean merely "not generic," because a set may be neither a Baire set nor a meager set, for example, a half-space $x_1 > 0$ in $\mathbb{R}^n$.

The Baire category theorem says that if the space is a complete metric space a Baire set is dense in it, but a meager set may also be dense; hence the distinction is not based on denseness. Furthermore, in case the space happens to be finite-dimensional a Baire set may have Lebesgue measure zero; hence the distinction is not based on measure, either. See Section 31A.G below.

If a property a is generic, the property "not a" is nongeneric; if a and b are generic, the property "a and b" is generic. Two contradictory properties cannot both be generic.

31.D Strongly Generic; Physical Interpretation

We shall not give a physical interpretation of genericity, except in the following special case: A property is called *strongly generic* if it occurs on a dense open set in the space, i.e., not merely on a countable intersection of such sets. (We note in passing that the intersection of finitely many dense open sets is again a dense open set.) If a property is strongly generic, if X is any system, and if ε_1 is any positive number, then, by a perturbation of X of magnitude $\leq \varepsilon_1$, we can obtain a system Y ($\|Y - X\| \leq \varepsilon_1$) which has the property. Furthermore, there is then another positive number ε_2 ($< \varepsilon_1$) such that the system retains the property under any arbitrary further perturbation of magnitude $\leq \varepsilon_2$. Nothing is said about how difficult it may be to find the first perturbation. (There may also be perturbations of X of magnitude $\leq \varepsilon_1$ that do *not* give the system the desired property.) But all sufficiently carefully prepared systems have the property; hence the property cannot be dismissed as one "not likely to appear in practice."

If a property is strongly nongeneric, that is, if it occurs on a nowhere dense set in the space, then, for any $\varepsilon_1 \geq 0$ there is a perturbation of norm $\leq \varepsilon_1$ that makes it disappear, and then there is an $\varepsilon_2 \geq 0$ such that no further perturbation of norm $\leq \varepsilon_2$ can make it reappear.

31.E Peixoto's Theorem

The theorem of Peixoto 1962 says that for dynamical systems on a compact 2-dimensional manifold, a particular strongly generic property is that every motion tends asymptotically to one of a finite number of fixed points and periodic orbits. Hence, in particular, quasi-periodic motions are nongeneric.

We state the theorem for the 2-torus; $\mathfrak{M} = T^2$. Let θ and φ be angle variables on T^2, so that the dynamical system is

$$\dot{\theta} = F(\theta, \varphi), \qquad \dot{\varphi} = G(\theta, \varphi),$$

where the functions F and G are periodic in both θ and φ with period 2π. Let $\mathfrak{B}$ be the Banach space of all pairs of periodic functions F and G of class C^1, with the norm given by

$$\max_{(\theta, \varphi)} \max\{|F|, |G|, |\partial_\theta F|, |\partial_\theta G|, |\partial_\varphi F|, |\partial_\varphi G|\},$$

so that the topology in $\mathfrak{B}$ is that of uniform C^1 convergence. Each point of $\mathfrak{B}$ represents a dynamical system. Then there is a dense open set in $\mathfrak{B}$ representing dynamical systems, for each of which

  i.   There are at most finitely many fixed points and finitely many closed orbits on T^2.
  ii. Each motion tends asymptotically to a fixed point or closed orbit. (This is intended to include stationary motions at the fixed points and periodic motions on the closed orbits.) The theorem actually says a little more than this; see Peixoto 1962, also Section 31.F below.

This is in sharp contrast with the Kolmogorov–Arnol'd–Moser theorem (see Moser 1973, Theorem 2.8), which says that for Hamiltonian systems, under certain assumptions, quasi-periodic motions are stable under almost all small perturbations. There is no contradiction, here, because the Hamiltonian systems are very special (for example, they are conservative) and are in a sense nongeneric among dynamical systems as a whole, so that the small perturbations referred to are, in this case, far from arbitrary, but are such that the system remains Hamiltonian.

31.F Other Examples of Generic and
Nongeneric Properties

Consider the space $C[0, 1]$ of continuous functions defined for $0 \leq x \leq 1$ with the usual norm $\|f\| = \sup|f(x)|$. For such functions, differentiability is a nongeneric property, and even differentiability at a single point, and even Lipschitz continuity at a single point; see Boas 1960, also Exercise 1 below.

The next two examples are part of a more complete version of Peixoto's theorem than the one given above. As stated, it is generic for a vector field on a compact 2-manifold to have a finite number of fixed points and periodic orbits. In a coordinate system with a fixed point at the origin, the vector field is

$$\mathbf{F}(\mathbf{x}) = A\mathbf{x} + \cdots,$$

where A is a 2×2 matrix. The theorem goes on to say that generically each such fixed point is *hyperbolic*, which means that no eigenvalue of A is pure imaginary; hence there are three possibilities: If both eigenvalues have negative real parts, the fixed point is *attracting*; if both have positive real parts, it is *repelling*; and if one has a positive real part and the other a negative one, it is a *saddle point*.

A further conclusion of Peixoto's theorem is that the existence of an orbit going from one saddle point to another is nongeneric; if there is such an orbit, for some vector field, the smallest perturbation can make it miss the second saddle point.

A conjecture that has not been proved, or even fully formulated, was made in Section 29.6. As was stated there, the existing completeness theorems on systems of eigenfunctions for hydrodynamic problems require that the generalized eigenfunctions, if any, be included; see equation (29.6-5). It is conjectured that in some suitable space of hydrodynamic systems, the existence of generalized eigenfunctions is nongeneric. An eigenvalue λ is said to have *index* 1 if there are no corresponding generalized eigenfunctions. It seems likely to be strongly generic for any eigenvalue λ_k to have index 1, but then only generic for them all to have index 1, since a countable collection of dense open sets is not necessarily open, but is in any case a Baire set.

EXERCISE

1.   For each positive integer n define a subset E_n of $C[0, 1]$ as follows: A function f is in E_n if there is an x_0 in $[0, 1 - 1/n]$ such that

$$\left| \frac{f(x_0 + h) - f(x_0)}{h} \right| \le n \quad \text{for } 0 < h < \frac{1}{n}. \tag{31.F-1}$$

Show first that E_n is a closed subset of $C[0, 1]$ by showing that if $f_k(x) \to f(x)$ uniformly, as $k \to \infty$, and if each f_k satisfies (31.F-1), then so does f. Show next that the complement of E_n is dense; show in fact that if f is any function in $C[0, 1]$, it can be made differentiable by a small perturbation, using a mollifier (see Volume I), then can be made to violate (31.F-1) by adding to it, if necessary, a further small perturbation having a sufficiently large derivative at $x = x_0$. Hence each E_n is a nowhere dense set, hence $\bigcup_{n=1}^{\infty} E_n$ is a meager set, and hence it is nongeneric for a function in $C[0, 1]$ to be Lipschitz continuous on the right at any point.

31.G Lack of Correspondence between Genericity and Lebesgue

1. Let X be the unit interval on the real line, as a metric space with the distance $(x, y) = |x - y|$. Let n be a positive integer. With each rational number p/q, where $0 < p < q$, assocaite the interval

$$\frac{p}{q} - \frac{1}{nq2^q} < x < \frac{p}{q} + \frac{1}{nq2^q},$$

and let S_n be the union of those intervals. Show that S_n is open and dense in X and has Lebesgue measure $\leq 4/n$. Conclude that a Baire set in X can have zero Lebesgue measure, and a meager set can have measure 1.

2. Let S be the intersection of the sets S_n. According to Exercise 1, S is a Baire set. Show that it is uncountable by considering numbers x in $[0, 1]$ having a binary representation

$$x = .a_1 \, 0 \, a_2 \, 0 \, 0 \ldots 0 \, a_3 \, 0 \, 0 \ldots 0 \, a_4 \, 0 \, 0 \ldots,$$

where each a_i is $= 0$ or 1. Show that if the number v_l of zeros between a_l and a_{l+1} increases rapidly enough with l, then x is in all the sets S_n. Show that the set of all such x is uncountable.

31.H Probability and Physics

Suppose that a topological space $\mathfrak{X}$ is a space of physical systems. When a physicist says that a certain property of such a system is very unlikely to appear, he means presumably that if a large number N of the systems is somehow selected from $\mathfrak{X}$, very few of them have the property in question, and, as N increases, the fraction of them that have the property tends to zero. Such a statement, if true, is unavoidably a probability statement. What is lacking from this description is a statement of how the N systems are selected from $\mathfrak{X}$. Since $\mathfrak{X}$ is generally infinite-dimensional, and hence has no Lebesgue measure, there is no possibility of the N systems being chosen at random from a uniform distribution in $\mathfrak{X}$, because no such distribution exists.

The method of selection has to be determined by considerations of physics, and cannot be decided by considerations of genericity alone.

There are of course many ways of defining probability measures in infinite-dimensional spaces; the Gaussian measures in a Hilbert space, described in Volume I, Section 13.11, are examples. Such a measure could be used for the selection of the N systems from $\mathfrak{X}$. The possibilities are presumably very numerous, and it may be necessary to appeal to many physical principles. For example, it would seem reasonable to require that the probability measure be positive on the open sets in $\mathfrak{X}$, for otherwise it would seem that the choice of the space $\mathfrak{X}$ was itself faulty; parts of it should be discarded. Then suppose there is a transformation in $\mathfrak{X}$ that represents some modification of the systems so as to alter their main properties without causing very "improbable" properties to become suddenly "probable"; then under such a transformation in $\mathfrak{X}$ sets of measure zero should go into sets of measure zero. In view of the physical interpretation given in 31.D, one requirement might be that a strongly nongeneric property should appear only on a set of measure zero.

All this is speculation at present, but it may indicate that genericity cannot replace probability in physics, but can perhaps be a guide to the appropriate choice of probability measures.

EXERCISES

1. Show that differentiation is nongeneric also in the Hilbert space $\mathfrak{H} = L^2[0, 1]$, by filling in the steps of the following outline of a proof: First define a linear manifold in $\mathfrak{H}$ by

$$\mathfrak{D} = \{\psi \in L^2 : \psi' \in L^2\}, \tag{31.H-1}$$

where the derivative is meant in the distribution sense (see Chapter 5 in Volume I). For any ψ in $\mathfrak{H}$, write

$$\psi(x) = \sum_{-\infty}^{\infty} \xi_k e^{2\pi i k x}. \tag{31.H-2}$$

Then the manifold $\mathfrak{D}$ can be characterized as

$$\mathfrak{D} = \{\psi \in L^2 : \sum |k\xi_k|^2 < \infty\}. \tag{31.H-3}$$

Define also the smaller manifold

$$\mathfrak{D}_M = \{\psi \in L^2 : \psi' \in L^2, \|\psi'\| \leq M\}$$

$$= \{\psi \in L^2 : 4\pi^2 \sum |k\xi_k|^2 \leq M^2\}, \tag{31.H-4}$$

for each $M = 1, 2, \ldots$. If it can be shown that each $\mathfrak{D}_M$ is nowhere dense, i.e., that its complement $C\mathfrak{D}_M$ is dense and open, it will follow that $\mathfrak{D} = \bigcup_{M=1}^{\infty} \mathfrak{D}_M$ is a meager set in $\mathfrak{H}$. Show first that $C\mathfrak{D}_M$ is dense by showing that any $\psi \in L^2$ that does not already violate the condition $\|\psi'\| \leq M$ can be made to do so by adding to it an arbitrarily small function with a large derivative. Then show that $C\mathfrak{D}_M$ is open as follows: Consider any ψ in $C\mathfrak{D}_M$, that is, any ψ such that either $\|\psi'\| = \infty$ or $\|\psi'\| = M + \delta$ for some $\delta > 0$. It must be shown that there is a neighborhood of that ψ that is contained in $C\mathfrak{D}_M$. Choose K such that

$$\left(\sum_{k=-K}^{K} 4\pi^2 |k\xi_k|^2\right)^{1/2} > M + \tfrac{1}{2}\delta,$$

and show that if χ is any function in L^2 such that $\|\chi\|$ is $< \delta/4K$, then $\|\psi' + \chi'\| > M + \tfrac{1}{4}\delta$, so that $C\mathfrak{D}_M$ is open.

2. Show that the same is true in the real Hilbert space $L^2[0, 1]$ by using the same argument as above but assuming throughout that $\xi_{-k} = \overline{\xi_k}$.

In order to introduce real coordinates and real basis functions, we write

$$\xi_0 = x_0, \qquad \xi_k = \frac{x_k + ix_{-k}}{\sqrt{2}} \qquad (k = 1, 2, \ldots), \qquad (31.\text{H-}5)$$

$$\varphi_0(x) \equiv 1, \qquad \varphi_k(x) = \sqrt{2} \cos 2\pi kx,$$
$$\varphi_{-k}(x) = \sqrt{2} \sin 2\pi kx. \qquad (31.\text{H-}6)$$

Then, from (31.H-2) with $\xi_{-k} = \bar{\xi}_k$, we find

$$\psi(x) = \sum_{-\infty}^{\infty} x_k \varphi_k(x), \qquad (31.\text{H-}7)$$

$$\psi'(x) = 2\pi \sum_{-\infty}^{\infty} kx_{-k} \varphi_k(x). \qquad (31.\text{H-}8)$$

Gaussian measures in a real Hilbert space $\mathfrak{H}$ were described in Section 13.11 of Volume I; the main points are as follows: If $\mathfrak{M}$ is any finite-dimensional subspace of $\mathfrak{H}$ and S is any Borel set in $\mathfrak{M}$, then the set

$$Z = S + \mathfrak{M}^{\perp},$$

i.e., the set of all points $x + y$, where $x \in S$ and $y \in \mathfrak{M}^{\perp}$, is called a *cylinder set*. To define a so-called Gaussian measure in $\mathfrak{H}$, we let B be a positive operator of the trace class, and we call $A = B^{-1}$. Then, given any cylinder set $Z = S + \mathfrak{M}^{\perp}$, where dim $\mathfrak{M} = m$, we let $\{\varphi_j\}_1^m$ be an orthonormal set in $\mathfrak{M}$, and we define an $m \times m$ matrix $A(\mathfrak{M})$ by $A(\mathfrak{M})_{jk} = (\varphi_j, A\varphi_k)$, $1 \leq j, k \leq m$. Then the probability $\mathbf{P}(Z)$ is taken as

$$\mathbf{P}(Z) = \frac{\sqrt{\det A(\mathfrak{M})}}{(2\pi)^{m/2}} \int_S \exp\left\{-\tfrac{1}{2} \sum_{j,k} x_j A(\mathfrak{M})_{jk} x_k\right\} dV. \qquad (31.\text{H-}9)$$

As stated in Section 13.11, this set function $\mathbf{P}(\cdot)$ can be extended to a unique probability measure defined on the σ-algebra $\mathfrak{A}$ generated by the cylinder sets.

We now apply these ideas to the real Hilbert space $\mathfrak{H} = L^2[0, 1]$, using the coordinates x_k and basis functions φ_k given by (31.H-5, 6). We define the operator A by the simple formula

$$A\psi(x) = \sum_{-\infty}^{\infty} a_k x_k \varphi_k,$$

where $\psi(x)$ is given by (31.H-7), and where the a_k are positive and $\sum (1/a_k) < \infty$, so that $B = A^{-1}$ is compact. We consider the cylinder sets

$$Z_{M,K} = \left\{\psi \in L^2 : 4\pi^2 \sum_{-K}^{K} |kx_k|^2 \leq M^2\right\},$$

so that

$$Z_{M,1} \supset Z_{M,2} \supset \cdots \supset Z_{M,K} \supset \cdots \supset \mathfrak{D}_M, \qquad (31.\text{H-}10)$$

where $\mathfrak{D}_M$ is given by (31.H-4). Then (31.H-9) takes the form

$$P(Z_{M,K}) = \prod_{-K}^{K} \sqrt{\frac{a_k}{2\pi}} \int_S \cdots \int \exp\left\{ -\tfrac{1}{2} \sum_{-K}^{K} a_k x_k^2 \right\} dx_{-K} \cdots dx_K, \quad (31.\text{H-}11)$$

where S is the ellipsoidal region

$$S: \sum_{-K}^{K} k^2 x_k^2 \leq \frac{M^2}{4\pi^2}. \quad (31.\text{H-}12)$$

EXERCISES

3.   By letting all the integration variables except the last in (31.H-11) go from $-\infty$ to $+\infty$, while the last one x_K is restricted to $|x_K| \leq M/(2\pi K)$, show that

$$P(Z_{M,K}) < \frac{M\sqrt{a_K}}{\sqrt{2\pi^3}\, K}.$$

4.   Now take $a_k = |k|^r$, where $r > 1$ (to make $\sum (1/a_k)$ converge). Show that if $1 < r < 2$, then $P(Z_{M,K}) \to 0$ as $K \to \infty$, so that from (31.H-10), $P(\mathfrak{D}_M) = 0$, and hence by countable additivity,

$$P(\mathfrak{D}) = P\left(\bigcup_{M=1}^{\infty} \mathfrak{D}_M \right) = 0.$$

This exercise shows that the meager set $\mathfrak{D}$ of differentiable functions in $L^2[0, 1]$ has Gaussian measure zero if $1 < r < 2$. For $r > 2$, $P(\mathfrak{D}) = 1$. Roughly speaking, when r is increased, the Gaussian measure is concentrated more and more toward the origin in $\mathfrak{H}$. The differentiable functions lie near the origin, according to (31.H-3), so that if the concentration of the probability measure toward the origin is large, the differentiable functions acquire positive probability, but for $r < 2$ they have zero probability.

In this case, therefore, the probability measure can be so chosen that the nongeneric property (differentiability) is improbable.

References

Abraham, R., and Robbin, J. (1967): *Transversal Mappings and Flows*. W. A. Benjamin, New York.

Adler, R., Bazin, M., and Schiffer, M. (1965): *Introduction to General Relativity*. McGraw-Hill, New York.

Barut, A. O., and Rączka, R. (1977): *The Theory of Group Representations and Applications*. PWN Polish Scientific Publishers, Warsaw.

Behrends, R. E., Dreitlein, J., Fronsdal, C., and Lee, W. (1962): Simple groups and strong interaction symmetries. *Rev. Mod. Phys.*, vol. **34**, pp. 1–40.

Bôcher, M. (1922): *Introduction to Higher Algebra*. The MacMillan Co., New York.

Boerner, H. (1955): *Darstellungen von Gruppen*. Springer-Verlag, Berlin, Heidelberg, New York.

Chevalley, C. (1946): *Theory of Lie Groups I*. Princeton Univ. Press, Princeton.

Curry, J. H. (1978): A generalized Lorenz system. *Comm. Math. Phys.*, vol. **60**, pp. 193–204.

Davey, A. (1962): The growth of Taylor vortices in flow between rotating cylinders. *J. Fluid Mech.*, vol. **14**, pp. 336–368.

Davey, A., DiPrima, R. C., and Stuart, J. T. (1968): On the instability of Taylor vortices. *J. Fluid Mech.*, vol. **31**, pp. 17–52.

Dirac, P. A. M. (1928): The quantum theory of the electron. *Proc. Roy. Soc. A*, vol. **117**, pp. 610–624.

Dirac, P. A. M. (1958): *The Principles of Quantum Mechanics*. Clarendon Press, Oxford.

Eagles, P. M. (1971): On stability of Taylor vortices by fifth-order amplitude expansion. *J. Fluid Mech.*, vol. **49**, pp. 529–550.

Eisenhart, L. P. (1926): *Riemannian Geometry*. Princeton Univ. Press, Princeton.

Eisenhart, L. P. (1933): *Continuous Groups of Transformations*. Princeton Univ. Press, Princeton.

Eisenhart, L. P. (1934): Separable systems of Staeckel. *Ann. Math.*, vol. **35**, pp. 284 ff.

Ellis, H. G. (1972): Ether flow through a drainhole: a particle model in general relativity. *J. Math. Phys.*, vol. **14**, pp. 104–118.

Feigenbaum, M. (1980): Universal behavior in nonlinear systems. *Los Alamos Science*, vol. **1**, pp. 4–27.

Finkelstein, D. (1958): Past-future asymmetry of the gravitational field of a point particle. *Phys. Rev.*, vol. **110**, pp. 965–967.

Flanders, H. (1963): *Differential Forms*. Academic Press, New York.

Gel'fand, I. M., Graev, M. I., and Vilenkin, N. Ya. (1966): *Generalized Functions, Vol. 5, Integral Geometry and Representation Theory*, (translated from the Russian) Academic Press, New York.

Gel'fand, I. M., Minlos, R. A., and Shapiro, Z. Ya. (1963): *Representations of the Rotation and Lorentz Groups and Their Applications*, (translated from the Russian) MacMillan (Pergamon), New York.

Gleason, A. (1952): Groups without small subgroups. *Ann. of Math.* vol. **56**, pp. 193–212.

Gollub, J. P., and Swinney, H. L. (1975): Onset of turbulence in a rotating fluid. *Phys. Rev. Letters*, vol. **35**, pp. 927–930.

Haar, A. (1933): Der Massbegriff in der Theorie der kontinuierlichen Gruppen. *Ann. Math.*, vol. **34**, pp. 147–169.

Hassard, Brian, D. (1980): Computation of invariant manifolds. To appear.

Hausner, M., and Schwartz, J. T. (1968): *Lie Groups and Lie Algebras: a Brief Description*. Gordon and Breach, New York.

Hocking, J. G., and Young, G. S. (1961): *Topology*. Addison Wesley, Reading, Mass.

Hopf, E. (1948): A mathematical example displaying the features of turbulence. *Comm. Pure Appl. Math.*, vol. **1**, pp. 303–322.

Kelley, Al (1967): *The Stable, Center-Stable, Center, Center-Unstable, and Unstable Manifolds*. Appendix C in Abraham and Robbin 1967.

Kerr, R. P. (1963): Gravitational field of a spinning mass as an example of algebraically special metrics. *Phys. Rev. Letters*, vol. **11**, pp. 237–238.

Krueger, E. R., Gross, A., and DiPrima, R. C. (1966): On the relative importance of Taylor-vortex and non-axisymmetric modes in flow between rotating cylinders. *J. Fluid Mech.*, vol. **24**, pp. 521–538.

Kruskal, M. D. (1960): Maximal extension of Schwarzschild metric. *Phys. Rev.*, vol. **119**, pp. 1743–1745.

Kurosh, A. G. (1956): *Theory of Groups I, II*. Chelsea Pub. Co., New York.

Ladyzhenṣkaya, O. (1969): *Mathematical Theory of Viscous Incompressible Flow*. Gordon and Breach, New York.

Ladyzhenskaya, O. (1975): Mathematical analysis of Navier-Stokes equations for incompressible liquids. *Annual Rev. of Fluid Mech.*, vol. **7**, pp. 249–272.

Landau, L. D., and Lifshitz, E. M. (1959): *Fluid Mechanics*. Pergamon Press, London.

Lanford, O. E. (1973): Bifurcation of periodic solutions into invariant tori: the work of Ruelle and Takens. in *Nonlinear Problems in the Physical Sciences and Biology*. Lecture Notes Vol. 322, Springer-Verlag, Berlin, Heidelberg, New York.

Lang, S. (1962): *Introduction to Differentiable Manifolds*. Wiley Interscience, New York.

Lorenz, E. N. (1981): Noisy periodicity and reverse bifurcation. To appear.

Lorenz, E. N. (1963): Deterministic nonperiodic flow. *Jour. Atmos. Sci.*, vol. **20**, pp. 130–141.

Marsden, J. E., and McCracken, M. (1976): *The Hopf Bifurcation and Its Applications*. Springer-Verlag, Berlin, Heidelberg, New York.

Miller, W. (1973): *Symmetry Groups and Their Applications*. Academic Press, New York.

Montgomery, D., and Zippin, L. (1952)—see next item.

Montgomery, D., and Zippin, L. (1955): *Topological Transformation Groups*. Interscience Publishers, New York.

Morse, P. M., and Feshbach, H. (1953): *Methods of Theoretical Physics I, II*. McGraw-Hill, New York.

Moser, J. (1973): *Stable and Random Motions in Dynamical Systems*. Princeton Univ. Press, Princeton.

Nachbin, L. (1965): *The Haar Integral*. Van Nostrand, Princeton.

Naimark, M. A. (1976): *The Theory of Group Representations*. Nauka, Moscow.

Naimark, M. A. (1959): On some cases of periodic motions depending on parameters. *Dokl. Akad. Nauk SSSR*, vol. **129**, pp. 736–739.

Pauli, W. (1927): Zur Quantenmechanik des magnetischen Elektrons. *Zeits. f. Physik*, vol. **43**, pp. 601–623.

Peixoto, M. (1962): Structural stability on two-dimensional manifolds *Topology*, vol. **1**, pp. 101–120.

Peter, F., and Weyl, H. (1927): Die Vollständigkeit der primitiven Darstellungen einer geschlossenen kontinuierlichen Gruppe. *Math. Ann.*, vol. **97**, pp. 737–755.

Redei, L. (1959): *Algebra*. Akademische Verlagsgesellschaft, Leipzig.

Richtmeyer, R. D., and Morton, K. W. (1967): *Difference Methods for Initial-Value Problems*. Wiley-Interscience, New York.

Richtmeyer, R. D. (1981): A study of the Lorenz attractor. *Advances in Mathematics*, to appear.

Riesz, F., and Sz. Nagy, B. (1953): *Leçons d' Analyse Fonctionnelle*. Akadémiai Kiadó, Budapest.

Robertson, H. P. (1927): Bemerkung über separierbare Systeme in der Wellenmechanik. *Math. Annalen*, vol. **98**, pp. 749 ff.

Ruelle, D., and Takens, F. (1971): On the nature of turbulence. *Comm. Math. Phys.*, vol. 20, pp. 167–192.

Russmann, H., and Zehnder, E. (1980): On a normal form of symmetric maps of [0, 1]. *Comm. Math. Phys.*, vol. **72**, pp. 49–53.

Sacker, R. (1964): (a bifurcation theorem) Thesis (unpublished), New York University.

Schiff, L. (1955): *Quantum Mechanics*. McGraw-Hill, New York.

Schur, I. (1905): Neue Begründung der Theorie der Gruppencharaktere. *Sitzungsber. preuss. Akad. Wiss.*, vol **1905**, pp. 406–432.

Sell, G. R. (1971): *Topological Dynamics and Ordinary Differential Equations*. Van Nostrand-Reinhold, London.

Siegel, C. L. (1956): *Vorlesungen über Himmelsmechanik*. Springer-Verlag, Berlin, Heidelberg, New York.

Siegel, C. L. and Moser, J. (1971): *Lectures on Celestial Mechanics*. Springer-Verlag, Berlin, Heidelberg, New York.

Smale, S. (1967): Differentiable dynamical systems. *Bull AMS*, vol. **73**, pp. 747–817.

Snyder, H. A. (1970): Waveforms in rotating Couette flow. *Intl. Jour. Nonlinear Mech.*, vol. **5**, pp. 659–685.

Sommerfeld, A. (1929). Das Molekül als symmetrische Kreisel, in *Atombau und Spektrallinien, wellenmechanischer Ergänzungsband*. Friedr. Vieweg u. Sohn AG, Braunschweig.

Sugiura, M. (1975): *Unitary Representations and Harmonic Analysis*. Halsted Press, New York.

Taylor, G. I. (1923): Stability of a viscous liquid contained between two rotating cyclinders. *Phil. Trans. Roy Soc. A*, vol. **223**, pp. 289–343.

Thomas, T. Y. (1961): *Concepts from Tensor Analysis and Differential Geometry*. Academic Press, New York.

Vilenkin, N. Ya. (1968): *Special Functions and the Theory of Group Representations*. AMS Translations, Providence, R.I.

Warner, G. (1972): *Harmonic Analysis on Semi-Simple Lie Groups I, II*. Springer Verlag.

Weyl, H. (1928): *Gruppentheorie und Quantenmechanik*. S. Hirzel Verlag, Leipzig.

Weyl, H. (1932). *The Theory of Groups and Quantum Mechanics*. E. P. Dutton and Co., New York.

Whitehead, J. H. C. (1932): Convex regions in the geometry of paths. *Quart. Jour. of Math.*, vol. **3**, pp. 33–42.

Wigner, E. (1931): *Gruppentheorie und ihre Anwendung auf die Quanenmechanik der Atomspektren*. Fr. Vieweg u. Sohn AG, Braunschweig.

Williams, R. F. (1977): The structure of Lorenz attractors. *Turbulence Seminar* (Univ. of Calif., Berkeley) 1976–77, pp. 94–112.

Index

Texts and Monographs in Physics

Edited by W. Beiglböck, M. Goldhaber, E. Lieb, and W. Thirring

Texts and Monographs in Physics includes books from any field of physics that might be used as basic texts for advanced training and higher education in physics, especially for lectures and seminars at the graduate level.

Polarized Electrons
by J. Kessler
1976. ix, 223p. 104 illus. cloth

The Theory of Photons and Electrons
The Relativistic Quantum Field Theory of
    Charged Particles with Spin One-Half
Second Expanded Edition
By J. Jauch and F. Rohrlich
1976. xix, 553p. 55 illus. cloth

Essential Relativity
Special, General, and Cosmological
Revised Second Edition
By W. Rindler
1980. xvii, 284p. 44 illus. paper

**Inverse Problems in Quantum Scattering
    Theory**
By K. Chadan and P. Sabatier
1977. xxii, 344p. 23 illus. cloth

Quantum Mechanics
By A. Böhm
1979. xvii, 521p. 105 illus. cloth

Relativistic Particle Physics
By H. Pilkuhn
1979. xii, 427p. 89 illus. cloth

**The Concepts and Logic of Classical
    Thermodynamics as a Theory of Heat
    Engines**
Rigourously Constructed upon the Foundation
    Laid by S. Carnot and F. Reech
By C. Truesdell and S. Bharatha
1977. xxii, 154p. 15 illus. cloth

Principles of Advanced Mathematical Physics
By R.D. Richtmyer
Volume I
1978. xv, 400p. 45 illus. cloth
Volume II
1981. xi, 322 p. 60 illus. cloth

Foundations of Theoretical Mechanics
By R.M. Santilli
Part I: The Inverse Problem in Newtonian
    Mechanics
1978. 288p. cloth
Part II: Generalizations of the Inverse Problem
    in Newtonian Mechanics
In preparation

**Advanced Quantum Theory and Its
    Applications Through Feynman Diagrams**
By M.D. Scadron
1979. xiv, 386p. 78 illus. cloth

**Operator Algebras and Quantum Statistical
    Mechanics**
By O. Bratteli and D.W. Robinson
Vol I: C* and W* Algebras. Symmetry Groups.
    Decomposition of States
1979. 500p. cloth.
Vol. II: Equilibrium States. Models in Quantum
    Statistical Mechanics
1981. 517p. approx. cloth

The Nuclear Many-Body Problem
By P. Ring and P. Schuck
1980. 716p. approx. 171 illus. cloth

Nuclear Reactions with Heavy Ions
By R. Bass
1980. 410p. 176 illus. 31 tables. cloth